Studies in Computational Intelligence

Volume 616

Series editor

Janusz Kacprzyk, Polish Academy of Sciences, Warsaw, Poland
e-mail: kacprzyk@ibspan.waw.pl

About this Series

The series "Studies in Computational Intelligence" (SCI) publishes new developments and advances in the various areas of computational intelligence—quickly and with a high quality. The intent is to cover the theory, applications, and design methods of computational intelligence, as embedded in the fields of engineering, computer science, physics and life sciences, as well as the methodologies behind them. The series contains monographs, lecture notes and edited volumes in computational intelligence spanning the areas of neural networks, connectionist systems, genetic algorithms, evolutionary computation, artificial intelligence, cellular automata, self-organizing systems, soft computing, fuzzy systems, and hybrid intelligent systems. Of particular value to both the contributors and the readership are the short publication timeframe and the worldwide distribution, which enable both wide and rapid dissemination of research output.

More information about this series at http://www.springer.com/series/7092

Paulo Novais · David Camacho
Cesar Analide · Amal El Fallah Seghrouchni
Costin Badica
Editors

Intelligent Distributed Computing IX

Proceedings of the 9th International
Symposium on Intelligent Distributed
Computing – IDC'2015, Guimarães,
Portugal, October 2015

 Springer

Editors
Paulo Novais
Departamento de Informática/Centro
  ALGORITMI, Escola de Engenharia
Universidade do Minho
Braga
Portugal

David Camacho
Computer Science Department
Universidad Autónoma De Madrid
Madrid
Spain

Cesar Analide
Departamento de Informática/Centro
  ALGORITMI, Escola de Engenharia
Universidade do Minho
Braga
Portugal

Amal El Fallah Seghrouchni
LIP6—University Pierre and Marie Curie
Paris Codex 05
France

Costin Badica
 Software Engineering Department, Faculty
  of Automatics, Computers and Electronics
University of Craiova
Craiova
Romania

ISSN 1860-949X   ISSN 1860-9503   (electronic)
Studies in Computational Intelligence
ISBN 978-3-319-25015-1   ISBN 978-3-319-25017-5   (eBook)
DOI 10.1007/978-3-319-25017-5

Library of Congress Control Number: 2015950052

Springer Cham Heidelberg New York Dordrecht London

Springer International Publishing AG Switzerland is part of Springer Science+Business Media
(www.springer.com)

Preface

The emergent field of Intelligent Distributed Computing focuses on the development of a new generation of intelligent distributed systems. It faces the challenges of adapting and combining research in the fields of Intelligent Computing and Distributed Computing. Intelligent Computing develops methods and technology ranging from classical artificial intelligence and computational intelligence to multi-agent systems and machine learning. The field of Distributed Computing develops methods and technology to build systems that are composed of interacting and collaborating components.

The 9th Intelligent Distributed Computing—IDC'2015 continues the tradition of the IDC Symposium Series that started as an initiative of two research groups from:

(i) Systems Research Institute, Polish Academy of Sciences, Warsaw, Poland;
(ii) Software Engineering Department of the University of Craiova, Craiova, Romania.

The IDC Symposium welcomes submissions of original papers on all aspects of intelligent distributed computing ranging from concepts and theoretical developments to advanced technologies and innovative applications. The symposium aims to bring together researchers and practitioners involved in all aspects of Intelligent Distributed Computing. IDC is interested in works that are relevant for both Distributed Computing and Intelligent Computing, with scientific merit in these areas.

This volume contains the proceedings of the 9th International Symposium on Intelligent Distributed Computing, IDC'2015. The symposium was hosted by the Intelligent Systems Lab (ISLab) from the ALGORITMI Center at the University of Minho, in Guimarães, Portugal, between the 7th and the 9th of October, 2015.

The IDC'2015 event comprised a main conference, with two special sessions, and two collocated workshops. The special sessions organized within the main conference were Energetic Sustainable Ambient Intelligence (ESAmI'2015) and Cognitive Models and Emotions Detection for Ambient Intelligence (COME-DAI'2015). The collocated events were the Workshop on Cyber Security and

Resilience of Large-Scale Systems (WSRL'2015) and International Workshop on Future Internet and Smart Networks (FI&SN'2015).

This book contains contributions from the main conference, with 2 invited, 22 regular, and 5 short papers, the ESAmI'2015 with 3 papers, the COMEDAI'2015 with 4 papers, the WSRL'2015 with 5 papers, and the FI&SN'2015 with 5 papers, one of them invited.

The IDC'2015 Symposium received 62 submissions from 17 countries (counting the country of each co-author for each paper submitted). Each submission was carefully reviewed by at least three members of the Program Committee. Acceptance and publication were judged based on the relevance to the symposium topics, clarity of presentation, originality and accuracy of results, and proposed solutions. Finally, 22 regular papers and 5 short papers were selected for presentation and included in this volume, resulting in a 35.48 % acceptance rate, counting only regular papers, and 43.55 % when including short papers.

The 46 contributions published in this book address many topics related to theory and applications of intelligent distributed computing including: Intelligent Distributed and High-Performance Architectures, Organization and Management, Intelligent Distributed Knowledge Representation and Processing, Networked Intelligence, and Intelligent Distributed Applications.

We would like to thank Janusz Kacprzyk, editor of Studies in Computational Intelligence series and member of the Steering Committee, for his continuous support and encouragement for the development of the IDC Symposium Series. Also, we would like to thank the IDC'2015, ESAmI'2015, COMEDAI'2015, WSRL'15, and FI&SN'2015 Program Committee members for their work in promoting the event and refereeing submissions. A special thanks to all colleagues who submitted their work to this event.

We deeply appreciate the efforts of our invited speakers Amílcar Cardoso and Francisco Fernandez de Vega and thank them for their interesting lectures.

Special thanks also go to the WSRL'15 organizers, Massimo Ficco and Salvatore D'Antonio, and to the FI&SN'2015 organizers, Alexandre Santos, Pascal Lorenz, and António Costa.

Finally, we appreciate the efforts of local organizers on behalf of ISLab from the ALGORITMI Centre, University of Minho, Guimarães, Portugal, for hosting and organizing these events.

Guimarães	Paulo Novais
Madrid	David Camacho
Guimarães	Cesar Analide
Paris	Amal El Fallah Seghrouchni
Craiova	Costin Badica
July 2015	

Organization

Organizer

Intelligent Systems Lab (ISLab)
ALGORITMI Center, University of Minho, Portugal

General Chairs

Paulo Novais, University of Minho, Portugal
Cesar Analide, University of Minho, Portugal

Steering Committee

Costin Badica, University of Craiova, Romania
David Camacho, Universidad Autonoma de Madrid, Spain
Filip Zavoral, Charles University Prague, Czech Republic
Frances Brazier, Delft University of Technology, The Netherlands
George A. Papadopoulos, University of Cyprus, Cyprus
Giancarlo Fortino, University of Calabria, Italy
Janusz Kacprzyk, Polish Academy of Sciences, Poland
Kees Nieuwenhuis, Thales Research & Technology, The Netherlands
Marcin Paprzycki, Polish Academy of Sciences, Poland
Michele Malgeri, University of Catania, Italy
Mohammad Essaaidi, Abdelmalek Essaadi University in Tetuan, Morocco
Paulo Novais, University of Minho, Portugal

Invited Speakers

Amílcar Cardoso, University of Coimbra, Portugal
Francisco Fernandez de Vega, University of Extremadura, Spain

Program Committee Chairs

Amal El Fallah Seghrouchni, LIP6—University Pierre and Marie Curie, France
David Camacho, Universidad Autonoma de Madrid, Spain
Paulo Novais, University of Minho, Portugal

Program Committee

Adina Magda, Florea, University Politehnica of Bucharest, Romania
Ajith, Abraham, Machine Intelligence Research Labs, USA
Alessandro Longheu, DIEEI—University of Catania, Italy
Amal El Fallah Seghrouchni, LIP6—University Pierre and Marie Curie, France
Amparo Alonso-Betanzos, University of A Coruña, Spain
Ana Madureira, Instituto Superior de Engenharia do Porto, Portugal
André C.P.L.F. de Carvalho, University of São Paulo, Brazil
Andrea Omicini, Alma Mater Studiorum–Università di Bologna, Italy
Anna Toporkova, National Research University Higher School of Economics, Russia
Antonio Fernández-Caballero, Universidad de Castilla-La Mancha, Spain
Antonio Gonzalez-Pardo, Basque Center for Applied Mathematics-TECNALIA, Spain
Antonio Liotta, Eindhoven University of Technology, The Netherlands
António Pereira, Instituto Politécnico de Leiria, Portugal
Barna Laszlo Iantovics, Petru Maior University of Târgu Mureş, Romania
Bertha Guijarro, University of A Coruña, Spain
Cesar Analide, University of Minho, Portugal
Corrado Santoro, University of Catania, Italy
Costin Badica, University of Craiova, Romania
Dana Petcu, West University of Timisoara, Romania
Dariusz Krol, Bournemouth University, UK
David Bednárek, Charles University Prague, Czech Republic
David Camacho, Universidad Autonoma de Madrid, Spain
David Fernandez Barrero, Universidad de Alcalá (UAH), Spain
David Obdrzalek, Charles University, Czech Republic
Doina Bein, The Pennsylvania State University, USA
Domenico Rosaci, University Mediterranea of Reggio Calabria, Italy
Dorian Cojocaru, University of Craiova, Romania

Dumitru Dan Burdescu, University of Craiova, Romania
Emilio Corchado, University of Salamanca, Spain
Fernando Otero, University of Kent, UK
Ficco Massimo, Second University of Naples, Italy
Filip Zavoral, Charles University Prague, Czech Republic
Florin Leon, Technical University "Gheorghe Asachi" of Iasi, Romania
Florin Pop, University Politehnica of Bucharest, Romania
Giacomo Cabri, Università di Modena e Reggio Emilia, Italy
Giancarlo Fortino, University of Calabria, Italy
Giandomenico Spezzano, University of Calabria, Italy
Giuseppe Di Fatta, University of Reading, UK
Giuseppe Mangioni, University of Catania, Italy
Goreti Marreiros, Polytechnic Institute of Porto, Portugal
Grzegorz J. Nalepa, AGH University of Science and Technology, Poland
Héctor D. Menéndez, University College of London, UK
Ichiro Satoh, National Institute of Informatics, Japan
Igor Kotenko, St. Petersburg Institute for Informatics and Automation of the
Russian Academy of Sciences, Russia
Inés Galván, Universidad Carlos III de Madrid, Spain
Ioan Salomie, Technical University of Cluj-Napoca, Romania
Jakub Yaghob, Charles University in Prague, Czech Republic
Jason Jung, Yeungnam University, South Korea
Javier Alfonso, University of Leon, Spain
Javier Bajo Pérez, Universidad Politécnica de Madrid, Spain
Javier Del Ser, Tecnalia Research & Innovation, Spain
Jen-Yao Chung, IBM, USA
Joel Rodrigues, Universidade da Beira Interior, Portugal
Jose Neves, Universidade do Minho, Portugal
José Machado, University of Minho, Portugal
Juan Pavón, Universidad Complutense de Madrid, Spain
Juan E. Tapiador, Universidad Carlos III de Madrid, Spain
Juan Manuel Corchado, University of Salamanca, Spain
Lars Braubach, University of Hamburg, Germany
Lucian Vintan, "Lucian Blaga" University of Sibiu, Romania
Luís Correia, University of Lisbon, Portugal
Maria Ganzha, University of Gdańsk, Poland
Maria D. R-Moreno, Universidad de Alcala, Spain
Marie-Pierre Gleizes, Université de Toulouse, France
Marjan Gushev, UKIM University St. Cyril and Methodius, Macedonia
Martijn Warnier, Delft University of Technology, The Netherlands
Michal Wozniak, Wroclaw University of Technology, Poland
Mirjana Ivanovic, University of Novi Sad, Serbia
Nick Bassiliades, Aristotle University of Thessaloniki, Greece
Nik Bessis, University of Derby, UK
Paul Davidsson, Malmö University, Sweden

Paulo Moura Oliveira, UTAD University, Portugal
Paulo Novais, University of Minho, Portugal
Pawel Pawlewski, Poznan University of Technology, Poland
Phan Cong-Vinh, NTT University, Vietnam
Radu-Emil Precup, Politehnica University of Timisoara, Romania
Rainer Unland, University of Duisburg-Essen, Germany
Razvan Andonie, Central Washington University, USA
Ricardo Aler, Universidad Carlos III, Spain
Ronaldo Menezes, Florida Institute Technology, USA
Safeeullah Soomro, Indus University, Pakistan
Salvador Abreu, Universidade de Évora, LISP/CRI, Portugal
Salvatore Venticinque, Second University of Naples, Italy
Shahram Rahimi, Southern Illinois University, USA
Stanimir Stoyanov, University of Plovdiv "Paisii Hilendarski", Bulgaria
Stefan-Gheorghe Pentiuc, University Stefan cel Mare Suceava, Romania
Vadim Ermolayev, Zaporizhzhya National University, Ukraine
Vicente Julian, Valencia University of Technology, Spain
Viviana Mascardi, University of GENOVA, IT, Italy
Weiming Shen, NRC, Canada

Organizing Committee Chairs

Paulo Novais, University of Minho, Portugal
Cesar Analide, University of Minho, Portugal

Local Organizing Committee

André Pimenta, University of Minho, Portugal
Angelo Costa, University of Minho, Portugal
Celestino Gonçalves, Polytechnic Institute of Guarda, Portugal
Davide Carneiro, University of Minho, Portugal
Fábio Silva, University of Minho, Portugal
Francisco Andrade, University of Minho, Portugal
Javier Alfonso-Cendón, University of Leon, Spain
João Carneiro, University of Minho, Portugal
João Ricardo Ramos, University of Minho, Portugal
José Machado, University of Minho, Portugal
Marco Gomes, University of Minho, Portugal
Sérgio Gonçalves, University of Vigo, Spain
Sorin Ilie, University of Craiova, Romania
Tiago Oliveira, University of Minho, Portugal

Special Session Chairs

Davide Carneiro, University of Minho, Portugal
Cesar Analide, University of Minho, Portugal

Publicity and Web Chairs

Angelo Costa, University of Minho, Portugal
André Pimenta, University of Minho, Portugal

Energetic Sustainable Ambient Intelligence (ESAmI'2015)

Session Chairs
Angelo Costa, University of Minho, Portugal
José Carlos Castillo, University Carlos III of Madrid, Spain
Fábio Silva, University of Minho, Portugal

Cognitive Models and Emotions Detection for Ambient Intelligence (COMEDAI'2015)

Session Chairs
Goreti Marreiros, Polytechnic Institute of Porto, Portugal
Davide Carneiro, University of Minho, Portugal
Ester Martinez-Martins, Jaume I University Carlos, Spain
André Pimenta, University of Minho, Portugal

2nd Workshop on Cyber Security and Resilience of Large-Scale Systems (WSRL'2015)

General Chairs
Massimo Ficco, Second University of Naples, Italy
Salvatore D'Antonio, University of Parthenope, Italy

Publicity Chair
Salvatore Venticinque, Second University of Naples, Italy

Steering Committee
Igor Kotenko, St. Petersburg Institute for Informatics and Automation of the Russian Academy of Sciences, Russia

Aniello Castiglione, University of Salerno, Italy
Michal Choras, University of Science and Technology, Poland
Marco Vallini, Politecnico di Torino, Italy
Luigi Coppolino, University Parthenope, Italy
Roberto Pietrantuono, University Federico II, Italy
Francesco Palmieri, Second University of Naples, Italy

International Workshop on Future Internet and Smart Networks, FI&SN'2015

Workshop Chairs
Alexandre Santos, University of Minho, Portugal
António Costa, University of Minho, Portugal
Pascal Lorenz, University of Haute Alsace, France

Invited Speaker
Fernando Boavida, University of Coimbra, Portugal

Workshop Program Committee
Adriano Moreira, University of Minho, Portugal
Alexandre Santos, University of Minho, Portugal
António Costa, University of Minho, Portugal
Bruno Dias, University of Minho, Portugal
Edmundo Monteiro, University of Coimbra, Portugal
Fatima Bendella, University of Oran, Algeria
Fernando Boavida, University of Coimbra, Portugal
Halina Tarasiuk, Warsaw University of Technology, Poland
Helena Rodrigues, University of Minho, Portugal
Jaime Lloret Mauri, Polytechnic University of Valencia, Spain
Joaquim Macedo, University of Minho, Portugal
Joel Rodrigues, University of Beira Interior, Portugal
Juan Carlos Burguillo, University of Vigo, Spain
Liane Tarouco, University Federal of Rio Grande Sul, Brazil
Luis Sabucedo, University of Vigo, Spain
M. João Nicolau, University of Minho, Portugal
Manuel Ricardo, University of Porto, Portugal
Marília Curado, University of Coimbra, Portugal
Mário Freire, University of Beira Interior, Portugal
Miguel Rio, University College London, UK
Pascal Lorenz, University of Haute Alsace, France
Paulo Carvalho, University of Minho, Portugal
Pedro Sousa, University of Minho, Portugal
Rui Aguiar, University of Aveiro, Portugal

Contents

Part I
Invited Papers

A Distributed Approach to Computational Creativity

Amílcar Cardoso, Pedro Martins, Filipe Assunção,
João Correia and Penousal Machado

Abstract Computational Creativity is an emerging field that studies and exploits the potential of computers to be more interventive and autonomous in creative processes. This paper presents a recently proposed distributed computational architecture that is based on a cognitive theory that intends to model human consciousness. In its nucleus lies a Global Workspace and a number of Generators that compete to have access to it. Two Generators being developed by our team are described, one based on the Conceptual Blending Theory, the other based on an evolutionary process.

1 Introduction

The use of software tools that allow us to express our creativity is becoming increasingly popular. They are being made available in all sorts of devices, allowing us to produce creative outputs like images, photos, texts, videos, and to share them with friends, colleagues, within social networks. However, most of the existing applications assume the mere role of *tools*, leaving the creative act entirely on the side of the user, who assumes the role of *creator*.

Computational Creativity (CC) is an emerging field that studies and exploits the potential of computers to be more interventive and autonomous in the creative processes [4]. One of the motivations behind the field is the belief that computational

A. Cardoso (✉) · P. Martins · F. Assunção · J. Correia · P. Machado
CISUC/DEI, University of Coimbra, Coimbra, Portugal
e-mail: amilcar@dei.uc.pt

P. Martins
e-mail: pjmm@dei.uc.pt

F. Assunção
e-mail: fga@student.dei.uc.pt

J. Correia
e-mail: jncor@dei.uc.pt

P. Machado
e-mail: machado@dei.uc.pt

© Springer International Publishing Switzerland 2016
P. Novais et al. (eds.), *Intelligent Distributed Computing IX*,
Studies in Computational Intelligence 616,
DOI 10.1007/978-3-319-25017-5_1

3

creative systems are potentially effective in a wide range of artistic, technical and scientific domains where innovation is a key issue.

The literature on CC illustrates that a wide range of computational techniques have been used to develop creative systems[1] including single-agent approaches (most of them) and some *a-life* and *swarm*-based approaches. However, multi-agent architectures allowing heterogeneous agents are scarce (an exception is the work of Eigenfeldt and Pasquier [5, 6]). Nonetheless, such architectures could allow the parallel exploration of heterogeneous strategies and techniques for computational creativity.

In this paper we present a distributed computational architecture that is being developed in the context of an European project. We will start in Sect. 2 by presenting a distributed computational architecture [22] inspired by a cognitive theory [3] that intends to model human consciousness. Then we focus on two *concept generator* modules that our team is developing for the architecture: in Sect. 3, we present a concept generator based on the Conceptual Blending Theory [7]; a bio-inspired concept generator is described in Sect. 4. Finally, in Sect. 5 we will draw some conclusions.

2 A Distributed Conceptual Architecture

Our team is currently involved in a FET[2] project on Computational Creativity called ConCreTe (Concept Creation Technology).[3] As a whole, the project investigates computational models for the representation and production of previously unseen concepts, and apply them in context of various forms of creativity (for example, design, narrative and poetry). The project explores the central framework of a computational cognitive architecture that space constraints only allow for a brief explanation here. For more details, we redirect the reader to [22].

In brief, the architecture intends to simulate human *spontaneous creativity*, i.e., the creative process described in [20] as the "illumination" ("Aha!" moment) resulting from a subconscious "incubation". It is based on Baars' Global Workspace Theory [3], adopting principles from Peter Gardenfors' Theory of Conceptual Spaces [8] and from Shannon's Information Theory [19]. In Baars' theory, the non-conscious mind can be seen as a large collection of expert generators (Fig. 1), processing data in parallel, and competing for access to a Global Workspace, which contains the information of which the organism is conscious at any given time. The Global Workspace is the only communication means for the generators and is always visible to all them, but has constraints on the information that it may contain at any time. A threshold based throttling mechanism regulates the access of the generators to the workspace.

---

[1] The Proceedings of the International Conferences on Computational Creativity and their ancestor workshops (http://computationalcreativity.net/home/conferences/).

[2] Future Emerging Technologies.

[3] http://www.conceptcreationtechnology.eu.

Fig. 1 Baars' cognitive model

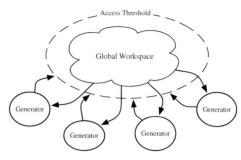

Fig. 2 Schematic diagram of the ConCreTe distributed cognitive architecture [22]

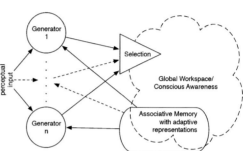

Our distributed cognitive architecture (Fig. 2) follows Baars' model: a number of generators act in parallel, like in a multi-agent architecture [15], and compete for access to a Global Workspace that resembles a Blackboard system [10] and simulates conscious awareness. A geometrical and multilevel knowledge representation scheme, which allows continuous representation of meaning, based on Gardenfors' theory, is being adopted and developed to support the blackboard architecture [22].

A selection mechanism regulates the access of the generators to the workspace using measures of probability and information content [21]. At each time, the generator that buffers more relevant output gains access to the workspace, which flushes its previous contents to an associative memory that keeps track of past information. The generators are fed by a continuous perceptual input and have access to the shared associative memory.

The generators may have different purposes and be heterogeneous regarding their internal structure. Within the context of this paper, we identify two basic kinds regarding function: *predictors* and *concept generators*. The first are intended to make predictions by scanning the input and matching it with the associative memory contents. Successful predictions are more likely to be selected by the throttling mechanism. Overall, these generators, together with the selection mechanism, the workspace and the shared memory contribute to a continuous unsupervised learning process that allows the simulation of conscious awareness of an entity when interacting with a dynamic environment.

The second kind of generators play an essential role within the creative process: they have access, like the others, to the perceptual input and to the shared memory, but continuously try to produce novel information with enough information content to be selected to the workspace and gain access the attention of the simulated mind, like in a "Aha!" moment. In the remaining of this document we will present two different concept generators that are being developed by our team: one is based on the Conceptual Blending Theory [7], the other is bio-inspired.

3 A Conceptual Blending Generator

Many important discoveries in history reportedly resulted from wandering in domains apparently not directly related to the target domain. We can find several theories in Psychology proposing explanatory models for this kind of phenomenon (e.g., the *Divergent Production* proposed by Joy Paul Guilford [9] and the *Bisociation* proposed by Mark Koestler [12]). These reasoning processes may occur within confined areas, or domains, of the space of concepts, where knowledge pieces are highly interconnected, but have much higher creative potential when occurring in cross-domain scenarios, i.e., when they connect previously unconnected areas.

The Conceptual Blending Generator is, as its name may suggest, a concept generator based on Concept(ual) Blending (CB) Theory [7]. CB theory relies on a framework that accounts for several cognitive phenomena related to the creation of ideas and meanings. Since CB theory provides an elaborate description of the concept integration process as well as a terminology and a set of consistent principles to be used in creativity modeling, CB theory has served as the basis for several computational creative systems.

A key element in the CB framework is the *mental space*, which is a temporary and partial structure of knowledge built for the purpose of local understanding and action. To describe the blending process, the CB framework makes use of a network of mental spaces, as depicted in Fig. 3. Two or more of these spaces correspond to

Fig. 3 The original four-space conceptual blending network [7]

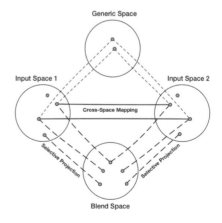

the input spaces, i.e., the content that will be blended. A *partial matching* between elements of the different input spaces is established. Note that such a correspondence is not arbitrary; a correspondence between elements from different mental spaces only occurs if they are perceived as similar in some respect. Such association between elements of the different input spaces is reflected in the *generic space*, a mental space that captures the conceptual structure shared by the input spaces. The next step in the blending process is typically referred to as *selective projection*. At this point, the conceptual structure encapsulated by the generic space defines which elements from the input spaces will be projected into a new mental space, called the *blend space*. Finally, an emergent structure emerges in the blend space by pattern completion or elaboration. The outcome of the whole blending process is therefore the blend space, a new mental space that keeps partial structures from the input spaces, combined with an emergent structure of its own.

The integration of input elements in the blend space is guided by optimality principles [7], which are responsible for generating the so-called "good blends", i.e., consistent blends which are more easily interpreted. Among these principles, there is the integration principle which states that the blend must be recognized as a unit. Another example is the unpacking principle, which requires that the blend alone must enable the "blend reader" to unpack the blend to reconstruct the inputs, the cross-space mapping, the generic space, and the network of connections between all these spaces. There is also the principle of relevance, which demands for the existence of a reason for the blend to occur. Other principles such as web, topology and pattern completion are responsible for managing the relationship between the input spaces and the blend.

3.1 The Framework

The CB generator is built on Divago [16], which is one of the most elaborate computational approaches to CB. The framework of the CB generator is composed of several modules (Fig. 4) that have access to a Knowledge Base, which, in the context of the overall cognitive architecture, may be shared by other generators. The architecture of Divago reflects the different stages of the CB mechanism, starting with the selection of the input knowledge, i.e., the choice of a pair of input spaces (domains) from the knowledge base. This selection is currently not informed by the dynamics of the Global Workspace, but we envisage that such possibility may increase the likelihood of selecting pieces of knowledge with greater relevance to the current state of the simulated mind.

The input spaces feed the Mapper module, which is responsible for finding analogy mappings between the input spaces using structural alignment. This operation looks for the largest isomorphic pair of sub-graphs contained in the input spaces.

For each mapping/alignment provided by the Mapper, the Blender module performs a selective projection into the blend space. This leads to the construction of the Blendoid, a graph structure that subsumes the set of all possible blends.

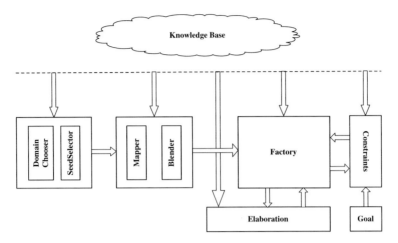

Fig. 4 CB generator architecture

The Factory module is responsible for exploring the space of all possible combinations of projections of the input spaces. It is based on a Genetic Algorithm (GA), which uses the Elaboration module to enrich blends with additional knowledge and the Constraints module to assess their quality. This module provides an implementation of the optimality principles (a set of principles that ensure a coherent and highly integrated blend [7]). When an adequate solution is found or a pre-defined number of iterations is attained, the Factory stops the execution of the GA and returns the best blend. The Constraints module acts, therefore, as the "fitness function" of the algorithm.

3.2 Generating Concepts: An Example

The Divago framework has been tested in several domains [17]. An early and interesting example that illustrates the potential of this architecture is the *Creature Generation Experiment*. In this example, Divago was used in a context of procedural content generation for a game environment. The role of Divago was to produce novel creatures from a set of existing ones. A 3D interpreter [18] was used to visualise the objects: it was able to convert outputs from Divago (in the form of *concept maps*), representing creatures, into Wavefront OBJ files that could then be rendered.

Divago was fed with the representations of three creatures whose 3D renderings are shown in Fig. 5: a werewolf, a dragon and a horse. These creatures were used as input spaces. Examples of hybrid creatures produced by Divago are shown in Fig. 6.

Fig. 5 Input creatures.
From *left* to *right* werewolf;
dragon; horse

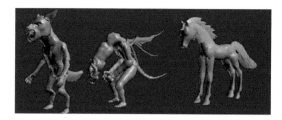

Fig. 6 Output creatures.
From *left* to *right*
horse-dragon;
horse-werewolf;
werewolf-dragon

4 A Bio-Inspired Concept Generator

The bio-inspired concept generator is based on an evolutionary engine that works on augmented grammars performing concept recombination and manipulation. One of the key ideas behind this work is to develop mutation and recombination operators that are sensible to the structures being manipulated and that are informed by background knowledge associated with the domain where the individuals are being evolved. This separation allows the development of general-purpose operators that can be easily adapted to different domains, and thus take advantage of domain-specific knowledge.

In order to apply the engine to other search spaces, one just needs to change the grammar that is used to form the candidate solutions to another one, capable of properly forming solutions that can solve the problem being tackled; in this case, the creation of aesthetic works, such as images and musical sequences.

In Sect. 4.1 we start by presenting, in a nutshell, the concepts that guided the creation of the engine, moving then to one applicational domain, in this case, the evolution of Context Free Design Grammars [11] (Sect. 4.2).

4.1 Overview of the Evolutionary Engine

The evolutionary engine used for this task is thoroughly discussed and tested in [1, 13] and is an extension of [14]. Therefore, we provide a brief overview focusing on the aspects that are needed for the understanding of the current work.

The aim of the engine is to evolve candidate solutions encoding grammatical formulations. In essence, a grammar is formed by a 4-tuple: (V, Σ, R, S) where: V is a set of non-terminal symbols; Σ is a set of terminal symbols; R is a set of production rules that map from V to $(V \cup \Sigma)^*$; S is the initial symbol. Because we allow the specification of parameters in the calls to terminal and non-terminal symbols, we say that the used grammars are augmented. Moreover, it is also possible to define more than one possibility for expanding a non-terminal symbol. When several production rules are applicable, one of them is randomly selected and the expansion proceeds.

In order to promote the evolution of this type of grammars across generations, two representations are used. On the one hand, individuals are encoded as directed graphs, where each node represents a production rule and the connections between them the flow of control and passing of parameters. On the other hand, we use a tree-based representation, where individuals stand for derivation trees of a pre-defined Backus-Naur Form grammar, adding, as such, another grammatical level to the framework. In this last representation scenario, internal nodes encode non-terminal symbols, whereas leaves represent terminal ones.

For the purposes of recombination and mutation, operators based on both representations were investigated. According to the results presented in [1], the best solutions are obtained when using tree mutation with tree and graph crossover. That is, it should be decided first whether mutation or crossover should be applied. If the decision is set towards crossover, it is required to choose which type of crossover should be employed. With regard to generalization, the tree initialization procedure is the one that is used.

4.2 Evolving Context Free Design Grammars

To assess the adequacy of the engine to properly evolve the desired outputs, we performed extensive experiments using Context Free Design Grammars (CFDGs) [11], which are capable of encoding images through a compact set of production rules. Results, which are detailed in [1, 13], show that guiding evolution with distinct automatic fitness assignment schemes promotes evolution, leading to the expected results. With the previous in mind, we tested a fitness function that combines several others, attaining results that bespeak characteristics from each one of the components being merged.

Taking into account the non-deterministic nature of CFDGs, i.e., if they are mapped several times to images, the result is often different, we introduce the concept of families [2], which is simplistically defined as a set of outputs which should have similar characteristics. For that, we perform the mappings from a grammar to an image a pre-defined number of times, assessing then the quality of the set, considering that the quality of each image should be maximized, the differences in quality should be minimized and that a proper degree of similarity should exist among them. An example of a family evolved using this formulation is shown in Fig. 7.

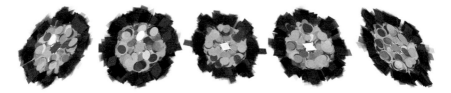

Fig. 7 Samples of a family produced by a single CFDG

5 Conclusions

We presented a distributed computational architecture based on the cognitive theory of Bernard Baars [3], which aims to model human consciousness. It consists of a number of Generators acting in parallel, like in a multi-agent architecture, competing for access to a Global Workspace that resembles a Blackboard system. A selection mechanism regulates the access of the generators to the workspace using measures of probability and information content. Besides providing a computational simulation environment for studying creativity processes, the architecture allows the exploration in parallel of heterogeneous strategies and techniques for concept creation. We identified two basic kinds of generators: *predictors* and *concept generators*. Two Concept Generators being developed by our team were described: the first, based on the Conceptual Blending Theory [7]; the second is a bio-inspired concept generator. We illustrated the output of the former by providing results from an experiment in procedural content generation for games. For the latter, we showed outputs from the evolution of Context Free Design Grammars.

This is an ongoing project and many aspects are to be deepened and polished in the next steps. Still untouched in this paper was the issue of evaluating the creative potential of the architecture. Evaluation of computational creativity is acknowledged as one of the most hard research problems faced by the community. The project is already developing work in a parallel thread to develop the methodologies and the tools that will allow such evaluation.

Acknowledgments This research is partially funded by project ConCreTe. The project ConCreTe acknowledges the financial support of the Future and Emerging Technologies (FET) Programme within the Seventh Framework Programme for Research of the European Commission, under FET grant number 611733.

References

1. Assunção, F.: Grammar based evolutionary design. Master's thesis, Department of Informatic Engineering, Faculty of Sciences and Technology, University of Coimbra, July 2015
2. Assunção, F., Correia, J., Martins, P., Machado, P.: Evolving families of shapes. In Proceedings of the Twenty-fourth International Joint Conference on Artificial Intelligence, pp. 4134–4135. AAAI Press, AAAI Press (2015)

3. Baars, B.J.: A Cognitive Theory of Consciousness. Cambridge University Press, New York (1988)
4. Cardoso, A., Veale, T., Wiggins, G.: Converging on the divergent: The history (and future) of the international joint workshops on computational creativity. AI Mag. **30**(3), 15–22 (2009)
5. Eigenfeldt, A., Pasquier, P.: Negotiated content: generative soundscape composition by autonomous musical agents in coming together: freesound. In: Proceedings of the ICCC-2011, 2nd International Conference on Computational Creativity, pp. 1–6, March 2011
6. Eigenfeldt, A., Pasquier, P.: Creative agents, curatorial agents, and human-agent interaction in coming together. In: Proceedings of Sound and Music Computing Conference, Copenhagen (2012)
7. Fauconnier, G., Turner, M.: The Way We Think. Basic Books, New York (2002)
8. Gärdenfors, P.: Conceptual Spaces: The Geometry of Thought. MIT Press, Cambridge (2000)
9. Guilford, J.P.: The Nature of Human Intelligence. McGraw-Hill, New York (1967)
10. Hayes-Roth, B.: Bb1: An architecture for blackboard systems that control, explain, and learn about their own behavior (1984)
11. Horigan, J., Lentczner, M.: Context Free (2014). http://www.contextfreeart.org/
12. Koestler, A.: The Act of Creation. Macmillan, New York (1964)
13. Machado, P., Correia, J., Assunção, F.: Graph-based evolutionary art. In: Gandomi, A., Alavi, A.H., Ryan, C. (eds.) Handbook of Genetic Programming Applications. Springer, Berlin (2015)
14. Machado, P., Nunes, H., Romero, J.: Graph-based evolution of visual languages. Applications of Evolutionary Computation, pp. 271–280. Springer, Berlin (2010)
15. Minsky, M.: The Society of Mind. Simon and Schuster Inc., New York (1985)
16. Pereira, F.C.: Creativity and AI: A Conceptual Blending approach. Ph.D. thesis, Department Engenharia Informática da FCTUC, Universidade de Coimbra, Portugal (2005)
17. Pereira, F.C., Cardoso, A.: Experiments with free concept generation in divago. Knowledge-Based Systems **19**(7), 459–470 (2006)
18. Ribeiro, P., Pereira, F.C., Marques, B., Leitao, B., Cardoso, A.: A model for creativity in creature generation. In 4th International Conference on Intelligent Games and Simulation (GAME-ON 2003) (2003)
19. Shannon, C.E.: ACM SIGMOBILE Mob. Comput. Commun. Rev. A mathematical theory of communication. **5**(1), 3–55 (2001)
20. Wallas, G.: The Art of Thought. Harcourt Brace, New York (1926)
21. Wiggins, G.A.: Crossing the theshold paradox: Modelling creative cognition in the global workspace. In Proceedings of the 3rd Int. Conference on Computational Creativity, ICCC-12, Dublin, Ireland (2012)
22. Wiggins, G.A., Forth, J.: Idyot: a computational theory of creativity as everyday reasoning from learned information. Computational Creativity Research: Towards Creative Machines, pp. 127–148. Springer (2015)

Evolutionary Algorithms: Perspectives on the Evolution of Parallel Models

F. Fernández de Vega

Abstract This chapter discusses the inherent parallel nature of evolutionary algorithms, and the role this parallelism can take when implementing them on different hardware architectures. We show the interest in studying ephemeral behaviors that distributed computing resources may feature and some EA's self-properties of interest, such as the fault-tolerant nature that helps to fight the *churn* phenomenon. Moreover, interactive versions of EAs, which require distributed computing systems, allow to incorporate human based knowledge within the algorithm at different levels, providing new means for improving their computing capabilities while also requiring a proper analysis of human behavior under an EA framework. A proper understanding of ephemeral properties of hardware resources, human behavior in interactive applications and intrinsic parallel behaviors of population based algorithms will lead to significant improvements.

1 Introduction

Although evolutionary algorithms [1], and other population based algorithms, have been successfully applied to solving a wide number of optimization problems, researchers typically apply sequential version of the algorithms. Several reasons explain this traditional approach to software development, including the learning curve required to properly apply parallel models and libraries, and the wide number of available software tools that were developed in the traditional sequential approach. Although things are slightly changing, the literature is still dominated by sequential EAs.

Nevertheless, parallelism was soon recognized as an intrinsic property of EAs that works in the background even when a sequential version of the algorithm is run. The schema theorem, proposed by Holland in the seventies, was in charge of

F. Fernández de Vega (✉)
Centro Universitario de Mérida, Universidad de Extremadura,
Sta. Teresa Jornet, 38, 06800 Mérida (Badajoz), Spain
e-mail: fcofdez@unex.es

© Springer International Publishing Switzerland 2016
P. Novais et al. (eds.), *Intelligent Distributed Computing IX*,
Studies in Computational Intelligence 616,
DOI 10.1007/978-3-319-25017-5_2

explaining this inherent parallel property [2]. Although the reasoning is of interest for understanding how EAs can build solutions to problems, it doesn't allow to speed up the behavior of the algorithm: researchers have happily relied on this explanation for its intrinsic parallel behavior until hard real life problems have been faced. Only then, researchers have resorted to parallelism, when days, weeks or even months are consumed until a proper solution is found [3].

This chapter reviews different parallel models that have been proposed, how they can be deployed on different hardware architectures, and focuses in new properties that have been studied in the last few years, involving non reliable hardware resources as well as human interaction with the algorithms, showing that work ahead may provide new means for improving the performance of the algorithm.

This chapter is organized as follows: In Sect. 2, an overview of available parallel models is provided; Sect. 3 discusses the role distributed users may have on the algorithm; Finally, Sect. 4 describe our conclusions and paths for future improvement of parallel EAs.

2 Parallel Models Have Evolved

Embarrassingly parallel models were firstly propose as a way to quickly embody parallelism within EAs [4]. The easier incarnation of parallelism allows to simultaneously evaluate a number of individuals when hardware resources are available. We must remind that the standard evolutionary loop includes the computationally expensive evaluation of a number of individuals from the population, candidate solution to the problem at hand, followed by the crossover process that give rise to the new generation of individuals. Thus, the master-slave based model doesn't change the main algorithm, in charge of selecting parents for the next generation and applying genetic operators, being the fitness function the only one requiring a change. Given that fitness evaluation is typically the most time consuming part of the algorithm, and how easily a sequential implementation of an EA can be parallelized using this model, it quickly attracted attention. Thus the simplest parallel EA has been deployed and run on networks of transputers [5], clusters and grids [6], and more recently on GPUs [7] and clouds [8], and has probably been the most frequently used version of a parallel EA.

Nevertheless, researchers soon devised new ways for improving convergence properties, adding new functions to parallel models that in the end implied a deep change in the underlying algorithm and produced a change in the process of searching for solutions. Instead of evaluating single individuals in parallel, researchers decided to run the main algorithm over a number of them -a subpopulation- within each of the processors available, thus resulting in the *Island Model* [9].

Each of the subpopulation run the standard algorithm in the island model, and a new step, the migration, allows selected individuals to be exchanged among subpopulations -islands- with a given frequency. Thus the researcher must set up the value of some new parameters: island size, frequency of migration, number of migrating individuals, selection operator in charge of selecting migrants, discharging policy allowing to maintain the size of islands once new individuals arrive, etc. All these new parameters have already been widely studied, and its influence on the convergence process exposed for different flavors of EAs, including Genetic Algorithms and Genetic Programming [10]. The conclusions points out the benefit of migrating individuals, which helps to improve diversity in the subpopulations, thus helping to find better solutions, regardless of the time saved thanks to the parallel hardware infrastructure employed.

Yet, the island model is not the only one available for improving convergence properties of EAs as well as speeding up the finding of solutions. The cellular model is another possibility [11]. In this case, individuals from the population are distributed on a grid, so that interaction required when genetic operators must be applied only occur within a previously established neighborhood. This means that one individual can only interact with surrounding ones, which changes the way chromosome information spreads along the population [9]. Several authors have applied successfully this model borrowed from the cellular automata literature, although the implementation details make it more difficult to be adopted by researchers.

An interesting difference among the available parallel models, regarding hardware resources to be used and their properties, can be noticed: While for the island model, each of the subpopulations can run semi-isolated within each of the processors employed, and only a migrating step is required after a number of generations, the embarrassingly parallel model requires fitness values computed to be returned to the master in charge of applying each of the genetic operators, and this implies the sending of fitness values at least once per generation, which may be of importance or not depending on the time required to compute fitness values: shorter time to compute fitness value means worse performance of the algorithm, given that the latency of communications has a larger impact. Similarly, this is also something that must be taken into account when using the cellular model, which requires communication among adjacent individuals from the topological point of view that are run on different processors every time a crossover operation must be applied. Summarizing, high latencies will significantly deteriorates the speedup of both cellular and embarrassingly parallel model, even preventing them to compete in certain situations with the sequential version, while it will not hinder island model to properly finding solutions in shorter times.

In any case, communication libraries had to be adopted by researchers when implementing parallel EAs, such as classic PVM or MPI [12], when using clusters of computers; other different approaches can be considered when resorting to internet and grid frameworks. Even in this latest hardware infrastructures, interesting software packages allow to quickly deploy any algorithm on an easy to build desktop grid system, such as that based in the BOINC framework [13], which has opened up a world of possibilities for EAs. As we will see below, a proper study of the dynamics of this model has allowed to develop in the last decade new proposals for parallel EAs that benefit from the properties of the underlying communication model: in the area of Grid computing, the well known desktop grid model has been employed to run massively parallel evolutionary algorithms applied to real-world problems; On the other hand, P2P models have allowed to implement new agent based EAs that change the standard dynamics of the algorithm. Both models have changed the way we understand the algorithm, and have shed light into some of the properties that the new parallel models have unveiled.

2.1 Desktop Grids and Shrinking Population

When referring to Desktop Grid Computing, we consider a particular case of Grid technology where all of the computing resources are homogeneous: desktop personal computers. Given that all of the computers are based on the same hardware architecture, and basically the same operating system, the grid system significantly simplifies the way parallel algorithms can be deployed on the network of computers: a single version of the algorithm must be implemented (linux—i386, for instance), instead of considering all of the potential hardware architectures and operating systems combinations that are present in a more traditional Grid infrastructure. Moreover, available software tools, such as BOINC [13], allow to easily manage the desktop grid infrastructure, allowing researchers to only focus on the Evolutionary Algorithm to be deployed. The basic desktop-grid model follows the master-slave approach, and is well suited to embarrassingly parallel EAs: typically desktop grids are deployed within institutions, and communication latencies are thus under control.

The simplicity of desktop grids, has allowed researchers to face hard real life problems: packages of individuals are distributed every generation along the available computing nodes, allowing researchers to manage large population sizes for real life problems requiring long fitness evaluation time [3]. Thus, the model was shown to perfectly work on desktop grids provided by the researchers. The surprise came when the model was applied using computing resources provided by volunteers under the well known volunteer computing model [14].

Volunteer computing is based on the desktop grid model, and desktop computers are provided by a number of volunteers connected to internet that are willing to contribute to a scientific project. Thus, the scientist is typically in charge of setting up the master node, where all of the computing tasks are established, and then, the volunteers connect to the server and agree to provide computing resources for each of the tasks. The model has worked fine for decades, being the *Seti@home* project one of the best known with several million volunteers providing resources [13]. Nevertheless, and given that resources are switched on and off according to volunteers' needs, nobody can assure the time a computing node will be alive, and whether a specific task submitted will be thus completed on time. The dynamic of the volunteer computing infrastructure is thus characterize by this well known *churn* phenomena, and scientists interested in profiting from volunteers must encode a number of fault-tolerant techniques if they want their project to finish properly [15]. But this inherent property, churn, was recently considered from a different perspective in the context of EAs, specifically from the point of view of Genetic Programming (GP) with interesting results.

2.1.1 Distributed GP and the Churn Phenomena

One of the main flavors of Evolutionary Algorithms is GP, popularized by Koza in the nineties [16] as a mean for automatic programming. One of the main features of GP, considered as a problem, is the *bloat* phenomenon: given that variable size chromosomes are employed in GP, the evolution dynamics make chromosomes to grow out of control, which implies an increase in memory consumption and usually time required to evaluate longer individuals. This behavior in GP has lead researchers to focus on chromosome growth [17], an although a number of techniques have already been propose to fight it, we think future research on the topic will show how this behavior may find a strong connection to improvements on the way GP is run in parallel systems: a natural load balancing technique could make use of individual differences to run them on different computing elements, as well as applying genetic operations as soon as individuals have been evaluated, thus favoring shorter computing-times, which typically implies smaller sizes. Thus parallel systems could naturally fight bloat. We must also bear in mind, the difficulty for properly running GP on GPUs, which has been an issue in latest years. Although some proposals have already been published, we still feel there is room for improvement, considering main differences among GP and other EAs.

Among the different techniques introduced in the literature for the last decades, the *plague* operator was proposed to remove progressively individuals from the population as a countermeasure for the bloat phenomenon, thus maintaining the amount of memory required to manage the population: individuals' growth is fought with a shrinking population [18]. Since then, different studies have shown the interest of considering variable size populations for GP and other EAs, which require a self-analyzing capability of the algorithm to know when the size must be changed. But a deeper analysis allowed to recently see the connection between this idea and the churn phenomenon in volunteer computing infrastructures: if instead of removing selected individuals, we consider churn phenomenon as the component in charge of randomly discarding individuals along the run of a GP experiment in a volunteer computing environment, we have a quite similar experiment, the only difference being the way individuals are selected.

In the last few years, a number of experiments have tested this approach showing that not only Genetic Programming, but also Genetic Algorithms are fault tolerant, and can cope with up to 30 % of population decrease without applying any particular fault tolerance technique. This has opened up the possibility of running distributed versions of the algorithm in non reliable distributed computing resources with results whose quality does not significantly deteriorates [19], boosting a line of research that focuses on self-properties of EAs in the context of parallel and distributed systems. The experiments have thus shown that other network topologies and communication models can also be employed within this context, such as Peer to Peer networks.

2.2 To Peer or Not to Peer

Peer to Peer models (P2P), have been recently studied in the context of EAs by Laredo et al. [20]. One of the main features of the model, is the lack of a central node, both in terms of hardware resources and in the main algorithm to be run. The model instead relies in a number of software agents with capability for establishing connections with surrounding agents, being them run on the same or different computing element.

P2P models require specific communication protocols, that allow agents to know where other agents are located, and from the algorithmic point of view, also requires changes when a task must be performed. If we consider EAs in a P2P context, we will see each of the individuals in the population as an agent. No central storage location for the population exist anymore, nor a single algorithm applying genetic operations to the individuals. Instead, each of the agents must include the capability to interact with other agents, individuals, so that they can crossover and create offspring. New software tools allow to deploy EAs using P2P models, and some of them rely on web browsers to run the genetic operations, including fitness evaluations [21]. The fault tolerance nature of these agent based models have also been studied reaching similar conclusions as with its volunteer model counterpart [19]. But one of the main interests now, is the possibility of using web browsers, and also user interaction, as the underlying system where the algorithms are run. The possibility of allowing users to interact with the algorithm through a web browser, in the context of P2P EAs but also when using the master-slave approach, and the churn properties featured are allowing to explore new properties for distributed EAs.

3 Interactive EAs, Ephemeral Properties and the Role of Users

Although the possibility for allowing users to interact with EAs was soon recognized as a means for fitness evaluation, similarly as how volunteer computing based projects invite users to collaborate by performing visual analysis of images [21], in the context of EAs, the interaction has been exclusively used as a way for aesthetic assessment in Evolutionary Art. Thus interactive EAs are directly related to Evolutionary Art and Design, and typically the interaction has been facilitated through web based applications in charge of displaying each of the *individuals* in the population, that are then rated by the users, so that fitness values provided are employed to apply selection, crossover, etc. Users are thus contributing not only with fitness values, but also with hardware resources to run the user interface, one of the main part of the algorithm, and are therefore prone to the same kind of problems that were previously described in the context of volunteer computing and P2P environments.

Only recently, new software tools have been developed trying to generalize the model allowing users to both run and/or interact with EAs through the web browser, such as Evospace and Evospace-i agent based software models that connect through web browsers and allow to face any kind of problem by means of Evolutionary Algorithms [8]. The dynamics of the underlying model feature some of the ephemeral properties that naturally arise in an agent based model, and have already been studied with satisfactory results [15]. We will focus now in one component of these latest distributed models that are increasingly attracting attention: the user.

3.1 Distributed Users

The fact that users collaborating with interactive EAs, deployed through the web, are part of the algorithm changes the way we understand distributed EAs. On the first hand, users may visit a website but their collaboration is not guaranteed: in order for the evolutionary algorithm to progress, users must get involved in the experiment. Similarly as with volunteer computing experiments, the scientists must properly provide information of interest that attract users attention. On the other hand, and given that usually web browser must remain open with the application running while the user is devoting attention to other tasks, the experiment must keep user interest to collaborate and donate both computing resources and their time. Finally, when repetitive actions are required by the user, some kind of reward may be necessary if we want to fight the problem of users fatigue. These are some open problems in the area, and although efforts are applied trying to model users interaction that may in the future reduce the number of times an action is required from the user, we still lack a general solution to that problem [22].

Also, the number of users to be involved in a given experiment and also the way they interact should be adjusted: although typically users are simply in charge of rating individuals, different possibilities could be also adopted, such as asking users to select the parents for a crossover operation, so that indirectly a fitness evaluation is performed every time new children is generated. As we will show below, this later approach has been recently adopted in the context of *unplugged evolutionary algorithms* [23], but it is not still a common method.

Therefore, the actual influence of users in interactive EA experiments still allow for deeper studies, and a number of questions remain to be explored: is it possible to allow users to perform other operations different from fitness evaluations? What are the main reasons that lead a user to apply a specific rate to a given individual? Is it useful to allow different users to rate the same individual? What is the situation in the context of evolutionary art? These and other questions are leading efforts in the area, and one of the most recently proposed approaches is the Unplugged Evolutionary Algorithm.

3.2 *Unplugged EAs*

The idea behind the model arise from the interactive version of the EA: human beings are in charge of performing fitness evaluations. The question is: is it possible to delegate *all* of the operations to human beings? In the context of Evolutionary Art, the idea of making artists to perform the whole evolutionary algorithms tries to analyze the creative model when applied by a team of artists: they apply every operation required for an evolutionary algorithm so that no computer is required in the experiment.

Thus, a team of five artists developed an artistic experiment: each of the artist was in charge of producing an artwork every week by applying any kind of crossover and mutation over two works of their colleagues produced the week before. This way, instead of explicitly asking for fitness values -rates- for each of the paintings, the artists introduce an indirect fitness evaluation: only the two preferred works are given best rates and selected as an inspiration source when producing offspring. After ten weeks of work, a collective art work was produced and interesting information on the operations applied were described within forms provided to artists [24].

The analysis of the work gives us some clues for a better understanding of the creative processes developed by human artists, such as information on the key elements when applying crossover or mutation. For instance, artists always perceive a story within each figurative work, that may lead mutation operations towards a new work. Figurative work is typically preferred, instead of abstract works that are usually the output of evolutionary art experiments [23]. Yet, is not easy to foresee how some of the concepts learnt can in the future be incorporated into software tools in charge of producing human-like art.

On the other hand, if we want to fully emulate the creative process developed by human artists, a possible way to future improvement should include studying audience response to the work, including audience understanding of the genetic operations developed. Given the need of an audience when an art work is exhibited, audiences should be somehow included in the Evolutionary Art loop, being part of experimental research, and artworks should be exhibited in art museums and galleries.

4 Conclusions

This chapter presents an overview of latest attempts to parallelize Evolutionary Algorithm considering different points of view. On the one hand we have reviewed the models that have arisen in the last decade, such as those based on agents making use of Desktop grids and P2P infrastructures; on the other hand we have seen new paths that are being explored when distributed users are included as part of the parallel versions of the algorithm, particularly when art and creativity are pursued.

This review has led us to a number of questions that show paths towards future improvements on the way we understand and apply parallel and distributed versions of the Evolutionary Algorithms, that may be summarized as follows: (i) a proper understanding of the dynamics of algorithms employing variable size chromosomes, such as GP, as well as employing self-properties that allow to be aware of individual-size dynamics may make it easier to profit from parallel infrastructures, including GPUs as well as those characterized by ephimeral properties, such as desktop grids. (ii) the proper understanding of users interaction dynamics in the context of unplugged evolutionary algorithms may provide clues to improving how distributed interactive evolutionary algorithms are applied when facing evolutionary art project. It can make it easier for scientists to atract users and also avoid users' fatigue, as well as provide a better understanding of creative process that helps in the future to improve computer assited creativity.

Acknowledgments This work is supported by EU Merie Curie actions, FP7-PEOPLE-2013-IRSES, Grant 612689 ACoBSEC; MINECO project EphemeCH (TIN2014-56494-C4-P) and Gobierno de Extremadura,Consejería de Economía-Comercio e Innovación y FEDER, proyect GRU10029.

References

1. Back, T.: Evolutionary Algorithms in Theory and Practice: Evolution Strategies, Evolutionary Programming, Genetic Algorithms. Oxford University Press, New York (1996)
2. Bertoni, A., Dorigo, M.: Implicit parallelism in genetic algorithms. Artif. Intell. **61**(2), 307–314 (1993)
3. González, D.L., de Vega, F.F., Trujillo, L., Olague, G., Araujo, L., Castillo, P., Sharman, K.: Increasing gp computing power for free via desktop grid computing and virtualization. In 17th Euromicro International Conference on Parallel, Distributed and Network-based Processing, IEEE, pp. 419–423 (2009)
4. Cantu-Paz, E.: Efficient and accurate parallel genetic algorithms. Springer (2000)
5. Andre, D., Koza, J.R.: Parallel genetic programming: a scalable implementation using the transputer network architecture. Advances in Genetic Programming, pp. 317–337. MIT Press, Cambridge (1996)
6. Lim, D., Ong, Y.S., Jin, Y., Sendhoff, B., Lee, B.S.: Efficient hierarchical parallel genetic algorithms using grid computing. Future Gener. Comput. Syst. **23**(4), 658–670 (2007)
7. Wong, M.L., Wong, T.T., Fok, K.L.: Parallel evolutionary algorithms on graphics processing unit. In The 2005 IEEE Congress on Evolutionary Computation, IEEE, vol. 3, pp. 2286–2293 September 2005
8. García-Valdez, M., Trujillo, L., Merelo, J.J., de Vega, F.F., Olague, G.: The EvoSpace model for pool-based evolutionary algorithms. J. Grid Comput., 1–21 (2014)
9. Tomassini, M.: Spatially Structured Evolutionary Algorithms. Springer, Berlin (2005)
10. Fernández, F., Tomassini, M., Vanneschi, L.: An empirical study of multipopulation genetic programming. Genet. Program. Evolvable Mach. **4**(1), 21–51 (2003)
11. Folino, G., Pizzuti, C., Spezzano, G.: A Cellular Genetic Programming Approach to Classification. In GECCO, pp. 1015–1020, July 1999
12. Fernández, F., Tomassini, M., Vanneschi, L., Bucher, L.: A distributed computing environment for genetic programming using MPI. Recent Advances in Parallel Virtual Machine and Message Passing Interface, pp. 322–329. Springer, Berlin (2000)

13. Anderson, D. P. Boinc: A system for public-resource computing and storage. In Grid Computing. 2004. Proceedings. Fifth IEEE/ACM International Workshop on (pp. 4–10). IEEE
14. Cole, N., Desell, T., González, D.L., de Vega, F.F., Magdon-Ismail, M., Newberg, H., Varela, C.: Evolutionary algorithms on volunteer computing platforms: the milkyway@ home project. Parallel and Distributed Computational Intelligence, pp. 63–90. Springer, Berlin (2010)
15. González, D.L., Laredo, J.L.J., de Vega, F.F., Guervós, J.J.M.: Characterizing fault-tolerance of genetic algorithms in desktop grid systems. Evolutionary Computation in Combinatorial Optimization, pp. 131–142. Springer, Berlin (2010)
16. Koza, J.R.: Genetic Programming: On the Programming of Computers by Means of Natural Selection, vol. 1. MIT press, Cambridge (1992)
17. Alfaro-Cid, E., Merelo, J.J., de Vega, F.F., Esparcia-Alcázar, A.I., Sharman, K.: Bloat control operators and diversity in genetic programming: a comparative study. Evol. Comput. 18(2), 305–332 (2010)
18. Fernandez, F., Vanneschi, L., Tomassini, M.: The effect of plagues in genetic programming: a study of variable-size populations. Genetic Programming, pp. 317–326. Springer, Berlin (2003)
19. Laredo, J.J., Bouvry, P., González, D.L., de Vega, F.F., Arenas, M.G., Merelo, J.J., Fernandes, C.M.: Designing robust volunteer-based evolutionary algorithms. Genet. Program. Evolvable Mach. 15(3), 221–244 (2014)
20. Laredo, J.L.J., Eiben, A.E., van Steen, M., Castillo, P.A., Mora, A.M., Merelo, J.J.: P2P evolutionary algorithms: A suitable approach for tackling large instances in hard optimization problems. Euro-Par 2008-Parallel Processing, pp. 622–631. Springer, Berlin (2008)
21. Secretan, J., Beato, N., D Ambrosio, D.B., Rodriguez, A., Campbell, A., Stanley, K.O.: Picbreeder: evolving pictures collaboratively online. In Proceedings of the SIGCHI Conference on Human Factors in Computing Systems, ACM, pp. 1759–1768 (2008)
22. Frade, M., Fernandez de Vega, F., Cotta, C. (2012). Automatic evolution of programs for procedural generation of terrains for video games: accessibility and edge length constraints
23. de Fernandez Vega, F., Cruz, C., Navarro, L., Hernández, P., Gallego, T., Espada, L.: Unplugging evolutionary algorithms: an experiment on human-algorithmic creativity. Genet. Program. Evolvable Mach. 15(4), 379–402 (2014)
24. Fernendez de Vega, F., Navarro, L., Cruz, C., Chavez, F., Espada, L., Hernandez, P., Gallego, T.: Unplugging evolutionary algorithms: on the sources of novelty and creativity. In IEEE Congress on Evolutionary Computation (CEC), pp. 2856–2863 (2013)

Part II
Agent-Based Systems

Agents and Ontologies for a Smart Management of Heterogeneous Data: The IndianaMas System

Daniela Briola

Abstract The IndianaMas system is a platform for storing, retrieving and analyzing images, manual sketches and multilingual texts about rock carvings: to manage heterogeneous data formats, languages and data sources, it adopts a multiagent architecture that makes easier coping with such a mixed environment, and exploits ontologies to have a clear, formal and self contained representation of the domain. These two architectural solutions make possible to have a very modular organization of the overall system and to manage in parallel different kinds of data, which are exposed thanks to a digital library. In this paper we present the architecture and the functionalities of the IndianaMas system, focusing on the coordination level between the agents and on the organization of structured data in the digital library, to show how the adopted architectural approaches and solutions really allow the management of such a complex system.

1 Introduction and State of the Art

The IndianaMas project (funded by the Italian Ministry for Education, University and Research) is a technology platform based on intelligent software agents for the digital preservation of rock carvings, to support domain experts in the creation of a repository of images and multilanguage texts about rock art that will support them in their studies about this topic [9]. As testbed, we chose the rock art of Mont Bego, in France, as described in [2].

The agent technology suites naturally the IndianaMas architecture, considering the need of each component to operate in a highly autonomous way, interacting and coordinating with the other components to share information and to reason about them in the most effective and parallel way. To obtain a real modular architecture, and to let all the involved modules act in a coherent way, we adopt an ontology as

D. Briola (✉)
DIBRIS, Genoa University, Via Dodecaneso 35, Genoa, Italy
e-mail: daniela.briola@unige.it

© Springer International Publishing Switzerland 2016
P. Novais et al. (eds.), *Intelligent Distributed Computing IX*,
Studies in Computational Intelligence 616,
DOI 10.1007/978-3-319-25017-5_3

25

core part of the system, so that the domain and every information that must be shared is formalized and not integrated directly in the source code.

The aim of this paper is to describe in detail the system architecture, focusing on the components interaction and on some specific organizational decisions we made, which let us fulfill the complex goals of coordinating, integrating and automatically classifying heterogeneous data and data sources.

Since ontologies are a powerful means to formally describe complex domains and can be used to develop Semantic Web applications, they are already widely adopted in the area of Cultural Heritage [6], where often large datasets exist but are hardly accessible by the community. For example we can cite [1], where a service-oriented architecture that explicitly includes a semantic layer which provides primitive services to the applications built on top of a digital library is presented, including a module called PIRATES that assists end users to retrieval relevant contents, or [10] that proposes an approach to ontology development that is user-centered and designed to facilitate access to digital cultural heritage materials, or [11], which presents the "MultimediaN E-Culture" system, whose architecture is fully based on open Web standards and uses explicit background knowledge in the form of ontologies/vocabularies. Another very important initiative with the aim of creating semantic web applications for cultural heritage is the "Europeana" project.[1]

Similarly, ontologies are often exploited in multiagent systems (MAS for short): for example, [5] presents "SEMMAS", an ontology-based framework for seamlessly integrating Intelligent Agents and Semantic Web Services, [8] describes an agent based integration approach using ontologies for merging and combining an internal Data Warehouse with external data gained from competitors web sites and from other relevant Web sources, and [12] proposes a method based on agent and ontologies to design knowledge management systems: in this method, an ontology is used to represent the knowledge and the message content for the messages exchanged among agents.

Anyway, it is difficult to find in literature projects integrating both agents and ontologies, implemented and based on real datasets instead of case studies, managing images and multilingual texts, to be compared with the whole IndianaMas system. At the best of our knowledge, we are not aware of systems really similar to ours: from this point of view, considering its architecture, the offered services and the domain it refers, IndianaMas is a novelty.

The paper is organized in the following way: Sect. 2 provides the architecture of the system, Sect. 3 focuses on the functionalities offered by the system and on the coordination between agents, giving some details about the implementation too, and Sect. 4 concludes the paper.

---

[1]http://www.europeana.eu, accessed June 2015.

2 IndianaMas Architecture

The IndianaMas system (Fig. 1) exploits the MultiAgent paradigm, and is organized in an holonic structure [7]: it is based on three subMASs (holons), each devoted to specific operations on different types of input (images, sketches or texts), each having an agent acting as interface toward the rest of the MAS. Following this model, it was possible to divide the burden of performing ad hoc complex, and time consuming, operations among separated groups of agents (the holons), each devoted to the management of different data types. The coordination among holons is appointed to some specific agents, external to the subMASs, and hinges on a common ontology, as described later.

Our design and implementation phases follow the general ASPECS process methodology [3], since it is particularly suited for holonic MASs, but without strictly adhering to it: our "System Requirements" phase covers quite all the steps foreseen by this phase in the methodology, as happens for the "Implementation and Deployment" and the "Agent Society Definition" phases. However, we only adopted this high level organization of the overall work, but then we did not implemented the methodology using its UML diagrams or the ontology formalization as foreseen in its specification, since we already had an ontology as the core part of the system, with a strict structure: we anyway adopted standard tools, languages and formal-

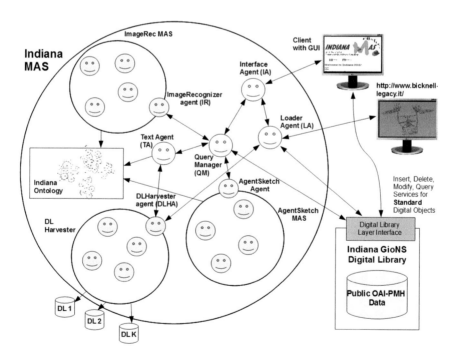

Fig. 1 The high level architecture of IndianaMas

ism (such as UML sequence diagrams, partially described in Sect. 3, OWL ontologies, described later on, Use Case descriptions, detailed in [9] and in the deliverables of the project), to exploit for each part of the system well known and accepted technologies.

The data managed by the system are stored in a digital library (DL) called *IndianaGioNS*, that makes them available to the community on the web as digital objects (DOs). Nowdays, another website, http://www.bicknell-legacy.it/, an outcoming of the IndianaMas project too, hosts some images from the *Bicknell legacy* (Bicknell is a famous archaeologist that spent many years on the Mount Bego at the beginning of 1900, drawing reliefs of petroglyphs), property of the Genoa university: the images and data on that website, plus many others still unpublished, will be inserted in *IndianaGioNS* too.

All the single agents and the subMASs act, interact and operate on the base of a shared ontology, the *Indiana Ontology*, so that every information regarding data inserted in the DL is coherent and semantically based on a common domain formalization. The ontology is a reference point for all the entities in the system, acting as a facilitator in the integration of data from different sources and of heterogeneous types.

The multiagent system itself is not reachable through the web, it is contacted locally using Java code, sending messages to the Interface Agent (IA), which is always up and running and manages the requests from the *IndianaGioNS* Web GUI to classify, with respect to the *Indiana Ontology*, images and texts, and to perform a similarity searches, as described in Sect. 3. The GUI autonomously manages the manual creation of new DOs.

For each new request of classification or search from the *IndianaGioNS* GUI, the IA starts a new Query Manager (QM) that manages the execution of the request and that sends the IA the result of the requested operation. Similarly, the QM starts new agents/subMASs, with respect to the type of request. The Loader Agent (LA) manages the requests to create DOs not coming from the GUI, or pro-actively manages the data collected by the DL Harvester subMAS. Later on in Sect. 3 we describe the logic process followed by IndianaMas to manage all these operations.

The ImageRecognizer subMAS (IR) is able to classify an input image (namely, to identify the carvings in the image) with respect to the *Indiana Ontology*, while the TextAgent (TA) classifies (namely, finds those carvings described/cited in the text) and extracts the Ages and Regions, again with respect to the ontology. The DL Harvester subMAS is in charge of autonomously searching the web to collect texts and metadata from other DLs: this subMAS retrieves information about external textual DOs (background operation, performed when IndianaMas is running) and stores them in a private folder that is then used by the LA to create new DOs, as described in Sect. 3.

Finally, we developed a middle layer to let the MAS interface with our DL *IndianaGioNS*, exploiting Rest calls: in this way the agents use only the ontology, ignoring the DL internal representation of DOs, and a clear separation between the MAS and the library is kept.

2.1 The Indiana Ontology

The *Indiana Ontology* results from the collaboration between computer scientists and archaeologists. The OWL^2 ontology is made up of three main parts: the first describes the classifications (related with petroglyphs shapes) and interpretations (related with petroglyphs meaning) of individual and aggregated petroglyphs, the second models the types of DOs managed by the IndianaMas framework, and the third models the actions that the agents can perform.

The *Indiana Ontology* has been presented in detail in [2], so in this paper we only give a brief description of it, reporting the details needed to understand the MAS, which is organized around the *Indiana Ontology*.

The petroglyphs (or carvings) can be classified into macro categories, which can be further refined into more detailed sub-categories.

The classes in this part of the ontology are used as "controlled vocabulary" for the classification of texts and images, as described in [2].

The IndianaMas framework manages two types of DOs: texts and images. Manual sketches are a specific type of images, as the Bicknell tracings. Texts may be multilingual and are classified according to the *Indiana Ontology*, whose concepts are expressed in English. DOs are represented as an OWL Class, with subclasses for Text and for the different image types, and have properties to store standard information like Title, Authors, Description and so on, and domain related information, like for example Geographical Area, Classification and Interpretation. Each DO has a list of physical files that will be loaded into the DL.

Lastly, we modeled in the ontology the actions that can be performed by the MAS: this part of the ontology, and its usage, is described in Sect. 3.

2.2 Mapping the Ontology into the Digital Library

Texts and Images managed by IndianaMas are stored in a DL called *IndianaGioNS*, which has been created and managed with *DSpace*.[3] As in any standard DL, each DO can be equipped with a set of metadata composed of couples (*term*, *value*). As described in [2], we mapped the OWL classes and properties to metadata and their values, exploiting the Dublin Core schema and a new ad hoc schema (called *IndianaMas*) for storing our domain related information, such as *classification* and *interpretation*.

For example, a DO can have one or more classifications represented by the OWL multiple property has_classifications. This property has been mapped into the metadata IndianaMas:classification, which can be added many times, one for each classification, and its value, corresponding to the instance of OWL class

[2] http://www.w3.org/TR/owl-features/.
[3] http://www.dspace.org/, accessed June 2015.

Calculated_Classification, will have the format: carving: *subclass of Indiana ontology* Classification; confidence: *Number*; source: *String*;

We use some specific metadata, from the schema *IndianaMas*, to describe the type of the DO, and we exploit the *dc.relation* metadata and the *IndianaMas.relation* metadata to represent the organization and the hierarchical structure of our DOs: in fact, above all for the Bicknell images (that are a set of manual tracings grouped into sheets in Rolls, each containing many carvings), we are managing figures containing many symbols, which sometimes can be grouped into scenes. We have to store different information if we are describing a single image, a scene or the overall image containing them all. Since a DL is inherently a flat structure, we had to find a way to represent this hierarchical structure, and we adopted the solution in Fig. 2:

- each image containing many symbols is mapped in a DO of type *MainDO*
- each symbol appearing in a *MainDO* is mapped in a new DO, of type *SubDO*, and has a metadata *dc.relation.ispartof* containing the identification Id of the *MainDO*. The *MainDO* has a metadata *dc.relation.haspart* for each *SubDO* it contains
- for each scene in the *MainDO*, a new DO of type *Scene* is created, with a metadata *IndianaMas.isScene* set to True and a metadata *IndianaMas.relation.hasScenePart* for each *SubDO* in the scene
- each *SubDO* in the scene has the metadata *IndianaMas.relation.isPartOfScene* containing the internal Id of the *Scene*
- The *Scene* has the metadata *dc.relation.ispartof* containing the *MainDO* Id, as the *MainDO* has a metadata *dc.relation.haspart* with the *Scene* Id.
- *MainDO*, *SubDO* and *Scene* types are modelled as metadata

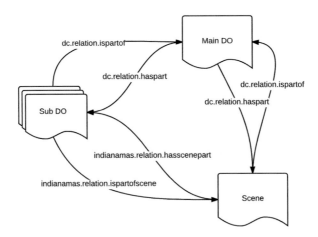

Fig. 2 Metadata used for representing the hierarchical structure of DOs

3 Functionalities

The IndianaMas system is a Mas developed with *JADE*[4] and consists of intelligent agents that perform many different activities: end users interact with the system through the *IndianaGioNS* DL, but the main operations over data are performed in background by agents. To contact the Mas from outside we exploit the *JadeGateway*[5] interface, so that we are able to directly send, and receive, messages to/from the Mas form the backend of *IndianaGioNS*.

The implementation of the *Indiana Ontology* done with *Protégé*[6] allowed us to automatically export the OWL ontology into a set of Java interfaces and Classes with a format that *JADE* is able to directly import and manage inside the Mas. The *Ontology BeanGenerator*[7] plugin generates a *Java* representation of the ontology respecting the *JADE* formalism, so that *JADE* agents can directly exchange messages based on the ontology.

To maintain a clear separation between the logical core of the application (the Mas) and the DL implemented with *DSpace*, we created a middle layer that offers to the world, and in our case to the Mas, a set of functions to interact with the DL: this layer wraps the available REST calls offered by *DSpace*. In this way, the Mas only interacts with the layer to insert, retrieve, update and search for DOs, represented using the *Java* representation mentioned above. Further details about this middle layer and the *Java* representation of the ontology can be found in [2].

We use the *Indiana Ontology* to model the agents' actions and their arguments: the results of these actions can only be concepts of the ontology mapped in *Java* or basic data types (int, String and so on). Figure 3 shows the portion of the *Indiana Ontology* representing the agents' actions with their input parameters. The actions results are not modeled in the Jade Agent_Action class (since we did not modify the Agent_Action class, we cannot model them explicitly): in our system, they are usually a list of DOs or a list of Calculated_Classifications (each containing a class from the Classification part of the ontology, a confidence and a source).

3.1 Agent Actions

The operations for classifying data and for searching for similar data in our DL are requested by the end user from *IndianaGioNS*. The MAS is always up and running, and the IA is always waiting for a new request, that is then forwarded to a new QM agent. In this way, each request is managed by a dedicated QM, which is in charge to create the agents needed to perform the operation: so, the QM will create a new

---

[4]http://jade.tilab.com/.

[5]http://jade.tilab.com/doc/api/jade/wrapper/gateway/JadeGateway.html.

[6]http://protege.stanford.edu/.

[7]http://protegewiki.stanford.edu/wiki/OntologyBeanGenerator.

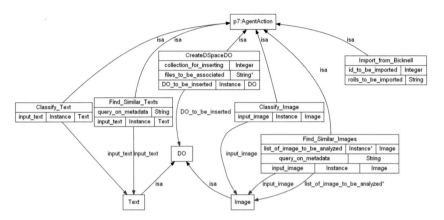

Fig. 3 The *Indiana Ontology* part representing agent actions. The name p7 is the name of an external ontology used by *BeanGenerator*, containing all the *JADE* classes needed for the ontology exportation

ImageRecognizer (IR) subMas for managing the operations regarding images and a new TextAgent (TA) for managing the operations regarding texts.

Classify Image/Text request. When a user is inserting a new Image DO in *IndianaGioNS*, he can ask the system for an automatic classification of the uploaded image. In this case, from the backend of the GUI, a request of type Classify_Image is sent to the IA in the MAS. The IA creates a new QM (named QM plus an internal identification number), specifying as input an object of type DO, which must be received in the previous step from the DL. The QM creates a new IR and forwards it this request: the IR processes the image found in the DO and answers with a list of objects of type Calculated_Classifications. The QM integrates these new metadata in the DO and forwards it to the IA that in turn forwards it to the DL backend, that will update the GUI.

If the request is Classify_Text, the protocol is the same but the QM creates a new TA and interacts with it instead of IR: in this case, the TA is able to automatically extract the Classifications (so returning a list of Calculated_Classifications) and Ages and Regions: these results are added by the QM as new metadata to the Text DO. This updated Text DO is sent back to the IA, as in the previous functionality.

Find Similar Images request. When a user is building an image similarity query (he is searching in the DL the Image DOs with an image similar to the one in input), the MAS is automatically involved. In this case, from the backend of the GUI, a Find_Similar_Images request is sent to the IA.

The IA creates a new agent of type QM, specifying as input a query string on the metadata, an object of type Image DO (both must be received in the previous step from *DSpace*), and an empty list of DOs. The QM creates a new IR and first of all asks it to classify the image in input (following the same steps described before). Then, exploiting the layer toward the DL that offers a search method, it requests the list of DOs with similar metadata and classifications (that is, with similar value in the

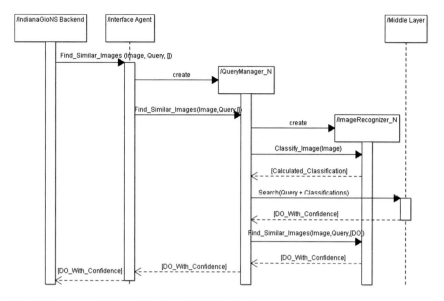

Fig. 4 The protocol followed for the task Find Similar Images

metadata *indianamas.classifications*, adding this filter to the query in input), and receives a list of DOs. Lastly, the QM asks the IR to perform the Find_Similar_ Images task, specifying as parameters the DO in input (the one with the image in input from the GUI) and the list of DOs just received from *DSpace*. IR makes a comparison between the image in input and all the images associated to the DO in the list, and returns an ordered list of DO_With_Confidence (similar DOs plus a confidence). This procedure lets the MAS avoid to compare the input image with all the images in *IndianaGioNS*, saving a lot of time for the user.

The QM forwards this list to the IA that in turn forwards it to the digital library backend, which will update the GUI. Figure 4 describes this flow.

Find Similar Texts request. Similarly, for the text similarity query (searching in the DL the Text DOs with a text similar to the one in input) a Find_Similar_Texts request is sent to the IA, which creates a new agent of type QM to manage it. The QM creates a new TA and first of all asks it to classify the text in input (following the same steps described before). Then, exploiting the layer toward the DL, it requests the list of DOs with similar metadata and classifications and receives a list of DO_With_Confidence.

When searching for similar texts, we are considering texts with similar contents (considering only metadata), not similar phrases, so in this case there is no need to contact again the TA since we use the DL for executing this search. The QM forwards this result to the IA that in turn forwards it back.

The CreateDSpaceDO and Import_From_Bicknell actions. Actually there are two ways to insert new DOs in *IndianaGioNS*: manually, thanks to the GUI interface (procedure that does not involve the MAS), or automatically, thanks to the Mas exploiting the action CreateDSpaceDO, which takes in input a DO and the collection where it must be inserted.

As said before, Genoa university owns the original Bicknell tracings, some of which are actually been published on the website http://www.bicknell-legacy.it/: the images and data presented on that site are being inserted in *IndianaGioNS* too, and the action Import_From_Bicknell was created to perform this task. The Mas is able to automatically transform the information stored in the web site, saved in a relational database, into a set of metadata coherent with the design of our DL. In this way, the Mas is actually a modular and distributed tool able to integrate and transform heterogeneous data stored in different repositories, with different schemas, in a transparent way to the user, which simply must specify which data has to be loaded.

The Loader Agent (LA) performs this operation: these images have been already classified before their publication, so there is no need to involve the IR to classify them. The import_From_Bicknell Action can have as parameter the identification number of the Roll to be imported (Bicknell tracings are a set of manual tracing on sheets, grouped into Rolls), or the single identification number of a tracing. The LA connects to the database and collects the information about the requested tracing or sheets in Roll. For each sheet, using the CreateDSpaceDO action, it creates a *MainDO* with the information, mapped into metadata, regarding it. Then, for each tracing in the sheet, it creates a new *SubDO* with its specific metadata, and if needed it groups them into *Scene*, creating the hierarchical structure described in Sect. 2.2.

The second way to automatically create a new DO in our DL is performed again by the LA, which in this case operates autonomously and pro-actively on the data collected by the DL Harvester subMas. The DL Harvester subMas is in charge of searching the web to collect information from other DLs: we assume to insert in *IndianaGioNS* only references to other already existing textual DOs, so this subMas retrieves information about external DOs (in background, when IndianaMas is idle) and stores them in a private folder. Then the LA autonomously picks up these data (saved in textual files, one for each DO), asks the TA to classify them and finally inserts new DOs in *IndianaGioNS* (performing the action CreateDSpaceDO), which shows the metadata retrieved from the web and a link to the original external DOs.

4 Conclusion

In this paper we described the functionalities of the IndianaMas system, focusing on the interactions among the intelligent agents performing the operations of data classification and management. Our system adopts a multiagent architecture to maximize the performances of the system, managing many concurrent requests, optimizing the agents allocation and speeding up the operations thanks to the agents creation (and destruction) on the fly.

The adoption of an ontology as the core of the system lets us integrate many different data with different information and structure and coming from many sources: this ontology-centric architecture helps to formalize the domain, to semantically describe our data and to coordinate the independent entities involved in the system. Lastly, the choice of a digital library as final data storage lets us present our data, and their related information, in a standard way, simplifying the sharing of the data stored in *IndianaGioNS*.

These features contribute to obtain a highly modular system, parametric on the ontology and able to manage heterogeneous data sources in a fast and clear way, answering many requests concurrently.

The system is implemented and we are only fine-tuning it, creating an appealing GUI, configuring the IR and TA to run as fast as possible testing them with domain related data, and we are inserting the thousands of Bicknell tracings. We are still searching for domain related DLs so that the LA can do its job: in the meanwhile, we are completing the DL Harvester subMAS. As soon as a considerable set of images and texts will be inserted in *IndianaGioNS* (in the next months), we will make the system available to the web community. Evaluating the system performances, some tests and results of IR and AgentSketch Mas have been described in [4, 9], and the TA is able to classify a text in few seconds (results to be published). Tests of the overall system are under execution, but from the first experiments the classification and searching operations are completed in less then 3 seconds.

References

1. Baruzzo, A., Casoto, P., Challapalli, P., Dattolo, A., Pudota, N., and Tasso, C.: Toward semantic digital libraries: exploiting web 2.0 and semantic services in cultural heritage. J. Digit. Inf. **10**(6) (2009)
2. Briola, D., Deufemia, V., Mascardi, V., Paolino, L., Bianchi, N.: Ontology-driven processing and management of digital rock art objects in indianamas. In: Proceedings of EuroMed 2014, LNCS, pp. 217–227. Springer (2014)
3. Cossentino, M., Gaud, N., Hilaire, V., Galland, S., Koukam, A.: ASPECS: an agent-oriented software process for engineering complex systems. Auton. Agent. Multi-Agent Syst. **20**(2), 260–304 (2010)
4. Deufemia, V., Paolino, L., de Lumley, H.: Petroglyph classification using the Image distortion model. In: Arnold, D., Kaminski, J., Niccolucci, F., Stork, A. (eds) VAST: International Symposium on Virtual Reality, Archaeology and Intelligent Cultural Heritage (2012)
5. García-Sánchez, F., Valencia-García, R., Martínez-Béjar, R., Fernández-Breis, J.T.: An ontology, intelligent agent-based framework for the provision of semantic web services. Expert Syst. Appl. **36**(2), 3167–3187 (2009)
6. Hendler, J., Hyvnen, E., Heath, T., Bizer, C.: Publishing and using cultural heritage linked data on the semantic web (2012)
7. Horling, Bryan, Lesser, Victor: A survey of Multi-Agent organizational paradigms. Knowl. Eng. Revl. **19**(4), 281–316 (2005)
8. Lavbič, D., Vasilecas, O., Rupnik, R.: Ontology-based Multi-Agent System to support business users and management. Technol. Econ. Dev. Econ. **16**(2), 327–347 (2010)
9. Mascardi, V., Briola, D., Locoro, A., Deufemia, V., Paolino, L., Bianchi, N., de Lumley, H., Grignani, D., Malafronte, D., Ricciarelli, A.: A holonic multi-agent system for sketch, image and text interpretation in the rock art domain. IJICIC **10**(1), 81–99 (2014)
10. Pattuelli, M.C.: Modeling a domain ontology for cultural heritage resources: a user-centered approach. J. Am. Soc. Inf. Sci. Technol. **62**(2), 314–342 (2011)

11. Schreiber, G., Amin, A., Aroyo, L., van Assem, M., de Boer, V., Hardman, L., et al.: Semantic annotation and search of cultural-heritage collections: the multimedian e-culture demonstrator. Web Semant. **6**(4), 243–249 (2008)
12. Yue, X., Song, M., Xin, Z.H.: An architecture of knowledge management system based on agent and ontology. J. China Univ. Posts Telecommun. **15**(4), 126–130 (2008)

Scrutable Multi-agent Hazard Rescue System

Andrei Mocanu and Costin Bădică

Abstract Time-efficient and competent intervention in environmental hazard situations is critical. Numerous experts on the field have to be coordinated and the most appropriate one(s) should be sent to the danger zone. A traditional system would require a lot of effort on the part of the people involved, but would also involve the risk of not taking the best decisions under pressure. We propose an automated system that incorporates multiple autonomous agents, negotiation and argumentation techniques that is not only capable of making rational selections, but which can also provide justifications for the decisions it takes. Making transparent decisions is fundamental for gaining the people's trust in this solution, but can also be used for learning purposes and for fine-tuning parameters. The system, which involves a command center and a dedicated smartphone app for the field experts, has been tested in a simulated experimental scenario around the area of Craiova, Romania.

Keywords Multi-agent system · Assumption based argumentation · Decision frameworks · JADE · Contract net protocol · Location-aware system · Android

1 Introduction

In a typical environmental hazard system the Dispatchers from a Command Center coordinate the Environmental Experts by communicating over the phone or through radio transmitters. As this can waste valuable time and is often difficult to maintain, we have developed a system [10] based on FIPA[1]-compliant [8] agent communication

---

[1]FIPA Specifications available at http://www.fipa.org.

A. Mocanu (✉) · C. Bădică
University of Craiova, Craiova, Romania
e-mail: mocanu.andrei@ucv.ro

C. Bădică
e-mail: badica_costin@software.ucv.ro

© Springer International Publishing Switzerland 2016
P. Novais et al. (eds.), *Intelligent Distributed Computing IX*,
Studies in Computational Intelligence 616,
DOI 10.1007/978-3-319-25017-5_4

[14], the Android mobile operating system and the Contract Net Protocol (CNP) [12] in order to effortlessly select the most appropriate Environmental Expert for the task according to its skill set and location.

A fundamental direction that we wanted to explore was to make rational decisions that the system is able to justify in real time. These justifications can be then inspected by humans who can critique the way the decisions are made and suggest alternative actions, thus contributing to the overall quality of the system. A self-documenting system can also be of great help to authorities in charge of monitoring it, so rather than inspecting cold figures, they would be looking at intuitive graphs and possibly natural language explanations. We are also confident that designing transparent software will gain the users' trust at a faster pace and will lead to a better adoption rate compared to traditional software.

2 Related Work

There has been a wide research interest in the area of crisis management. The ALADDIN project [1] is a multidisciplinary project concerned with developing techniques and architectures for multi-agent systems in uncertain and dynamic conditions. While the scope of their work is vast, it differs from ours as they consider agents that are able to perform rescue and evacuation missions themselves, while we consider software agents that simply act on behalf of human field experts by participating in negotiation protocols and performing tasks such as logging messages from the Command Center or fetching directions to the danger zone. The authors also explore decision making for agents, but they use other techniques such as Reinforcement Learning, and do not explore the possibility of agents giving explanations for their actions, which is what we aim to achieve.

In [2], the authors explore dynamic negotiation protocol selection and self-configuration in the context of disaster management. Although this work is similar in trying to find the appropriate measure imposed by a critical event, it is not concerned with designing and implementing a hazard management system.

The authors of [4] describe a method for making autonomous systems transparent which involves argumentation and natural language generation. The authors do not apply their research in the field of disaster management.

Finally, argumentation has been applied to numerous application domains including medicine [6, 11], law [16] and social networks [15], but to the best of our knowledge it has not been applied in hazard management.

3 Background

Making transparent decisions relies on Assumption based Argumentation (ABA) [5] and decision frameworks with ABA.

Decision frameworks [6] are tuples $<D, A, G, T_{DA}, T_{GA}, P^S>$:

- a (finite) set of decisions $D = \{d_1,..., d_n\}$, $n > 0$;
- a (finite) set of attributes $A = \{a_1,..., a_m\}$, $m > 0$;
- a (finite) set of goals $G = \{g_1,...,g_l\}$, $l > 0$;
- a partial order over sets of goals P^S representing the preference ranking of sets of goals;
- two tables: T_{DA} (size n × m), and T_{GA} (size l × m);

We will use the notation $T[x, y]$ for a cell in row x and column y in table T.

– For all $d_i \in D$, $a_j \in A$, $T_{DA}[d_i, a_j]$ is either:

  1, meaning d_i has attribute a_j;
  0, meaning d_i does not have attribute a_j;
  u, if the information is unknown;

– For all $g_k \in G$, $a_j \in A$, $T_{GA}[g_k, a_j]$ is either:

  1, meaning g_k is satisfied by attribute a_j;
  0, meaning g_k is not satisfied by attribute a_j;
  u, if the information is unknown;

Given a decision framework $DF = <D,A,G,T_{DA},T_{GA}>$, a decision $d_i \in D$ meets a goal $g_k \in G$ wrt DF, iff there exists an attribute $a_k \in A$, such that $T_{DA}[d_i,a_k] = 1$ and $T_{GA}[g_j,a_k] = 1$. $\gamma(d) = S$, where $d \in D$, $S \not\subseteq G$, denotes the *set of goals met by d*.

Given a decision framework $edf = <D, A, G, T_{DA}, T_{GA}, P^S>$ the *most preferred set* decisions [7] are the decisions meeting the more preferred sets of goals that no other decisions meet, formally defined as follows. For every $d \in D$, d is the most preferred set decision iff the following holds true for all $d' \in D\backslash\{d'\}$: for all $s \in S$, if $s \not\subseteq \gamma(d')$, then there exists $s' \in S$ such that:

– $s' > s \in P^S$;
– $s \not\subseteq \gamma(d)$;
– $s' \not\subseteq \gamma(d')$;

Assumption-based Argumentation (ABA) *frameworks* are tuples $<L, R, A, C>$:

- $<L, R>$ is a deductive system, with L the *language* and R a set of *rules* of the form $\beta_0 \leftarrow \beta_1, \ldots, \beta_m (m \geq 0, \beta_i \in L)$;
- $A \subseteq L$ is a (non-empty) set, referred to as *assumptions*;
- C is a total mapping from A into $2^L - \{\{\}\}$, where each $\beta \in C(\alpha)$ is a *contrary* of α, for $\alpha \in A$.

Given a rule ρ of the form $\beta_0 \leftarrow \beta_1, \ldots, \beta_m$, β_0 is referred to as the *head* (denoted $Head(\rho) = \beta_0$) and $\beta_1, \ldots, \beta_m$ as the *body* (denoted $Body(\rho) = \{\beta_1, \ldots, \beta_m\}$). We focus on *flat* ABA frameworks, with no assumption is the head of a rule.

In ABA, *arguments* are deductions of claims using rules and supported by sets of assumptions, and *attacks* are directed at the assumptions in the support of arguments. Informally, following [5]:

- *an argument for (the claim) $\beta \in L$ supported by $A \subseteq A$ ($A \vdash \beta$ in short) is a* (finite) tree with nodes labeled by sentences in L or by τ,[2] the root labeled by β, leaves either τ or assumptions in A, and non-leaves β' with, as children, the elements of the body of some rule with head β';
- *an argument $A_1 \vdash \beta_1$ attacks an argument $A_2 \vdash \beta_2$ iff β_1 is a contrary of one of* the assumptions in A_2.

Attacks between (sets of) arguments in ABA correspond to attacks between sets of assumptions, where *a set of assumptions A attacks a set of assumptions A' iff an* argument supported by a subset of A attacks an argument supported by a subset of A'.

With argument and attack defined for a given $F = <L, R, A, C>$, standard argumentation semantics can be applied in ABA [5], e.g.: *a set of assumptions is admissible (in F) iff it does not attack itself and it attacks all $A \subseteq A$ that attack it; an argument $A \vdash \beta$ is admissible (in F) supported by $A' \subseteq A$ iff $A \subseteq A'$ and A' is admissible (in F); a sentence is admissible (in F) iff it is the claim of an argument* that is admissible supported (in F) by some $A \subseteq A$.

As shown in [6, 7, 9], ABA can be used to model decision making problems and compute "good" decisions. The ABA framework for computing the *most preferred set decisions* in a decision framework is defined as $AF = <L, R, A, C>$ for which:

- R is such that:

  for all $d_k \in D$, $isD(d_k)$;
  for all $g_j \in G$, $isG(g_j)$;
  for all $a_i \in A$, $isA(a_i)$;
  for all $g_j \in G$, $s_t \in S$,

  if $g_j \in s_t$ then $inSet(g_j, s_t)$;
  for all $s_t, s_r \in S$,

  if $s_t > s_r \in P^S$, then $prefer(g_j, s_t)$;
  for $k = 1,..,n$; $j = 1,..,m$

---

[2]$\tau \notin L$ represents "true" and stands for the empty body of rules.

if $T_{DA}[k, i] = 1$ then $hasAttr(d_k, a_i)$;

for $j = 1,..m; i = 1,..,l$

if $T_{GA}[j,i]=1$ then $satBy(g_j, a_i)$;

$met(X,Y)$ $hasAttr(X,Z), satBy(Y,Z), isD(X), isG(Y), isA(Z);$
$notMetSet(X,S)$ $notMet(X,Y), inSet(Y,S);$
$notSel(X)$ $metSet(X_1,S), notMetSet(X,S), notMetBetter(X,X_1,S);$
$metBetter(X,X_1,S)$ $metSet(X,S_1), notMetSet(X_1,S_1), prefer(S_1,S);$

- A is such that:

for all $d_k \in D$, $sel(d_k);$
for all $d_k \in D$ and $s_t \in S$, $metSet(d_k,s_t);$
for all $d_k \in D$ and $g_j \in G$, $notMet(d_k,g_j);$
for all $d_k, d_r \in D$, $d_k \neq d_r$ and $g_j \in G$, $notMetBetter(d_k, d_r, g_j);$

- C is such that:

$C(sel(X))=\{notSel(X)\};$
$C(notMet(X,Y))=\{met(X,Y)\};$
$C(metSet(X,S))=\{notMetSet(X,S)\};$
$C(notMetBetter(X,X1,S))=\{metBetter(X,X1,S)\}.$

4 System Design

Our initial system [10] in Fig. 1 comprises agent and non-agent software components. Agent components are marked with the letter 'A' on this figure. At the center of our architecture lie the servers which connect to the database management system, in our case MySQL. The servers are also configured to run the JADE platform [3] with the Main and Auxiliary Containers hosting a number of agents, being distributed on the available computing nodes.

The Dispatcher agent is responsible for triggering the negotiation in order to select the best Environmental Expert for completing the task. The Environmental Expert agents connected to JADE's Containers can be in one of two interchangeable states: Fixed (when the Expert user is at his/her workstation) and Mobile (when the Expert user is on the field staying connected through his/her Android enabled device). The latter enables the participation of mobile skilled people in the experts' selection process facilitated by automated negotiation protocols. Finally, we use Google Maps and Google Directions API for map visualization and navigational capabilities.

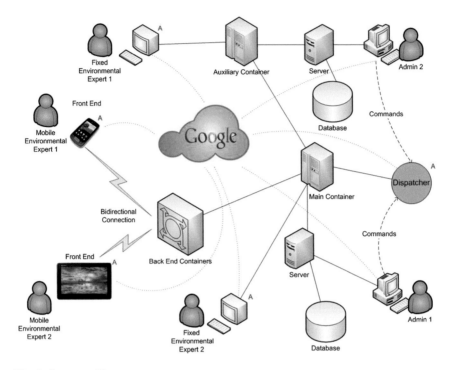

Fig. 1 System architecture

4.1 Utility Function Decision

In our original system [10] the Dispatcher agent uses the utility function u: X→ [0, 1] defined by Eq. 1 to evaluate the offer x of each Expert agent and to select the most appropriate one(s) that provides him/her maximum utility:

$$
u(x) = \begin{cases} \dfrac{\sum\limits_{i=1}^{6} \min(x_i^2, h_i^2)}{\sum\limits_{i=1}^{6} h_i^2} \cdot w + \dfrac{d}{x_{time}} \cdot (1-w), & \exists h_i > 0 \\[4mm] w + \dfrac{d}{x_{time}} \cdot (1-w), & otherwise \end{cases} \tag{1}
$$

Parameter d has a calibration role and its value is dynamically determined by the Dispatcher based on the offers received from the Experts. If x_{time}^a is the value provided by Expert agent a and if A is the set of Expert agents that submitted bids during a certain negotiation then the Dispatcher computes d as: $d = \min_{a \in A} x_{time}^a$. Parameter w represents the weight of the expertise component and was set to 0.5 in our scenario.

Another important mathematical equation, the haversine formula (Eq. 2, [13]), allowed us to calculate the distance between two points expressed by their geographical coordinates. The haversine formula, which is quite popular in navigation, returns the flight distance between two points.

$$d = 2r \cdot \ \arcsin(\sqrt{\sin^2(\frac{\alpha_2 - \alpha_1}{2}) + \cos(\beta_1)\cos(\beta_2)\sin^2(\frac{\beta_2 - \beta_1}{2})}) \qquad (2)$$

Replacing r in the formula with the Earth radius which is 6378 km we can successfully calculate the flight distance between two geographical points expressed in latitude and longitude.

4.2   Assumption Based Argumentation Decision Making

Although selecting the best Expert agent for the job based solely on utility is quite straightforward and quick it would be useful to explain the decisions taken by our hazard management system.

We are able to justify the selections made by the Dispatcher agent by converting our initial problem into an ABA framework to which we can apply decision making semantics, more specifically reasoning about preferences over sets of goals.

The potential decisions in our framework are going to be the Expert agents that responded to the initial CFP message. We discretize both the response times and levels of expertise of the Expert agents into a fixed number of attributes respectively, in our scenario we have chosen three. Thus, the Expert agent with the fastest response time along with other agents whose response times relative to the former's obtain a score of at least 70 % gain the attribute *timeOver70*. In the same manner, the agents who have a relative score between 40% and 70% get the attribute *timeBetween40And70*, while agents who score below 40 % get the attribute *timeBelow40*. Analogously, the agents with the best expertise level for the task along with those with a relative score of at least 70 % gain the attribute *expertiseOver70*, those with a relative score between 40 and 70% gain the attribute *expertiseBetween40And70*, while agents who have a score less than 40 % get the attribute *expertiseBelow40*.

The goals that the agents need to satisfy are represented by different levels of response times and expertise scores. For our scenario we have chosen the same number of corresponding goals for the previously defined attributes. We introduce the goals: *bestTime, goodTime, badTime, bestExpertise, goodExpertise, badExpertise* and their relation to the attributes is given by the following empty rules which are going to appear in all generated ABA frameworks corresponding to each instance of the Contract Net Protocol.

Finally, we need to introduce the sets of goals and the preference relationships between them. For generating the sets of goals one possibility is to do a Cartesian product between the set of goals referring to response time and the set of goals referring to expertise levels. In this way, for our scenario we generate nine distinct sets of goals named: *bestTimeBestExpertise, bestTimeGoodExpertise, bestTimeBadExpertise,*

Fig. 2 Goal sets preference
ranking

*goodTimeBestExpertise, goodTimeGoodExpertise, goodTimeBadExpertise, bad-
TimeBestExpertise, badTimeGoodExpertise, badTimeBadExpertise.*

In our setup, we wanted to make getting to the hazard site slightly more important
than the expertise level of the Expert agent involved. This lead to the preference
relationships: *bestTimeBestExpertise > bestTimeGoodExpertise > goodTimeBestExpertise > goodTimeGoodExpertise > bestTimeBadExpertise > goodTimeBadExpertise > badTimeBestExpertise > badTimeGoodExpertise > badTimeBadExpertise* as
can be seen graphically in Fig. 2.

In an ideal case we would like the Dispatcher agent to select an Expert agent that
satisfies the set of goals *bestTimebestExpertise*, but should this not be possible it will try
to select an agent that satisfies the next best set of goals *bestTimeGoodExpertise*, etc.

For a simplified scenario having just two possible choices of Expert agents
(*agent0* and *agent1*) we generate an ABA framework in order to select the most
appropriate agent. To aid understanding, in this sample run agent0 has the quickest
response time of 3 min, while *agent1* has a response time of 5 min. This means
agent0 will have the *timeOver70* attribute, while *agent1* will have a relative time
score of 3/5 = 60% meaning that it will gain the *timeBetween40And70* attribute. In
the same scenario, *agent0* will have an expertise level of 0.6 while *agent1* will have
an expertise level of 0.9, thus *agent0* will have the *expertiseBetween40And70*
attribute while *agent1* will have the *expertiseAbove70* attribute. There is no agent
which can simultaneously satisfy goals *bestTime* and *bestExpertise*, thus satisfying
the set of goals *bestTimebestExpertise,* and thus the Dispatcher needs to look for a
lower ranked set of goals in the preference list. The second most preferred set of
goals *bestTimeGoodExpertise* is satisfied by *agent0* who meets both the goals
bestTime and *goodExpertise*. By computing admissible solutions for the generated
ABA framework using the Proxdd solver we are able to produce the dispute derivation graph which is partially illustrated in Fig. 3. By inspecting it we can see that
the claim *sel(agent0)* successfully defends against all attacks (e.g. the attack stating
that *agent1* meets the set of goals *bestTimebestExpertise* that is a more preferred set
of goals than *bestTimeGoodExpertise* is refuted by the proponent who points out
that *agent1* does not meet that set of goals because it does not meet the goal
bestTime contained in that set).

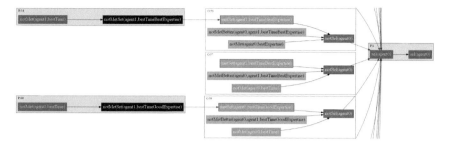

Fig. 3 Dispute derivation graph

5 Experiments

Since not only the number of Expert agents in the system is important, but also the geographical area that they cover, we can deduce that a useful measure of the system is best expressed by the number of Expert agents per square kilometer (density) that can be efficiently handled by the system. In our experiment we considered a 400 square kilometer area centered in Craiova, Dolj, Romania (10 km in any direction) and we constrained the Expert agents to this area.

For testing our argumentation based decision making discretization, we wanted to discover how the attributes of agents will be distributed for the three different densities of experts we had in our initial experiment. We ran a number of 100 iterations for 10 agents on the field, 20 iterations for 50 agents on the field and 10 iterations for 100 agents on the field, thus yielding 1000 separate computations for time and expertise for each of the three scenarios.

The outcome of these measurements is detailed in Fig. 4. As can be seen from the graph, the number of agents that satisfy the *bestTime* goal is always a minority and the phenomenon tends to be more pronounced as the density of agents increases. With the exception of the reduced density case where the numbers tend to be very close, the number of agents that satisfy the *goodTime* goal is 3–4 times greater than the number of agents that satisfy the *bestTime* goal. Over 65 % of

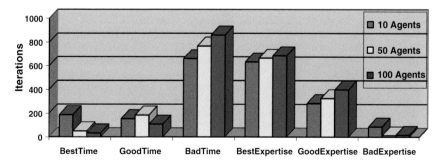

Fig. 4 Distribution of time and expertise attributes

agents fall into the *badTime* category in all 3 scenarios, with an 86 % figure in the last scenario. From this we can deduce that having an efficient relative response time is difficult to attain.

The same cannot be said entirely about the expertise level of the agents. Although the expertise levels of the agents and of the hazards are randomly generated in this scenario, by examining the graph we observe that more agents tend to sit in the *bestExpertise* category, followed by the *goodExpertise* category. The *badExpertise* category is represented by a small minority of agents.

6   Conclusions

In this paper we have described an autonomous system that involves Android mobile users in negotiation protocols for task assignment in environmental hazard scenarios. The interchangeable agent state of the Environmental Experts, switching seamlessly from mobile to fixed mode, translates to permanent connectivity and a higher degree of flexibility for the system users.

Furthermore, we have equipped our system with the ability to make transparent decisions using decision frameworks relying on assumption based argumentation. To the best of our knowledge, this is the first time that argumentation makes its way into a system of such complexity designed for environmental hazard management. Such a scrutable system manifests several benefits such as self-documentation, easy introspection, improved maintainability and higher trustworthiness. Simulation runs have been conducted around the area of Craiova, Romania, producing encouraging results.

Acknowledgement   This work was supported by the strategic grant POSDRU/ 159/1.5/S/133255, Project ID 133255 (2014), co-financed by the European Social Fund within the Sectorial Operational Program Human Resources Development 2007 - 2013.

References

 1. Adams, N.M., Field, M., Gelenbe, E., Hand, D.J., Jennings, N.R., Leslie, D.S., et al.: Intelligent agents for disaster management. In Proceedings of the IARP/EURON Workshop on Robotics for Risky Interventions and Environmental Surveillance (RISE) (2008)
 2. Bădică, A., Bădică, C., Ilie, S., Muscar, A., Scafeș, M.: Dynamic selection of negotiation protocol in multi-agent systems for disaster management. In: Computational Collective Intelligence. Technologies and Applications, vol. 6923. Springer, Berlin (2011). ISBN: 978-3-642-23937-3
 3. Bellifemine, F., Caire, G., Greenwood, D.: Developing multi-agent systems with JADE. John Wiley & Sons (2007). ISBN: 0470057475
 4. Caminada, M.W.A., Kutlak, R., Oren, N., Vasconcelos, W.W.: Scrutable plan enactment via argumentation and natural language generation. In: Proceedings of the 13th International

Conference on Autonomous Agents and Multiagent Systems (AAMAS 2014), Paris, France (2014)
5. Dung, P.M., Kowalski, R.A., Toni, F.: Assumption-based argumentation. In: Argumentation in Artificial Intelligence, pp. 199–218. Springer (2009)
6. Fan, X., Craven, R., Singer, R., Toni, F., Williams, M.: Assumption-based argumentation for decision-making with preferences: a medical case study. In: Proceedings of CLIMA (2013)
7. Fan, X., Toni, F.: Decision making with assumption-based argumentation. In: Second International Workshop on the Theory and Applications of Formal Argumentation (TAFA 2013), Beijing, China, 3–5 Aug 2013
8. Gotta, D., Trucco, T., Ughetti, M., Semeria, S., Cucè, C., Porcino, A.: Jade Android Add-on Guide (2011). http://jade.tilab.com/doc/tutorials/JADE_ANDROID_Guide.pdf
9. Matt, P.A., Toni, F., Vaccari, J.: Dominant decisions by argumentation agents. In: Proceedings of ArgMAS. Springer (2009)
10. Mocanu, A., Ilie, S., Badica, C., Ubiquitous multi-agent environmental hazard management. In: 14th International Symposium on Symbolic and Numeric Algorithms for Scientific Computing (SYNASC), pp. 513–521. 26–29 Sept 2012
11. Mocanu, A, Fan, X., Toni, F., Williams, M., Chen, J.: RecoMedic: recommending medical literature through argumentation. In: 4th International Workshop on Combinations of Intelligent Methods and Applications (CIMA 2014). Limassol, Cyprus, 10–11 Nov 2014
12. Smith, R.G.: The contract net protocol: high-level communication and control in a distributed problem solver. In: IEEE Trans. Comput. **C-29**(12), 1104–1113
13. Weisstein, E.W.: "Haversine" from mathworld—a wolfram web resource. http://mathworld.wolfram.com/Haversine.html
14. Wooldridge, M.: An Introduction to Multiagent Systems. Wiley, Chichester (2002)
15. Yaglikc, N., Torroni, P., Microdebates App for Android: a Tool for participating in argumentative online debates using a handheld device. In: IEEE 26th International Conference on Tools with Artificial Intelligence (ICTAI). 10–12 Nov 2014
16. Zhong, Q., Fan, X., Toni, F., Luo, X. Explaining best decisions via argumentation. In: European Conference on Social Intelligence (ECSI 2014). Barcelona, Spain, 3–5 Nov 2014

From Virtual to Real, Human Interaction as a Validation Process for IVEs

J.A. Rincon, Emilia Garcia, V. Julian and C. Carrascosa

Abstract This paper presents a development process for intelligent virtual environments (IVEs) considering the immersion of real objects and human beings in virtual simulations. This development process includes a meta-model for designing, a set of tools to facilitate the implementation and a simulation platform. The paper presents a case study that makes use of this development process for designing adapted spaces for people with physical disabilities.

1 Introduction

An Intelligent Virtual Environment (IVE) is a virtual environment that simulates a physical world inhabited by autonomous intelligent entities [1]. Today, this kind of applications are between the most demanded ones, not only for being the key for multi-user games such as *World Of Warcraft*[1] (with more than 7 million of users in

---

[1] http://eu.battle.net/wow.

J.A. Rincon (✉) · E. Garcia · V. Julian · C. Carrascosa
Departamento de Sistemas Informáticos y Computación (DSIC),
Universitat Politècnica de València, Camino de Vera s/n, Valencia, Spain
e-mail: jrincon@dsic.upv.es

E. Garcia
e-mail: mgarcia@dsic.upv.es

V. Julian
e-mail: vinglada@dsic.upv.es

C. Carrascosa
e-mail: carrasco@dsic.upv.es

[1] http://eu.battle.net/wow.

© Springer International Publishing Switzerland 2016
P. Novais et al. (eds.), *Intelligent Distributed Computing IX*,
Studies in Computational Intelligence 616,
DOI 10.1007/978-3-319-25017-5_5

49

2013)[2] but also for immersive social networks such as *Second Life*[3] (with 36 million accounts created in its 10 years of history).[4] These kinds of IVEs are addressed to a huge number of simultaneous entities (human or not), so they must be supported by highly scalable software. This software has also to be able to adapt to changes, and to incorporate human requirements and needs. Technology, currently used to develop this kind of products, lacks of elements that facilitate the self-adaptation of the system and the immersion of human entities. Traditionally, this kind of applications use the client/server paradigm. However, due to their features, a distributed approach such as multi-agent systems (MAS) seems to fit better. Its allow developing components that will evolve in an autonomous way and coordinated with the own environment's evolution. Current approaches that use Multi Agent Systems as a paradigm for developing IVEs have some unresolved issues, among which stand out: low generality and then reusability; weak support for handling full open and dynamic environments. This work tackles these issues by developing a methodology and a framework for the development and simulation of IVEs. Besides, this paper also tackles the issue of immersing human beings into IVE. When a human being is completely immersed into a system of this kind, he/she can interact with the system in a natural way. Moreover, agents immersed in the system can learn about human actions adapting its behaviors and taking decisions about future situations. Examples of these systems can be domotic scenarios, production lines in an industry, entertainment industry, ...In order to achieve this kind of immersion, an IVE development process has been specified over the *JaCalIVE* framework [2], specially designed for the execution and adaptation of IVEs, allowing an easy integration of human beings in the MAS. This framework can be downloaded from this url: http://jacalive.gti-ia.dsic.upv.es.

We also present a case study that makes use of this development process and it is oriented for the design of adapted spaces for people with physical disabilities. In this kind of problems, designs must take into account the special need of each position in order to allow people to move freely through the environment. Designers need to find a balance in which both people with or without disabilities can coexist in the same environment. Typically, problems of this kind can be solved using simulation toolkits. These tools allow designers, engineers and other professional to get a preview of their design before it is built. The main challenge of these tools is creating simulation environments which can be achieved by obtaining a high level of immersion by human being and it is in this type of problems where the use of tools such as the proposed in this work can play an important role. Our proposal immersed human beings into the human immersion within the virtual environment and thus create suitable environments for anyone (with or without physical disabilities). The pro-

---

[2]http://www.statista.com/statistics/276601/number-of-world-of-warcraft-subscribers-by-quarter/

[3]http://www.secondlife.com.

[4]http://massively.joystiq.com/2013/06/20/second-life-readies-for-10th-anniversary-celebrates-a-million-a/

posed approach allows the developer to observe how are the possible reactions to his design, being contrasted with a human who interacts within the virtual environment. The rest of the document is structured as follows: Sect. 2 presents our proposed IVE development process based on the *JaCalIVE* framework. Section 3 presents the case study commented above. Final Sect. 4 presents the main conclusions of this paper and our current and future work.

2 IVE Development Process

In this section we present an extension to the IVE development process presented in [2]. This process was devised according to the MAM5 meta-model [3] that defines an IVE in terms of Agents and Artifacts. Over the last few years, simulation toolkits have acquired a high importance as a validation toolkit for designers, engineers or any other professional to obtain a prototype visualization of his design before building it. But to develop a simulation toolkit enough to be useful it has also an associated cost. This cost higher if this simulation has to be re-designed or re-implemented to take into account changes. A new design and simulation toolkit has been developed to tackle with this problem, allowing the designer to visualize and to interact with his design without building it, and to make modifications in it in an easy and low-cost way.

In order to produce realistic simulations, there are three important elements: (1) Artificial Intelligence (AI) techniques, (2) physics simulation, and (3) linking AI with real world devices and high quality graphics. All these elements allow creating IVEs with a high degree of realism. The AI techniques allow creating behaviors to the entities (agents) inside the IVE. Physics simulation helps to create and to manage the static or dynamic constraints of the objects composing the virtual environment by means of physics simulation engines. Connecting the agents to the physical devices gives them the possibility to perceive and act over the real world. This link helps the developer to observe some behaviours, relations, and interactions that can be difficult to be observed in the simulation. Lastly, allowing the developer to observe what is happening inside the IVE by means of Virtual Reality or Augmented Reality devices enables him to check the validity of his designs in a close to reality way. Using all these toolkits allow the developers not only to check the validity of their designs, but also to easily make changes with a low cost associated.

The design method here presented is divided into three main steps (Virtual, Physical and Human) as can be observed in Fig. 1. Each one of these steps is composed of three phases: modelling, implementation and simulation. The designer can change his design and go back to any of the previous phases and steps at any moment. More specifically, these steps are defined as follows:

- *Step 1: Virtual*: The purpose of this step is to develop a virtual environment. This virtual environment can be executed, redesign and executed again easily and with

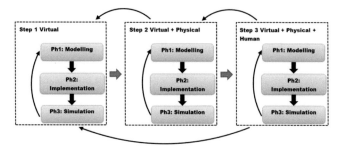

Fig. 1 IVE development processes

a low cost. It can be used to make an easy and quick review of the design, helping the designer to detect failures in the design or non achieved requirements.

- *Step 2: Virtual + Physical*: The purpose of this step is connecting the simulation environment with the real world, i.e. linking the simulation with objects in the real world. This linking helps to check if all that appears valid in the simulated environment is so. There may exist some properties, or details that have been abstracted in the simulation and that can make the design not valid. These failures can be easily detected including some physical world in the system, as can be, for instance, some physical agents, robots, sensors or effectors.

  To carry out this step the simulation, designed in the previous step, must be revised in order to include real physical objects. Some of the previous virtual elements will be substituted with real objects, that will move an behave in real life as if there were in the virtual environment. For example a real robot integrated into a simulation, could interact with virtual entities. It would move in real life and its representation in the virtual world would move as well. Moreover, since it is bounded by the limitation of the virtual environment it would not be able to move forward in real life if in the virtual world there was, for example, a wall. The immersion of real physical objects into the simulation can be used to verify that actually the virtual objects behave as the real ones, and also to test physical objects in different scenarios that can be virtually created.

- *Step 3: Virtual + Physical + Human*: The goal of this step is to integrate human beings into the simulation. Some of these virtual agents will be substituted with human beings immersed in the simulation, thanks to specific devices such as virtual reality glasses. The immersion of human beings into the simulation can be useful to verify that actually the virtual objects behave as the real ones, and also to train human beings in different scenarios that can be created virtually.

Next sections explain in detail each phase of the proposed process.

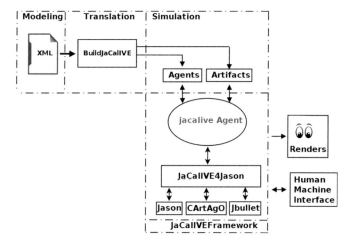

Fig. 2 Phases of each IVE development process step

2.1 Phase 1: Modelling

The first phase of any step is modelling the Virtual Environment in terms of the Extended MAM5 meta-model. This meta-model is based on A&A [4, 5]. It specifies the system in terms of agents and artifacts that can be situated in a workspace. More details about the meta-model can be found in the next subsection. As is summarized in Fig. 2, the designed virtual environment is specified in a XML file. *JaCalIVE*[5] provides a XSD schema that represents the MAM5 meta-model in order to facilitate the specification of the design in XML.

MAM5 [3] is addressed to be used by an IVE designer, that wants to design an IVE based on a multi-agent system. The MAM5 meta-model has been extended in order to include the immersion of real objects and human beings into the virtual environment. In the environment, it must be distinguished between the virtual environment and the real one. Concretely, we distinguish between the abstractions related to a virtual representation, (that we have called Virtually Physical Situated because they have a Virtually Physical representation in the Virtual Environment), and the ones that are not situated in such virtual environment, (and that we have called Non Virtually Physical Situated because they don't have a representation in the Virtual Environment). MAM5 classifies the entities in the design into two different sets. The first one is related to all the entities that do not have any physical representation in the IVE (Non Virtually Physical Situated), whilst the second one is formed by all the entities having a representation inside the IVE (Virtually Physical Situated). Inside the former set the main entities are: (1) *Agents* representing the intelligent part of the autonomous pro-active entities in the system, (2) *Artifacts* representing the basic bricks used to define the environment, and (3) *Workspaces* representing the concep-

---

[5]The *JaCalIVE* toolkit can be downloaded from: http://jacalive.gti-ia.dsic.upv.es.

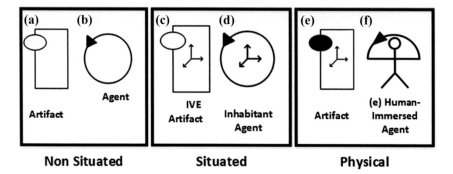

Fig. 3 MAM5 graphical icons

tual containers of agents and artifacts. In a similar way, inside the second set the main entities are: (1) *Inhabitant Agents* representing agents that are physically situated in the virtual environment, (2) *IVE Artifacts* representing artifacts that are physically situated in the virtual environment, and (3) *IVE Workspaces* representing the virtual place and the laws defining and governing such places. It has to be underlined that there is one special kind of Inhabitant Agents, that are the *Human-Immersed Agents*. It is important to distinguish these kind of agents at modelling time because the designer must take into account the different interactions the user is going to make and the different set of actions available for him (that may differ from the rest of Inhabitant Agents). In the same way there is a special kind of IVE Artifacts that are the *Physical Artifacts*. They represent the real objects that are immersed in the virtual environment. Moreover, the MAM5 meta-model also incorporates the specification of physical restrictions and properties of the environment by means of the specification of *IVE Laws* governing the different *IVE Workspaces*. Finally, MAM5 provides a XSD version of its meta-model in order to facilitate the modulation of IVE. Besides, it provides a set of icons, based on the ones used in the A&A model, that allows graphically representing the modelled IVEs (see Fig. 3).

2.2 Phase 2: Implementation

The second phase of any step is the implementation of the necessary code to simulate the model specified in the previous phase. As is shown in Fig. 2, the *JaCalIVE* framework provides the *BuildJaCalIVE* tool. This tool automatically generates code templates from the XML model. So, one *JASON*[6]. agent is generated for each agent defined in the XML, along with the corresponding *CArtAgO*[7]. files to implement the workspaces and artifacts defined. The developer has to fulfill these templates,

---

[6]http://jason.sourceforge.net/wp/

[7]http://cartago.sourceforge.net/

however, this automatic code generation facilitates considerably the implementation phase.

2.3 Phase 3: Simulation

The third and last phase of any step is the simulation of the modeled and implemented IVE. To support the execution of this simulation phase, the *JaCalIVE* Framework integrates the following toolkits:

1. **JASON**: It is the toolkit in charge of the creation and management of the agents populating the IVE. *JASON* will support the development of the agents' behaviours, plans, action and communication.
2. **CArtAgO**: It is the toolkit that gives support to the *workspaces* inhabited by all the artifacts modeled in the XML.
3. **Jbullet**[8].: It is the toolkit in charge of the physical simulation and of controlling all the physical restrictions inside the environment such as gravity, friction, and collision detection.

The *JaCalIVE* Framework is independent of the *render* engine used to visualize the simulation. *JaCalIVE* offers a basic *render* with which you can visualize the simulation where the entities are represented with basic primitives. Besides, *JaCalIVE* offers an API so that the developer can implement its own *render*.

3 Case Study: Virtual Furniture Distribution to Disabled People

The immersion of human beings and physical objects into virtual simulations can be useful in a wide range of domain applications (such as learning [6], training sociological skills [7], or team training [8]). In this section we introduce a case study based on the design of furniture distribution in classrooms with disabled people. When designing the furniture distribution of a classroom, it is important to take into account the static issues such as the maximum number of students that this classroom is allowed to have and the visibility of the blackboard. Besides it is important to considerer the dynamics of the class such as how many students get in and out at the same time, the necessary time to evacuate the class, the mobility restrictions of the students, and the possibility of every student to approach the blackboard. Moreover, it is necessary to follow the government laws of each region regarding the special requirements for disabled people. Therefore, it is necessary to provide a mechanism that allow to distribute the furniture taking into account the special needs of every

---

[8]http://jbullet.advel.cz/

Fig. 4 Render 3D

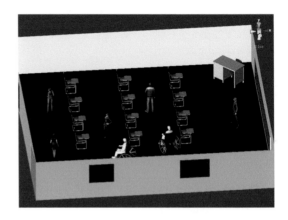

student. In the rest of the section we present how the IVE development process, presented in Sect. 2, can be used to design or redesign the furniture distribution of a classroom. Using this modeling and simulating framework, designers can try different approaches and detect unappropriated designs. They are able to try different designs changing the number of students and their mobility restrictions.

3.1 Step 1. Virtual

Following the process detailed in Sect. 2, the first step is the virtual simulation of the scenario following the three phases defined in the process: (1) modeling, (2) implementing, and (3) simulating.

First the environment is modeled by using the MAM5 meta-model. Figure 5 shows a graphical representation of three different designs. In this case, the Step1 representation only employs virtual agents and artifacts in the design. The designed model is composed of one IVE workspace that is inhabited by 7 students and one professor represented by 8 agents. Two of this students use a wheelchair. This workspace has also 10 desks and 10 chairs, and a blackboard that are represented by artifacts. As an example of physical restriction the figure shows the gravity and the friction that would be necessary to calculate the wheelchair movement. Secondly the model is translated automatically into code templates using the *JaCalIVE* framework. These templates were fulfilled by the developers in order to provide the specific movement restrictions of each student. Third and last, the simulation is executed. *JaCalIVE* provides a basic visualization render, however in order to provide more realism a specific *Unity 3D*[9]. render was created for this case study. Figure 4 shows a snapshot of a simulation. Thanks to this simulation the behavior of the students (agents) can be observed and it can be detected whether there is any conflict.

---

[9]http://unity3d.com/unity

Moreover the position, configuration and design of the furniture can be easily modified in order to test with which configuration students can move more smoothly, using less time and without conflicts. These changes were done easily using the *JACALIVE* framework. Once the most appropriate design was choose, we followed to the next step.

3.2 Step 2. Virtual + Physical

In this step furniture configuration designed in the first step is tested by incorporating into the simulation two real electric wheelchair. The Step 2 representation of Fig. 5 shows the graphical model which includes these physical artifacts. Each wheelchair is connected with the virtual simulation by means of a mini computer (Fig. 6). The wheelchair will move in real life and will also have its representation in the virtual world. As a result, it is possible to check that the real life wheelchair is actually able to move as the virtual one. Moreover, it is possible to detect conflicts and crashes that were not detected in the virtual world. For example, the real wheelchair may be not as precise as the virtual one, so wider corridors were needed. Once a conflict is detected the design has to be changed, and before continuing doing tests with the real chair, a 100 % virtual simulation must be executed, i.e. it is recommendable to go back (step one) in the development process.

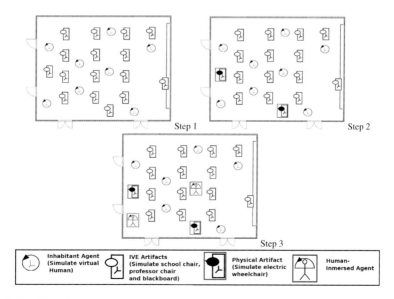

Fig. 5 Model to simulate

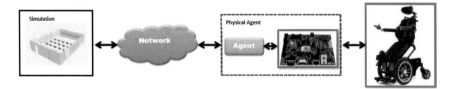

Fig. 6 Connection between the real robot and the virtual one

3.3 *Step 3. Virtual + Physical + Human*

The last step is to add human interaction into the IVE. The Step 3 representation in
Fig. 5 shows graphically that in the design we have change one virtual agent for one
human-immersed agent. Thanks to mobile virtual simulation devices such as *Oculus
3D*[10] and *Sulon 3D*,[11]. human beings can be integrated into the virtual simulation.
In this case study this step is very important because disabled people can test how
comfortable are their movements into the classroom and detect conflicts that were
not identified in the previous steps.

4 Conclusions and Future Work

This paper presents a development process for intelligent virtual environment IVE
considering the immersion of real objects and human beings in the virtual simulations. This IVE development process allows modeling, implementing and simulating
IVEs and their interactions with real objects and human beings.

Despite the increasing interest in simulation, current IVEs have been traditionally
developed for a specific purpose and domain. In this paper we present a set of tools
that allows developing IVEs for different purposes and domains in a homogeneous
way. Besides this, the approach simplifies the design and redesign of IVEs reducing
the designing time, and increasing the scalability and modularity of the IVEs. This
development process has been tested by the development of an IVE for designing
the furniture of spaces taking into account the special needs of different users.

As future work, regarding the case study we plan to integrate a connection with
real wheel chairs, since at this time the tests were performed connecting the system
with small robotic platforms. Moreover, we plan to add emotions into our meta-
model and framework. Our goal is to simulate emotions into our virtual agents and
how these emotions will affect the agent behaviors. Also we plan to perceive human
beings emotions during the simulation. In order to improve the immersion of these
human beings in the virtual environment.

---

[10]BuildJaCalIVE2.

[11]http://sulontechnologies.com/

Acknowledgments This work is partially supported by the MINECO/FEDER TIN2012-36586-C03-01, CSD2007-00022, COST Action IC0801, FP7-294931 and the FPI grant AP2013-01276 awarded to Jaime-Andres Rincon.

References

1. Aylett, R., Luck, M.: Applying artificial intelligence to virtual reality: intelligent virtual environments. Appl. Artif. Intell. **14**, 3–32 (2000)
2. Rincon, J.A., Garcia, E., Julian, V., Carrascosa, C.: Developing adaptive agents situated in intelligent virtual environments. In: International Conference on Hybrid Artificial Intelligence Systems, pp. 98–109. Springer (2014)
3. Barella, A., Ricci, A., Boissier, O., Carrascosa, C.: MAM5: Multi-agent model for intelligent virtual environments. In: 10th European Workshop on Multi-Agent Systems (EUMAS 2012), p. 1630 (2012)
4. Ricci, A., Viroli, M., Omicini, A.: The a&a programming model and technology for developing agent environments in MAS. In: Programming Multi-Agent Systems, p. 89106. Springer (2008)
5. Ricci, A., Viroli, M., Omicini, A.: Give agents their artifacts: the a&a approach for engineering working environments in MAS. In: Proceedings of the 6th international joint conference on Autonomous agents and multiagent systems, p. 150 (2007)
6. Piccoli, G., Ahmad, R., Ives, B.: Web-based virtual learning environments: a research framework and a preliminary assessment of effectiveness in basic it skills training. MIS Q. 401–426 (2001)
7. Blascovich, J., Loomis, J., Beall, A.C., Swinth, K.R., Hoyt, C.L., Bailenson, J.N.: Immersive virtual environment technology as a methodological tool for social psychology. Psychol. Inq. **13**(2), 103–124 (2002)
8. Thomas, W.: Mastaglio and Robert Callahan. a large-scale complex virtual environment for team training. Computer **28**(7), 49–56 (1995)

MAESTROS: Multi-Agent Simulation of Rework in Open Source Software

Thiago R.P.M. Rúbio, Henrique Lopes Cardoso
and Eugénio da Costa Oliveira

Abstract Rework Management in software development is a challenging and complex issue. Defined as the effort spent to re-do some work, rework implies big costs given the fact that the time spent on rework does not count to the improvement of the project. Predicting and controlling rework causes is a valuable asset for companies, which maintain closed policies on choosing team members and assigning activities to developers. However, a trending growth in development consists in Open Source Software (OSS) projects. This is a totally new and diverse environment, in the sense that not only the projects but also their resources, e.g., developers change dynamically. There is no guarantee that developers will follow the same methodologies and quality policies as in a traditional and closed project. In such world, identifying rework causes is a necessary step to reduce project costs and to help project managers to better define their strategies. We observed that in real OSS projects there are no fixed team, but instead, developers assume some kind of auction in which the activities are assigned to the most interested and less-cost developer. This lead us to think that a more complex auctioning mechanism should not only model the task allocation problem, but also consider some other factors related to rework causes. By doing this, we could optimise the task allocation, improving the development of the project and reducing rework. In this paper we presented MAESTROS, a Multi-Agent System that implements an auction mechanism for simulating task allocation in OSS. Experiments were conducted to measure costs and rework with different project characteristics. We analysed the impact of introducing a Q-learning reinforcement algorithm on reducing costs and rework. Our findings correspond to

T.R.P.M. Rúbio (✉) · H.L. Cardoso · E. da Costa Oliveira
LIACC / DEI, Faculdade de Engenharia, Universidade Do Porto,
Rua Dr. Roberto Frias, 4200-465 Porto, Portugal
e-mail: reis.thiago@fe.up.pt

H.L. Cardoso
e-mail: hlc@fe.up.pt

E. da Costa Oliveira
e-mail: eco@fe.up.pt

© Springer International Publishing Switzerland 2016   61
P. Novais et al. (eds.), *Intelligent Distributed Computing IX*,
Studies in Computational Intelligence 616,
DOI 10.1007/978-3-319-25017-5_6

a reduction of 31 % in costs and 11 % in rework when compared with the simple approach. Improvements to MAESTROS include real projects data analysis and a real-time mechanism to support Project Management decisions.

1 Introduction

Software development is facing a big change. Traditionally, software development companies have closed policies on choosing developer teams and rigorous control over the task allocation between them. In the last few years we have seen a massive adoption of Open Source Software (OSS) [1], based on free code in which any other developer can contribute by free will. Open source developers could be anywhere in the world, have learnt different techniques and ways of working. There is no guarantee that they will follow the same strategies and quality policies when compared with a traditional and closed project.

Resolving problems that were not solved consistently or bugs created by developer mistakes is called *Rework*, an additional and not planned work represented by the effort cost (in time and money) spent in order to resolve a problem with a requirement that was previously considered solved. Rework occur in both closed and open projects, but closed projects have many ways to early identify and control rework causes that open projects do not. To be able to manage rework in OSS is a necessary and important step to reduce project costs and to help project managers to better define their strategies to software improvement [2].

When considering open projects, a big problem is how to identify the main causes of rework. Developers introduce rework due to lack of specification, missing verifications or even unplanned new properties. In the literature, rework is considered a manifestation of the lack of communication between developers and a cause of stressed or uncommitted personnel [3]. Finding rework causes could help us to discover how developer's behaviours affect projects.

We observed that frequently activities are assigned to developers that are interested. There are no fixed team and the assignment of the activities rely only on developer's cost (mostly in time or even monetary) of development. Actually, this situation resembles a simple auction mechanism in which the activities are allocated to the less costly developers. A more appropriate mechanism would consider also other impacting factors such as developer experience and its past results on achieving activities completion with success. This motivates our work: we aimed at simulating this task allocation process using a MAS in order to understand, analyse and propose some optimisation that could help to reduce rework.

In this paper we present a multi-agent system for simulating the task allocation process in open source software. We seek the best opportunities: low development cost with minimum rework. We have evaluated the performance of our model in terms of the number of re-incident activities and their rework cost together with project final cost. Results show that our approach can get close-to-budget projects final cost, even with some rework present. Further developments of our system could be a good ally to project managers.

The rest of this paper is structured as follows: In Sect. 2 we present the Rework problem in Open Source Software. Section 3 discusses the project management process workflow. In Sect. 4 we describe the architectural design of our system. Section 5 presents the experimental evaluation of the model. We discuss the findings of this work and point lines of future research in Sect. 6.

2 Related Work

Open Source Software (OSS) is a trend in software development. Since late 1990, an uncountable number of projects have grown in this environment proving it can be successful and profitable [4]. Research interest in OSS is much diversified [5]. The open nature of the software creates a great difficulty in managing resources, planning and delivering projects. As mentioned by Raja et al. [6], resource allocation and budgeting in OSS is even a harder challenge. The cost of the development is an important factor for the success of a project, making the search for reducing rework an important matter. In Sect. 1, we introduced rework as the effort and consequently monetary cost of trying to fix something that was already considered a solved problem. Rework is, in fact, a big problem in software engineering, consuming big part of the project budget (40 % up to 70 %) [7]. Rework could be explained by human problems in project management like communication, formation and work conditions [3, 7] and the Industry believes that great part of rework could be early identified and avoided, but until now not much attention has been paid in studying rework.

Previous works have characterised the relation between rework and developers actions regarding their expertise and work profiles. Rbio et al. [8] divided developers into members and volunteers, in which the first are recognised by Project Owners because of their knowledge on some specific project or are permanent members of a development team. Volunteers, on the other hand are developers that might contribute spontaneously by their interest on the project. Members are usually have a lower probability of generating rework (about 10 %), while volunteers have a higher chance (30 %). By other side, the development cost of a member is known to be more than volunteers work [8, 9]. The other actor interested in this process is the Project Manager (PM). Although Project Managers work under different methodologies, their basic task is to distribute projects activities and manage the assignment of tasks to the available developers [10].

In traditional development the tasks are imposed, opposing to more flexible methodologies, where developers are able to discuss or vote their willingness for working in some task [11]. As referred in Sect. 1, process of task assignment in OSS is somewhere between this two: the Project Manager tries to choose the developer that best fits to some activity by its reputation, cost and availability [12].

Each actor in this system has its own decisions: developers must decide whether they are interested on developing a specific activity, determine how their assigned tasks will be accomplished, and even decide on which bids to propose. On the other hand, managers also make decisions in their effort on trying to reduce reworks and costs, while maximising the number of activities successfully concluded. This underlying autonomy of the actors involved lead us to an agent-based approach. Each software agent represents one actor in the process, behaving accordingly to its own goals. In multi-agent systems the autonomous entities (agents) can decide whether or not to accomplish some task and can deal with self and community goals [13].

The importance of mapping developers as agents relies on their free will to contribute to projects and the relationships built from their interactions. In fact, although many investigations about open source deal with some properties of OSS, like the actors roles in [14], very few discussions in the literature model the development process in this environment as a multi-agent system [15, 16].

Simulation is a good approach in this case, where the difficulty of gathering and analysing data on-line with real open source projects is high. Since the platforms restrain the access to data and most projects decisions are private, analysing the task allocation in real-time is hard and complex. The simulation, in the other side, needs to represent well the behaviours and the mechanisms used by agents to coordinate their actions. Once the abstracted characteristics represent significantly the behaviour of the actors, a multi-agent simulation could be used to investigate the impact of changing various characteristics of the project, for instance, how would PM behave when the number of available developers grow or even how to automate the negotiations about costs where the agents represent the interests of the real developers.

We do not intend to create a new method for optimising task allocation, but rather applying automated scheduling and negotiation intelligent techniques to help on the improvement of the decisions taken by project managers in open source projects. our investigation contributes by creating a first attempt to model the task assignment process in OSS with a MAS. Moreover, using learning strategies could lead to reduce the rework on this kind of projects.

As an introductory work, this opens opportunities to future works in this area and widens the applicability of MAS to a growing and rich environment.

3 Model Conceptualisation

The OSS development process described on the previous sections give us an insight about how to model actors and their behaviours. Transcribing agents goals, actions and decisions could help us to create a simulation model that represent how task assignment work on OSS projects [15].

Rubio et al. [8], analysed real data from big real open source projects such as Apache projects.[1] For simplicity, we are going to consider only the types of developers described in their work: members and volunteers. Regardless of its type, a developer can finish its tasks successfully or not. When the conclusion of an activity

---

[1] Apache Software Foundation—http://www.apache.org/.

fails, the Project Manager must reassign it causing an increase of cost, the so-called rework. When the current set of activities is successfully concluded, the Project Manager reports it to the Project Owner (PO), the person that knows the next milestones on the project or new set of activities to deliver to the development team. We characterise three main agents in the OSS environment: (1) Project Manager (PM), (2) Developer and (3) Project Owner. Developers are also divided into two subgroups: (a) Members and (b) Volunteers.

The Project Manager has the knowledge about a set of activities and the estimated cost for completing each one of them. The Project Manager must gather information about the interested developers and assign the activities to the best opportunities, considering risk, cost and other factors. Similar to an auction protocol, the PM wants to "sell" the activities to developers, who "bid" with the cost of development.

PM is, then, responsible to select the best developer for each activity. The complete process is represented in Fig. 1 and starts with the PM checking if there are to-do activities and selects one of them (based on priority or cost, for example). The information about the activity is spread among the developers who are not currently working and in case they are interested on developing such activity, they must send a proposal (cost of work) to the PM. For members this is mandatory and they must always present proposals. Volunteers could refuse to work on some activities. The PM evaluates the proposals received and if there are none, the activity goes back to the to-do list and the process starts again. On the other hand, if there are proposals, the PM awards the winner and sends the refusing message to the other bidders. At this time the winner developer starts the development process. This developer will not participate in other auctions until current work is done. The PM repeats the cycle with other activities until there are no more activities or available developers. Finally, when the developer finishes its work, it becomes available again by notifying the PM, who is responsible to check whether the activity was successfully completed or not. In case the developer has failed, the activity is put back into the to-do list. When the developers work is well done, the activity is put in the completed list. The process finishes when all the activities are completed.

In order to simulate this environment and analyse how the assignment process could be optimised in terms of the rework introduced and process completion, we

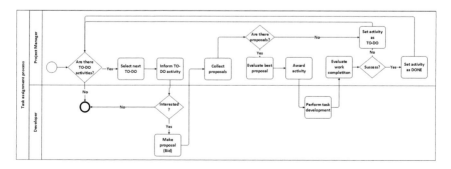

Fig. 1 Task assignment process

have proposed a multi-agent simulation. We analyse the problem contemplating the decisions taken by the Process Manager and propose the use of a reinforcement learning algorithm in order to reduce rework. When compared to a simple approach that only consider the available bids, the Q-Learning approach uses a more complex reasoning mechanism, by considering also the experience and past results from developers. Section 4 presents our model characteristics and architecture.

4 MAESTROS Architecture

We have analysed how Project Managers can distribute project tasks between available developers. As explained, work in on-line real projects is a complex matter. In order to study how the behaviours of the agents and how project manager decisions affect project results, we have created a simulation system that implements the process described on Sect. 3. Although there are many agent-based simulation platforms available, we wanted to create a flexible system that allows us to expand our work in the future, connecting the simulation with the real on-line project. Thus, we have constructed MAESTROS (**M**ulti **A**g**E**nt **S**imulation of **R**ework on **O**pen **S**ource Software).

Developed in Java, MAESTROS use JADE[2] multi-agent framework [17] in its core. JADE allows us to define agent behaviours that map actors reactions in different steps of the task assignment process.

We assume that the development cost of an activity is directly proportional to the time in which a developer is working on it. Basically, this simplifies the modelling of the inner development process of one activity into a time-frame cost problem. In our system, the cost of an activity is the portion of the time the developer will be occupied working on it. We do not simulate the working process described by one activity like coding generation or other kind of documentation by developers. Although this may seem a simplistic model, the focus here is whether the choices made could be optimised in order to reduce rework.

MAESTROS consists in three main parts: (1) Communication layer; (2) Negotiation mechanism and (3) Decision strategy. Regarding to the Communication layer, MAESTROS agents communicate through messages, sending messages with the desired content and the semantics of the information. For this, JADE We have used FIPA ACL messages [18], native on JADE platform. ACL messages uses standardised *performatives*, a special field on the message that contextualises the required action.

In the Negotiation Mechanism we model how the task allocation occurs. As explained, our model is a very simple auction mechanism. The Contract Net Protocol [19] is a good strategy that fits just our needs. The algorithm consists on one round of bidding in order to select the winner that will get the item or service auctioned. Activities could be seen as the services the Project Manager wants to sell

---

[2]http://jade.tilab.com/.

Fig. 2 Communication on
the assignment process cycle

and developers are the buyers that bid (make a proposal) in terms of working cost of
an activity. In fact, the Project Manager just needs one round to decide who will be
responsible for executing the activity, assuring the Contract Net as a good strategy
to MAESTROS.

Our implementation of the protocol follows Fig. 2, starting when the Project Manager identifies a to-do activity and want it to be developed. He makes an announce-
ment, sending a message with the performative CFP (Call For Proposals) and the
receivers (available developers) may answer with a PROPOSE message containing
the cost for the work. After evaluating the proposals, the winner developer is notified
with a ACCEPT-PROPOSAL message and the auction losers receive a REJECT-
PROPOSAL. Here we see clearly the establishment of a contract between the PM
and the winner of the auction and it starts working in order to get the activity done.
Finally, when the task is done developers send the result of the job with the activity
concluded with an INFORM message to the Project Manager. When all activities are
finished, the PM sends an INFORM with the completed set of activities and waits
for receiving more. The Project Owner, in turn, answers with another INFORM containing a new set of activities to re-start the development cycle.

Finally, we have to discuss the strategies for deciding the winners in the auction
mechanism. The final cost of a project summing up the rework cases could exceed
the total budget if the Project Manager does not take this into account. Thus, the reasoning process of deciding which proposal is the best at a specific time could follow
many strategies, from choosing the cheapest one to personal choices based on previous experience. Many development teams consider only the cost of development
as decision criteria. This is not a good decision because most of the time, cheap bids
have higher probability of rework and reassigned activities could even imply higher
costs in the future. We call this the simple case. In the other hand, more complex
strategies could be developed, considering many other factors to compose a decision. In this case, a more advanced technique is required. Since we are dealing with
agents, we decided to improve the criteria used by the PM and give it some learning mechanism in order to observe past results and try to predict the best opportunity to follow. We have opted to use the Q-Learning [20], a reinforcement learning
algorithm that tries to find and select an optimal policy function for choosing specific actions given the current state. This function represents the rules that the agent

will follow when giving some state-action pair. After constructing the function, the optimal path is given by selecting the action with the highest value in each state. We take advantage from Q-learning strength since it does not need a previous model of the environment.

$$Q_{t+1}(s_t, a_t) = \underbrace{Q_t(s_t, a_t)}_{\text{old value}} + \underbrace{\alpha}_{\text{learning rate}} \times \Bigg[\overbrace{\underbrace{R_{t+1}}_{\text{reward}} + \underbrace{\gamma}_{\text{discount factor}} \underbrace{\max_a Q_t(s_{t+1}, a)}_{\text{estimate of optimal future value}}}^{\text{learnt value}} - \underbrace{Q_t(s_t, a_t)}_{\text{old value}} \Bigg] \tag{1}$$

The algorithm works following Eq. (1), where the function calculates the Quality (Q) of a state-action combination. Our Q-learning states are defined as the current states of the project, regarding all the process of selecting developers and the result of their work impacting project cost and rework. The reward of choosing a developer is given by the difference between the budget and project's total cost at that given time, weighted by the probability of rework, seen in Eq. (2).

$$R(t) = (Budget(t) - FinalCost(t)) \times (1 - rework) \tag{2}$$

The learning rate α determines to what extent the newly acquired information will override the old information. A factor of 0 would make the agent not learn anything, while a factor of 1 would make it consider only the most recent information. On the other hand, the discount factor γ determines the importance of future rewards. A factor of 0 will make the agent or short-sighted, only considering current results, while a factor near 1 will make it strive for a long-term high reward.

We have conducted a set of experiments to compare the simple case without optimisation and others trying to find the optimal values of these parameters. Varying α and γ from 0.1 to 1.0 we analysed the influence on the number of members and volunteers chosen and on costs. Due to space limitations we only reference here the achieved optimal values of $\alpha = 0.3$ and $\gamma = 0.7$. Our conclusions were that with these values, the number of auctions was closer to the number of activities and mean numbers of selected members and volunteers were approximately equal. On the other hand, this configuration leaded us to minimal rework and final project costs. In all our experiments reported in Sect. 5 we use this optimal parameter values for α and γ.

4.1 Rework Visualisation

MAESTROS we have the opportunity to inspect all living agents in the environment and see their properties, together with the complete control of the communication flow between agents, given by JADE platform. In contrast, JADE is not a simulation tool and facing the lack of graphical visualisation is a limitation. We have developed then a graphical interface for analysing projects characteristics during simulations

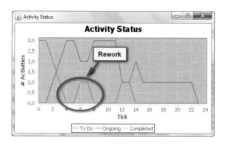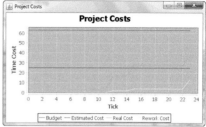

Fig. 3 MAESTROS Status chart (*left*) and Costs chart (*right*)

using JFreeCharts. This module allows us to visualise project's states during development. In Fig. 3 we see two important charts presented on MAESTROS visualisation module: Activity Status Chart and the Costs Chart. The Activity Status Chart shows how activities states change over the time and how rework manifestations, shown when the number of to-do activities increases, indicating that some activities re-entered on the stack for development, indicating clear rework cases. Meanwhile, the Costs Chart shows how costs evolve in terms of budget, estimated cost of the project, rework and final cost. The estimated cost is an important metric since it can be defined as the sum of all activities estimated costs and interpreted as a measure of the optimal minimum cost of the project.

5 Simulation Experiments

We wanted to test MAESTROS capabilities on simulating the OSS environment and verify our model. For that, we have designed two experimental scenarios: (1) Task assignment performance; (2) Impact of project characteristics (activities, budget and developers).

First Scenario—Analysing task assignment
We have designed a default workload that consists on simulating the development of a simple set of activities many times. In a project consisting on 5 activities and 10 available developers (5 members and 5 volunteers) on the environment, the project manager should lead the development trying to keep the final cost closer to the estimated cost and trying not to overpass the budget. The budget is set as 200 % of the estimated cost and to be more realistic, activities have a variable estimated cost (simulating different degrees of difficulty on the tasks) which is randomly set between predetermined values of 5 and 15. Members are allowed to bid between 8 and 15 with 0.1 rework probability (lower) and volunteers bid between 1 and 15 with a 0.3 rework probability (higher), according to literature.

The experiment consists on running the simulation 100 consecutive times and check if the learning mechanism really helps to reduce project's final cost while trying to get lower rework cost. Once each run is independent, the estimated cost,

Fig. 4 Performance over
100 runs

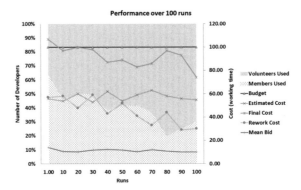

final cost and number of auctions may be different. In each run, MAESTROS assigns tasks to developers and records all costs.

The results of the simulation are condensed in Fig. 4. We have overlapped the graphical information of choosing members or volunteers with the project cost results. The shadowed area correspond to the percentage of developers that won the auctions through project development. It seems that at first Project Managers actions were more erratic, leading to a higher final cost that extrapolated both budget and estimated cost. We could see clearly that through the runs the final cost seemed to decrease under the budget and get closer to the optimal cost.

We have no doubt that MAESTROS was able to reduce the rework on this experiment. Comparing the first and the last runs on Table 1 we see the reduction on costs: final cost was reduced in 31 % and rework cost in 47 %. The winner bid also decreased a 29 %. Comparatively, the number of winner members was reduced in 57 % facing an increasing of 75 % on the number of volunteer winners. The number of auctions seemed to be more stable, suffering a small increase of 9 %.

Second Scenario—Different project characteristics
In the second scenario our experiments were focused on verifying how the system performed with different projects characteristics. We have setup and experimented different workloads. In each experiment, focusing in just one property we have varied

Table 1 Performance Improvements

	1st run	100th run	Improv. (%)
Final Cost	107.00	74.00	−31
Rework Cost	57.00	30.00	−47
Mean Bid	14.00	10.00	−29
Members	7.00	3.00	−57
Volunteers	4.00	7.00	75
#Auctions	11.00	12.00	9

its values to check the outcomes after 100 runs. The analysed properties were: **(1) Number of Activities:** Varying from projects with 1 simple activity and ending with 100 activities; **(2) Budget:** Fixing lower project budgets starting from 200 % and ending with only the estimated cost; We have followed the same characteristics described on the first scenario for the costs range of activities and bids, as also used the same probabilities of rework.

The results from running the system 100 times with different number of activities show that this is an important factor that influences projects costs. As seen in Fig. 5, MAESTROS managed to get the final cost under the budget and rework rate very stable, despite the increasing project final cost. When analysing the budget we wanted to see if the learning mechanism in MAESTROS could give us better final costs, trying not to exceed it and looking for the optimal value (estimated cost). In Fig. 6 we see that a broader margin to rework has lead to higher costs, almost 250 % over the estimated, but reducing the budget leaded to decreasing rework and the final cost. Results indicate that at best, MAESTROS achieved a rework cost of approximately 11 % of the final cost.

Fig. 5 Varying activities number

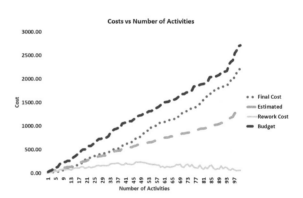

Fig. 6 Varying the budget

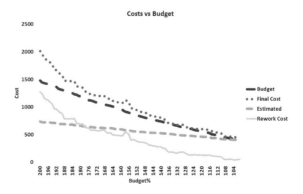

6 Conclusions and Future Work

MAESTROS is the first multi-agent approach that study how rework affects Open Source Software projects. This mechanism for activity assignment proved to reduce rework and final projects costs and could be used as an auxiliary tool in real on-line OSS platforms. Our experiments show that multi-agent systems are a good tool to model and simulate a software engineering environment, namely the Open Source Software. MAESTROS has its importance related to the lack of tools and simulations about rework on software development and could help to improve management decisions in this kind of management.

One of MAESTROS most important contributions is that it could help to reduce rework in real projects if the system parameters are fine-tuned with real projects characteristics. MAESTROS could reduce the final cost to a near optimal value (estimated cost of the project) and in comparison to the literature, where the value of 20 % is accepted as a usual rework cost, in MAESTROS we have managed to achieve a rework cost on the order or 11 % of the final cost in average.

We envisage to expand MAESTROS capabilities by providing access to real world projects data from known OSS development environments. We aim to create a good database about open source projects and development teams characteristics and costs. Other future works may include the expansion to model developers work, namely code generation and documentation of activities.

Acknowledgments This work has been funded through a IBRASIL Grant. IBRASIL is a Full Doctorate programme selected under Erasmus Mundus, Action 2 STRAND 1, Lot 16 and coordinated by University of Lille.

References

1. Gaff, B.M., Ploussios, G.J.: Open source software. Computer **45**(6), 9–11 (2012)
2. Software risk management. Springer, Berlin (1989)
3. Chua, B.B., Verner, J.: Examining requirements change rework effort: a study. arXiv preprint arXiv:1007.5126 (2010)
4. Sen, R., Singh, S.S., Borle, S.: Open source software success: measures and analysis. Decis. Support Syst. **52**(2), 364–372 (2012)
5. Raymond, E.: The cathedral and the bazaar. Knowl. Technol. Policy **12**(3), 23–49 (1999)
6. Raja, U., Tretter, M.J.: Defining and evaluating a measure of open source project survivability. IEEE Trans. Softw. Eng. **38**(1), 163–174 (2012)
7. Zahra, S., Nazir, A., Khalid, A., Raana, A., Majeed, M.N.: Performing inquisitive study of pm traits desirable for project progress. Int. J. Mod. Educ. Comput. Sci. (IJMECS) **6**(2) 41 (2014)
8. Rúbio, T.R., Gulo, C.A.: Characterizing developers rework on github open source projects. In: Proceedings of the 10th Doctoral Symposium in Informatics Engineering, FEUP Edicoes - Faculty of Engineering, University of Porto
9. Robles, G., González-Barahona, J.M., Cervigón, C., Capiluppi, A., Izquierdo-Cortázar, D.: Estimating development effort in free/open source software projects by mining software repositories: a case study of openstack. In: Proceedings of the 11th Working Conference on Mining Software Repositories, pp. 222–231. ACM (2014)

10. Crowston, K., Li, Q., Wei, K., Eseryel, U.Y., Howison, J.: Self-organization of teams for free/libre open source software development. Inf. Softw. Technol. **49**(6), 564–575 (2007)
11. Larman, C.: Agile and iterative development: a manager's guide. Addison-Wesley Professional, Boston (2004)
12. Warsta, J., Abrahamsson, P.: Is open source software development essentially an agile method. In: Proceedings of the 3rd Workshop on Open Source Software Engineering, Citeseer, pp. 143–147 (2003)
13. Wooldridge, M., Jennings, N.R.: Intelligent agents: theory and practice. Knowl. Eng. Rev. **10**(02), 115–152 (1995)
14. Feller, J., Fitzgerald, B.: A framework analysis of the open source software development paradigm. In: Proceedings of the twenty first international conference on Information systems, Association for Information Systems, pp. 58–69 (2000)
15. Koch, S.. Free/open source software development. IGI Global, Hershey (2005)
16. Madey, G., Freeh, V., Tynan, R.: Agent-based modeling of open source using swarm. In: Proceedings of the AMCIS 2002, p. 201 (2002)
17. Bellifemine, F., Bergenti, F., Caire, G., Poggi, A.: Jadea java agent development framework. In: Multi-Agent Programming, pp. 125–147. Springer, New York (2005)
18. Fipa, A.: Fipa acl message structure specification. Foundation for Intelligent Physical Agents. http://www.fipa.org/specs/fipa00061/SC00061G.html (2002). Accessed 30 Jun 2004
19. Smith, R.: The contract net protocol: Highlevel communication and control in a distributed problem solver. IEEE Trans. Comput. C **29**, 12 (1980)
20. Watkins, C.J., Dayan, P.: Q-learning. Mach. Learn. **8**(3–4), 279–292 (1992)

A CBR Approach to Allocate Computational Resources Within a Cloud Platform

Fernando De la Prieta, Javier Bajo and Juan M. Corchado

Abstract Cloud Computing paradigm continues growing very quickly. The underlying computational infrastructure has to cope with this increase on the demand and the high number of end-users. To do so, platforms usually use mathematical models to allocate the computational resource among the offered services to the end-user. Although these mathematical models are valid and they are widely extended, they can be improved by means of use intelligent techniques. Thus, this study proposes an innovative approach based on an agent-based system that integrated a case-based reasoning system. This system is able to dynamically allocate resources over a Cloud Computing platform.

1 Introduction

The technology industry and the scientific community have taken great strides in recent years toward implementing the Cloud Computing (CC) technological paradigm. This has resulted in a rapid growth of both private and public platforms [12, 17, 25, 28] aimed to provide innovative solutions that can resolve the current needs of the CC paradigm.

The marketing model used in the CC paradigm is innovative, as it is based on a pay-as-you-go concept [2], in which users must negotiate and previously establish a Service Level Agreement (SLA) in order to access services [1]. Once this contract

F. De la Prieta (✉) · J.M. Corchado
Department of Computer Science and Automation Control,
University of Salamanca, Plaza de la Merced s/n, 37008 Salamanca, Spain
e-mail: fer@usal.es

J. Bajo
Department of Artificial Intelligence, Technical, University of Madrid,
Bloque 2, Despacho 2101, Campus Montegancedo,
Boadilla del Monte, Madrid 28660, Spain
e-mail: jbajo@fi.upm.es

© Springer International Publishing Switzerland 2016
P. Novais et al. (eds.), *Intelligent Distributed Computing IX*,
Studies in Computational Intelligence 616
DOI 10.1007/978-3-319-25017-5_7

75

for computing goods has been established, both the users (through regular payments) and the CC system (by maintaining the service) are obligated to follow through with their agreement. In this regard, novelty is determined by the innovative spectrum of underlying technology (virtualization, service farms, web services, etc.), which have recently reached the point of allowing the services to be offered with the same level of quality, regardless of existing user demand [16, 26, 31]. These new possibilities at a technological level lead to the birth of a new concept, elasticity [9], which is based on the just-in-time production method [13].

Existing research in the state of the art is based on methods that use centralized algorithms based on mathematical and heuristic models [15, 19, 30], neither of which can ensure the efficiency of the system, or even its availability, in the event of a system failure.

Given these shortcomings, it is necessary to study new techniques that allow for the evolution of existing models with regard to elasticity of services. This study proposes the use of models derived from Artificial Intelligence (AI), since fron an internal point of view, a CC is characterized by its massive distribution, hetero-geneity, and high level of uncertainty, which is precisely where the application AI holds great potential. The inclusion of proactive, self-adaptation and learning capabilities, among others, is key for the evolution of these elastic management algorithms for computational resources. As a result, agents and multiagent systems [29] (MAS) were selected among all the available AI techniques because of their distributed nature and ability to work in environments such as CC systems, whose characteristics would clearly identify them as open systems.

Using this MAS-based approach, the framework of this study proposes a dynamic and self-adapting model for the distribution of computational resources in a CC environment. This model is based on the learning capabilities provided by a case-based reasoning (CBR) [10] system, an approach which has not previously been used in this type of distributed environment. These reasoning systems develop a reasoning model similar to that of humans, using past experiences to solve a specific problem.

This work is organized as follows: the following section provides a description of the context of and related approaches, Sect. 3 proposes a solution based on multiagent systems, while the evaluation and validation of these systems are pre-sented in Sect. 4. Finally, the last section presents the conclusions of the research.

2 Resource Allocation in Cloud Computing Platforms

In a CC environment, the hardware infrastructure is virtualized [7, 8], which means that there is an abstraction layer between the real hardware infrastructure and the computing nodes. Each of the services is actually deployed in the computing nodes of this abstraction layer (referred to as virtual machines). In turn, the services are generally distributed among various computational nodes, which is why their needs to be a work balance system that can distribute the requests among the various computational nodes attending the services.

The use of virtualization greatly simplifies the management of computational resources at the infrastructure level, making is possible to dynamically create or eliminate virtual machines on demand or even migrate a virtual machine from one physical server to another in execution time, without needing to stop or pause the machine. Therefore, and given the capabilities offered by virtualization technology, the problem, while complex, is actually simple in itself, since it is only based on the efficient redistribution of physical (real) resources among the different computational (virtual) nodes.

In current literature, the distribution of resources is viewed from two points of view [5]:

- **QoS-aware based, or market oriented** [5]. This first group is associated with a client-oriented distribution of resources model which attempts to minimize computational risks in order to distribute the computational resources according to the SLA reached, and following the pay-per-use economic model. According to this model, the management techniques for the computational resources aim to adhere to these agreements at all time, thus providing the quality of service that was requested and consequently expected by the end user. The state of the art includes studies in line with this approach by means of mathematical models [18, 23, 27].
- **Energy-aware based** [5]. In this second approach, the distribution of resources takes place by taking into account both the pre-established SLA and the energy consumption, which assumes compliance with both. There are fewer studies in the state of the art with this approach as compared to the first, although they are more novel. This includes a variety of techniques are also based on mathematical models [3, 15, 19].

In light of these studies in the current state of the art, it is necessary to propose a model for the distribution of computational resources which would take energy consumption into account. The aim is to reduce the energy consumption required to satisfy the SLA that have been established with the platform users. The present study follows a completely different approach based on optimization techniques and AI, which allows for the distribution of resources by following a distributed and scalable model, thus allowing the system to learn over an extended period of time.

As noted above, the CC computational paradigm has grown strongly in recent years; its development has led to the advancement of a large number of platforms, both public and private. A MAS framework based on VO has been selected to deal with these obstacles. Although one may initially consider these two distributed systems (MAS and CC) to be incompatible, a detailed analysis demonstrates that they are in fact not only complementary, but share considerable synergy between them. First of all, CC environments can cover the computational needs for persistence of information and the computing potential that MAS require for different applications such as data mining, management of complex services, etc. Additionally, MAS can be used to create a much more efficient, scalable and adaptable design for the CC environment than what is currently available. Finally, the use of MAS in the framework of the design for CC systems provides this paradigm with new characteristics such as learning or intelligence, which makes it possible to develop much more advanced computational environments in all aspects (intelligent services, interoperability among platforms, efficient distribution of resources, etc.). The number of studies that can be found on the state of the art relating CC with agent technology is actually quite low. However, this tendency is changing and it is becoming increasingly common to find studies and applications focused on this field. Despite the limited number of studies on the matter, **Agent-based Cloud computing**, or the **Agent-based Cloud platform**, is becoming a common concept, mentioned by various authors in recent years [4, 6, 14, 20–22, 24].

3 Proposed Architecture

Taking into account the needs and shortcomings detected in the review of the state of the art, this study proposes a new model of allocating resources based on a CBR approach and guided by a multiagent architecture especially designed for the management of CC environments. This section will describe the key components that allow extending the operation of the elastic algorithms for the distribution of resources proposed within this work.

To begin, we would like to note that since the proposed MAS is a distributed system by nature, each of the agents that work in the distribution of resources can be located throughout the entire CC environment. That is, the CC system is monitored and controlled in a distributed manner. This distributed monitoring model makes it possible to instantly adapt existing resources to the CC environment according to demand for each service, which in turn meets the dual objective of complying with the established SLA agreements and reducing energy consumption.

Figure 1 presents an overview of the agent-based architecture The following agents are directly related to the monitoring and control of the hardware:

- **Local Monitor**. In charge of gathering data related to the state of the local resources for each physical server, including the physical machine as well as the different virtual machine it hosts.
- **Local Manager**. In charge of controlling the computational resources of the physical machine. In other words, responsible for initiating or turning off virtual machines according to the previously configured service templates.
- **Global Manager**. The primary agent in charge of decision making with regard to the distribution of computational resources. In order to perform this task, the agent uses a CBR-BDI model, which will be explained in detail in the following section. As a means of support for making decisions regarding the distribution of resources, this agent uses a partial knowledge base (provided by the Local Monitor) and the ability to modify local resources in each machine (provided by the Local Manager).

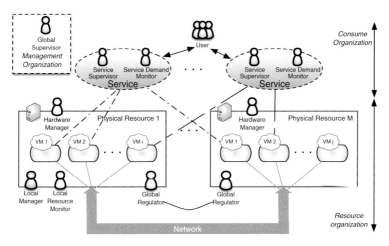

Fig. 1 Agent-based architecture for a Cloud Computing Platform

The redistribution of resources at a macro level is performed by the *Global Manager* agents, which have greater authority than the *Local Manager* agent and can inform them of the need to start up a new machine with a specific service and specific characteristics. At the end of this process, a new virtual machine (*VM*), with specific characteristics of Memory and virtual cpus will be instantiated in order to meet the current demand.

The *Global Manager* is a highly specialized agent that implements a CBR-BDI [10, 11] deliberative architecture. As a result, the reasoning process in each physical node is based on past experience gained from storing similar cases. The case memory is central to the entire CC system; the system's global knowledge can be shared by each of its members, in this case the *Global Manager* agent. Given that this memory can grow exponentially as a maintenance strategy, a high-speed schema-less database is used to provide fast access to the stored data, based on MongoDB.[1]

The Global Manager initiates the process by defining the concept of case $C = \{P, S(P), E\}$ where:

- P corresponds to the problem description, which has a matrix-matched representation associated to the instantiation of the use of resources.
- $S(P)$ is associated with the solution to the problem: $S(P) = \{M, vcpu\}$ in terms of memory and *vcp*u.
- The efficiency (E) is measured from two perspectives:

  - the degree of efficiency of the proposed solution within the physical server where the virtual machine has been instantiated. This degree of efficiency is

---

[1]http://www.mongodb.org.

proposed by the *Local Monitor* agent according to the usage rates of the processor and the allocated memory.

- The degree of efficiency from the point of view of the service. The degree of efficiency measures the number of additional nodes required by the service.

The CBR (*Case-Based Reasoning*) process is initiated and retrieves similar cases from the case memory. The most similar cases are selected according to the following steps: (i) Select the cases from the physical machines with similar characteristics; (ii) a vector is configured for each case that contains the same number of virtual machines that are in the case; and (iii) the cases selected from this subset are those that previously used the same service that is now requesting resources, and during a period of time similar to the current case.

A solution to the problem, which is based on the retrieved cases, will be prepared during the reuse phase:

- If the case base does not contain a previous similar case, the solution to the problem will be associated to the minimum resources determined at the level where the service is instantiated.
- If, on the other hand, similar cases are retrieved, the solution to the problem will be the closest case multiplied by the case efficiency:
- If the values assigned to the previous solution are greater than the values assumed by the machine, due to the fact that there are not many resources available, the result of the case will be the maximum amount of resources available in the machine.

Once the solution to the case has been calculated, the new node will be instantiated and its use evaluated from a micro and macro perspective, thus providing the value of efficiency for the solution. Finally, during the final state of the proposed CBR cycle, the case and its corresponding efficiency will be stored for future use.

4   Evaluation

The evaluation and validation of the model for this study will be done through a CC platform developed within the scope of the research carried out by the BISITE research group,[2] and will include different computational services at the hardware and software level. This CC platform was deployed in the HPC environment of the BISITE research group and composed of 15 latest generation machines that support virtualization in the hardware with the use of Intel-VT technology and the KVM virtualization system.

During the experiment, 10 threads that query to specific methods of the service (*GetSize* and *GetFolderContent*) are launched every three seconds, to a maximum

---

[2]http://bisite.usal.es.

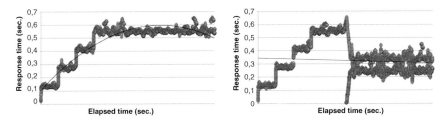

Fig. 2 Experiment 1: readjustment of the infrastructure resources for method (*left* no adaptation; *right* adaptation)

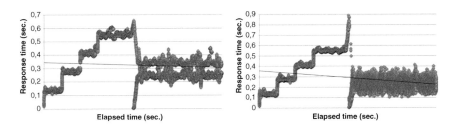

Fig. 3 Experiment 1: readjustment of the infrastructure resources for method (*left* experiment 1; *right* informed adaptation)

of 40 threads. The process starts once the agent-based architecture detects a decrease in performance, at which time it directly executes the adaptation process. The *Global Manager* agent for each of the physical machines that host the service nodes. We should recall that the *Global Manager* agent is a specialized agent that uses a CBR-BDI reasoning process [10] in charge of the distribution of resources at a macro level. Once they receive the initial alert, these agents resend the alert message to the remaining *Global Manager* agents in the CC system.

Each individual Global Manager hosted by each physical machine carry out in parallel the process described in the previous section. Thus *n* solutions are proposed. The agent-based architecture reactively selects the node that offers the most resources at the virtual machine level.

The results in terms of QoS can be seen in Fig. 2, which also show an increase in the quality level after the adaptation has been completed.

The case study was repeated numerous times, which made it possible to store a good number of past experiences in the case memory. However, as presented in Fig. 3 when there are many cases in the memory and a number of past experiences similar to the current problem, the adaptation results are actually better because the QoS level is lower.

5 Conclusions

This study initially set forth to be one of the first MAS approaches to fall within the framework of control and monitoring systems in a CC environment. The study proposed a new architectural model based on a MAS with a clearly integrative character. A series of algorithms for the distribution of computational resources in a CC environment were developed, evaluated and validated. Its biggest innovation centers on the system's dynamic ability to automatically adapt according to demand and learn from previous experiences.

This new model has demonstrated that a control and monitoring system in a CC environment can be designed with MAS. The inherently distributed nature of MAS makes it possible to implement elastic algorithms for services by following a distributed strategy. The distribution of responsibilities within the scope of this type of algorithm makes it possible not only to make decisions where the problems actually arise, but to distribute the computing capability required to reach a solution among different instances of the CC environment.

This approach also ensures independence of the decision-making process in software layers where the various actions are executed. There is no doubt that a change in the capabilities offered by the underlying technology will also require changes to be made in the proposed reasoning models, as with any approach with a traditional design. Given the definitions of roles at a high level, if the technology proposes new capabilities, the adaptation in the proposed architecture will consist of modifying the individual or individuals that perform specific tasks or have a role within the MAS.

Finally, This approach can maximize the degree of efficiency of the proposed solutions with regard to previous solutions, which in turn progressively improves the response and the system's capability since it is capable of learning. Moreover, this learning ability is important in an uncertain environment such as the CC system. If the context or environment of the CC platform changes at any given time, the adaptation model will evolve in turn, adapting the proposed solutions in order to maximize the efficiency of the given solution.

Acknowledgments This work has been supported by the MICINN project TIN2012-36586-C03-03.

References

1. Alhamad, M.; Dillon, T.S., Chang, E.: Conceptual SLA framework for cloud computing (2010)
2. Armbrust, M., et al.: A view of cloud computing. Commun. ACM **53**(4), 50–58 (2010)
3. Beloglazov, A. Abawajy, J. Buyya, R.: Energy-aware resource allocation heuristics for efficient management of data centers for cloud computing. Future Gener. Comput. Syst. **28**(5), 755–768 (2012)

4. Braubach, L., Jander, K. Pokahr, A.: A middleware for managing non-functional requirements in cloud PaaS. In: IEEE International Conference on Cloud and Autonomic Computing (ICCAC), pp. 83–92 (2014)
5. Buyya, R., Beloglazov, A., Abawajy, J.: Energy-efficient management of data center resources for cloud computing: a vision, architectural elements, and open challenges. Preprint arXiv: 1006.0308 (2010)
6. Cao, B.-Q., Li, B. Xia, Q.-M.: A service-oriented QoS-assured and multi-agent cloud computing architecture. In: Cloud Computing, pp. 644–649. Springer, Berlin (2009)
7. Che, J., et al.: A synthetical performance evaluation of OpenVZ, Xen and KVM. In: IEEE 2010 Asia-Pacific Services Computing Conference, pp. 587–594. IEEE (2010)
8. Chen, W., et al.: A novel hardware assisted full virtualization technique. In: The 9th International Conference for Young Computer Scientists, pp. 1292–1297. IEEE (2008)
9. Chiu, D.: Elasticity in the cloud. Crossroads 16(3), 3–4 (2010)
10. Corchado, J.M., et al.: Replanning mechanism for deliberative agents in dynamic changing environments. Comput. Intell. 24(2), 77–107 (2008)
11. Corchado, J.M., Laza, R.: Constructing deliberative agents with case-based reasoning technology. Int. J. Intell. Syst. 18(12), 1227–1241 (2003)
12. Fisher, P., Pant, R., Edberg, J.: Cloud Computing: Assessing Azure, Amazon EC2, Google App Engine and Hadoop for it Decision Making and Developer Career Growth. Apress, New York (2010)
13. Hutchins, D.:Just in time. Gower Publishing Ltd., London (1999)
14. Kang, J. Sim, K.M.: Cloudle: an ontology-enhanced cloud service search engine. In: Web Information Systems Engineering–WISE 2010 Workshops. Springer, Berlin, Heidelberg, pp. 416–427 (2011)
15. Kusic, D., et al.: Power and performance management of virtualized computing environments via lookahead control. Cluster Comput. 12(1), 1–15 (2009)
16. Liu, F., et al.: NIST Cloud Computing Reference Architecture, vol. 500, p. 292. NIST Special Publication (2011)
17. Luo, J.-Z., et al.: Cloud computing: architecture and key technologies. J. China Inst. Commun. 32(7), 3–21 (2011)
18. van Nguyen H., Dang Tran, F. Menaud, J.-M.: Autonomic virtual resource management for service hosting platforms. In: Proceedings of the 2009 ICSE Workshop on Software Engineering Challenges of Cloud Computing, pp. 1–8. IEEE Computer Society (2009)
19. Raghavendra, R., et al.: No power struggles: coordinated multi-level power management for the data center. In: ACM SIGARCH Computer Architecture News, pp. 48–59. ACM, (2008)
20. Sim, K.M.: Agent-based cloud computing. IEEE Trans. Serv. Comput. 5(4), 564–577 (2012)
21. Talia, D.: Cloud computing and software agents: towards cloud intelligent services. In: WOA 2011, pp. 2–6
22. Talia, D.: Clouds meet agents: toward intelligent cloud services. IEEE Internet Comput. 16(2), 78–81 (2012)
23. Van Hien N., Tran, F.D., Menaud, J.-M.: SLA-aware virtual resource management for cloud infrastructures. In: Ninth IEEE International Conference on Computer and Information Technology, CIT'09, pp. 357–362. IEEE (2009)
24. Venticinque, S., et al.: A cloud agency for SLA negotiation and management. In:Euro-Par 2010 Parallel Processing Workshops, pp. 587–594. Springer, Berlin, Heidelberg (2011)
25. Von Laszewski, G., et al.: Comparison of multiple cloud frameworks. In: 2012 IEEE 5th International Conference on Cloud Computing (CLOUD), pp. 734–741. IEEE (2012)
26. Wang, L., et al.: Cloud computing: a perspective study. New Gener. Comput. 28(2), 137–146 (2010)
27. Wei, G., et al.: A game-theoretic method of fair resource allocation for cloud computing services. J. Supercomput. 54(2), 252–269 (2010)
28. Wen, X., et al.: Comparison of open-source cloud management platforms: OpenStack and OpenNebula. In: 2012 9th International Conference on Fuzzy Systems and Knowledge Discovery (FSKD), pp. 2457–2461. IEEE (2012)

29. Wooldridge, M., Jennings., N.R.: Intelligent agents: theory and practice. Knowl. Eng. Rev. 10, (02), 115–152 (1995)
30. You, X., et al.: RAS-M: resource allocation strategy based on market mechanism in cloud computing. In: Fourth ChinaGrid Annual Conference, ChinaGrid'09, pp. 256–263. IEEE (2009)
31. Zhang, Q., Cheng, L., Boutaba, R.: Cloud computing: state-of-the-art and research challenges. J. Internet Serv. Appl. 1(1), 7–18 (2010)

A Multi-agent Strategic Planning System Based on Blackboard

J. Luis Dalmau-Espert, Faraón Llorens-Largo
and Rafael Molina-Carmona

Abstract In the last years, the organizations have faced deep changes in their environments that have led them to a new complex and uncertain world. More participatory, flexible and distributed structures are needed to address and reduce this uncertainty and complexity. This new form of governance involves changes in the Strategic Planning process to meet the new situation. A new agile, collaborative, integrated, and automated architecture of a Multi-Agent Strategic Planning System is presented. It is based on a blackboard and an ontology, and it lets the participating experts (human or not) cooperate and interact to ensure better decisions.

1 Introduction

The global economy, the technological dynamism and the growing need of knowledge, have led the organizations to a new complex and uncertain environment [17]. Technologies can support the smart organizations to change the traditional paradigm (centralized, reactive, inefficient) for Strategic Planning (SP) [12].

SP is defined as the process by which stakeholders of the firm analyze the internal and external environments for the purpose of formulating strategies and allocating resources to develop a competitive advantage in an industry that allows for the successful achievement of organizational goals [3]. Nowadays, the organizations approach the SP process using, in the best-case scenario, several tools and software technologies for some of the process steps, to achieve a certain degree of

J.L. Dalmau-Espert (✉) · F. Llorens-Largo · R. Molina-Carmona
Group "Informática Industrial E Inteligencia Artificial", University of Alicante,
Alicante, Spain
e-mail: jldalmau@dccia.ua.es

F. Llorens-Largo
e-mail: faraon@dccia.ua.es

R. Molina-Carmona
e-mail: rmolina@dccia.ua.es

© Springer International Publishing Switzerland 2016
P. Novais et al. (eds.), *Intelligent Distributed Computing IX*,
Studies in Computational Intelligence 616,
DOI 10.1007/978-3-319-25017-5_8

85

automation and formalization. This is the case of tools for competitive intelligence [13], business intelligence [9] or knowledge management [7]. However, they are not integrated with the SP process and, often, they just provide reports and documents designed with different criteria, unable to accelerate the subsequent revisions. There are three key issues that should be addressed within the SP process:

- Formally define a conceptual framework and a common vocabulary that serve to represent the information/knowledge (from the internal and external environment and from each participant) and to have a unified vision of the facts and a fluent communication and cooperation of the stakeholders.
- Formalize the SP process itself to determine the steps that make up this process [8], the information/knowledge type that is used and who is involved.
- Adapt the process to a shared, participatory and collaborative vision. The complexity and uncertainty are reduced and the agility and flexibility are increased.

Agents are strongly related to dynamic, uncertain and distributed problems due to its features of autonomy, collaboration, learning, induction and reasoning [11, 20]. A Multi-Agent System (MAS) is a set of interacting agents within an environment that, beyond their individual objectives, pursue a common goal [18]. MASs are widely used in decision support [20], planning and management [1], knowledge and search problems [6]. MASs support heterogeneous and distributed information sources and cooperation among agents with distributed logic. This is a key characteristic in dynamic systems where agile changes are needed.

Blackboard system is a problem-solving model suitable for tackling difficult, ill-structured problems [2]. It consists of three main components: the knowledge sources (KS) (independent computational modules equivalent to a domain expert), the blackboard (a global data repository containing data and partial solutions, where all interaction is made) and the control component (making decisions about the problem solving and the resources). A blackboard is a kind of MAS [16].

In this paper, a multi-agent strategic planning system based is proposed. It redefines the SP process in a formal, semi-automatic, integrated, agile, collaborative, concurrent and distributed manner. A blackboard is used because:

- The blackboard offers a workspace where agents can indirectly communicate and there is no need to establish any relation among them. This feature increases the system flexibility and agility to add new SP needs or requirements.
- In a blackboard system only a KS can be active at the same time. In the SP process it is necessary more than one agent can be active in a concurrent manner to carry out tasks that have no conflict. So, we use a hybrid approach [2].
- The role of the facilitator [12] within the SP process, whose principal function is to control the needs, steps, time and conflicts during the process, has an equivalent in the control component agent associated to the blackboard system.

In this system, the agents are the experts or stakeholders of the SP process; the blackboard offers a public workspace to participate, share, and collaborate; and the control component is an agent who controls changes on the blackboard. As a common language for information exchange, an application ontology is proposed.

In Sect. 2 the SP process and its ontology are presented. The Multi-Agent for SP is defined in Sect. 3. Finally conclusions and future works are presented.

2 Ontology for SP Process

The traditional SP process is performed in five main steps [8]: selection of the corporate mission and major corporate goals; analysis of the external competitive environment to identify opportunities and threats; analysis of the internal operating environment to identify strengths and weaknesses; selection of the strategies that, built on strengths and correcting weaknesses, take advantage of opportunities and oppose threats; and implementation of the strategy [12, 17].

An application ontology O_m (Fig. 1) is proposed to describe the concepts of the SP [4]. The design of O_m is based on the SP model of Llorens [10] and includes:

- The formalization of all concepts of this model and involved in the process and the properties and relationships between them.
- The formalization of existing tasks/steps and the order to perform them in the SP process, obtained from the dependencies among the concepts.
- The formalization of the Stakeholders as agents that carry out a particular task and the relation between each agent and the concept, which it references to.

The concepts are organized into levels based on a hierarchical structure (hierarchical taxonomy), which determines the inheritance of properties between a parent concept and its child concept. All concepts inherit Object properties.

The relations among concepts establish the dependences among them and the dependences among SP process tasks that generate them [14]. For practical

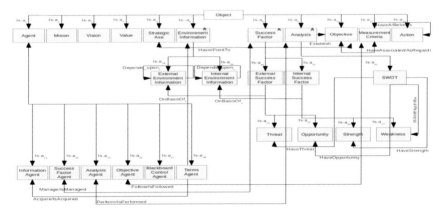

Fig. 1 Ontology for SP process

purposes, when performing the SP process, the definition of this ontology will allow that agents have a common language to formalize facts and knowledge that can be exchanged. Each instance of a concept is a particular fact. The concepts are classified as Passive (domain facts that agents perceive or conclude) and Active (entities which carry out tasks and can conclude and exchange passive concepts).

3   The Multi-agent Strategic Planning System

In general, a MAS is n-tuple **MAS** = (E, O, A, R, O_p, LoU) [5, 15, 19] where:

- E is the internal and external environment of the organization and it is defined as a set of infinite measurable variables [5]. E is considered inaccessible, no deterministic, dynamic and continuous [15]. The continuous evolution of the environment E is defined as a set of discrete states $E = \{e, e', \ldots\}$ where e is the environment state in a specific instant (when the SP process begins) [19].
- O is a finite set of O_m passive concept instances that represents the information and knowledge about the domain (SP) and the solution in the instant e. O contains the evolution of the solution while the SP process is being carried out.
- A is a finite set of O_m active concept instances that represents the participating agents. Each agent manages (create/modify/delete) a passive concept instance.
- R is a finite set of O_m relation instances existing, on one hand, among O_m passive concept instances (representing dependencies among passive concepts) and, on the other hand, among O_m passive and O_m active concept instances (representing which agent manages these passive concept instances).
- O_p is the finite set of operations typified as tasks that have to be made by the agents (stakeholders) in a SP process. LoU is the set of, so-called, laws of the universe, which are common for the environment E.

SP process is defined by a set of passive concept instances (SP information) and a set of active concept instances (agent's knowledge for generating this information). All passive concept instances created by an agent for representing its results must satisfy the syntax, relations, rules and axioms defined by ontology O_m.

As the ontology O_m, the proposed MAS architecture (Fig. 2) has a hierarchical organization where each level has an agent cluster that carries out a specific step of the SP process. An agent cluster is a group of agents which have identical or similar tasks and knowledge level. One of the reasons for such approach is to emphasize the communication among agents in the cluster. Information agents cluster is at the lowest level and it contains the agents who analyze the environment and extract information about it. This information is represented with instances of external/internal environment concepts. The next level is assigned to the success factor agent cluster who analyzes the information obtained at the lower level and establishes what external/internal environment concept instances may be considered success factors. It classifies these success factors in Strengths, Weaknesses, Opportunities and Threats (SWOT) creating instances of the related concept in O_m. At the next level,

Fig. 2 MAS architecture for
SP process

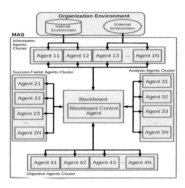

the analysis agent cluster creates an instance of the SWOT concept (a matrix with the strengths, weaknesses, opportunities and threats) and other agents (human or not) complete this matrix as a part of the SWOT analysis. At the highest level, an objective agent cluster determines the strategic objectives and checks out the necessary measurement criteria to reach them. Each agent is implemented with the most suitable and reliable architecture for its assigned task.

The communication mechanism among agents (regardless of its level) is based on the blackboard paradigm. So there is no direct communication between agents. The blackboard is divided into as many levels as agent clusters, in order to delimit the space in which each agent can manage instances (create, modify and delete). Each agent, as result of its work, creates instances of passive concepts in its corresponding blackboard level that contains formal and machine-readable information. Any other agent waiting for this information will receive a confirmation when it is ready and will query the blackboard for the instances that are necessary to complete its task. The agents can coordinate and cooperate through the blackboard.

The communication and coordination among agents are completed through the corresponding collaboration agreement, strategy, coordination and evaluation mechanisms that are associated to the Blackboard Control Agent (BCA). It is the brain system who has the facilitator role in the SP process. BCA directly controls the blackboard and indirectly the other agents. Each agent has an Agent Interface within its architecture, which is used to communicate to the BCA what information is needed to complete its work. The BCA registers the Concept Name of the needed information and the Agent Identifier in the Preconditions Workspace. There is a Tracing Module inside de BCA, constantly controlling content changes in the blackboard. When a change is produced in the blackboard and the Decision Module concludes that it matches one of the needs registered in the preconditions workspace, the decision module informs the Communication Module. Then, the Output Agent Interface communicates the agents that the information is available.

The decision module implements the heuristic function to coordinate the agents, eliminate conflicts and guarantee that the blackboard has the adequate solution space. This way, new agents can be added with no system changes.

4   Conclusions and Future Work

A Multi-Agent Strategic Planning System has been presented as a formal, semi-automated, agile and flexible approach to carry out the SP process. An ontology is used to formalize concepts, dependencies among them, process steps and stakeholders (agents) who participate within SP. This ontology provides a common vocabulary, which the Stakeholders can use to communicate and interact. They share information and knowledge through a blackboard in a cooperative and a collaborative manner controlled by the BCA who ensures the problem solution. As future work we are developing a visual tool, which lets the user define, formalize and execute the SP process, including active and passive concepts of O_m.

References

1. Aguilar, J., Rivas, F., Cerrada, M.: Multiagents systems for planning and management of the production factors in automation. Journal Ciencia e Ingeniería **31**(1), 13–24 (2010)
2. Corkill, D.: Collaborating software. blackboard and multi-agent systems and the future. In: Proceedings of the International Lisp Conference, New York (2003)
3. Cox, M.Z., Daspit, J., McLaughlin, E., Jones III, R.J.: Strategic management: is it an academic discipline? J. Bus. Strateg. **29**(1), 27–28 (2012)
4. Dalmau-Espert, J.L., Llorens-Largo, F., Molina-Carmona, R.: An ontology for formalizing and automating the strategic planning process. In: Proceedings of the 7th International Conference on Information, Process, and Knowledge Management, eKNOW, Lisbon (2015)
5. Ferber, J.: Multi-agent Systems: An Introduction to Distributed Artificial Intelligence. Adison-Wesley, London (1999)
6. Florin, L.: Design of a multiagent system for solving search problems. J. Eng. Stud. Res. **16** (3), 51–64 (2010)
7. Gruber, T.R.: A translation approach to portable ontology specifications. Knowl. Acquis. **5**(2), 199–220 (1993)
8. Hill, C.W.L., Jones, G.R.: Strategic Management: An Integrated Approach, South-Western College Pub, 10th edn. ISBN: 978–1111825843 (2012)
9. Howson, C.: Successful Business Intelligence: Unlock the Value of BI & Big Data, 2nd edn. McGraw Hill, Emeryville (2013)
10. Llorens, F.: Plan Estratégico de la Universidad de Alicante (Horizonte 2012) (2007). http://web.ua.es/es/peua/horizonte-2012.html
11. Llorens F., Rizo, R., Pujol, M.: An architecture model of intelligent agents system to approach problems of distributed knowledge. In: 5th FLINS Conference, pp. 237–243. Gent (2002)
12. Olivera Rodríguez, C.A.: Strategic Exercise: Facilitator Guide. Matanzas (2011)
13. Paradies, S., Zillner, S., Skubacz, M.: Towards collaborative strategy content management using ontologies. In: International Semantic Web Conference. Washington, DC (2009)
14. Prescott, J.E., Miller, S.H.: Proven Strategies in Competitive Intelligence: Lessons from the Trenches. Wiley, New York (2001)
15. Russell, S.J., Norvig, P.: Artificial Intelligence A Modern Approach, 3rd edn. Prentice Hall, Englewood Cliffs (2009)
16. Studer, R., Benjamins, V.R., Fensel, D.: Knowledge engineering: principles and methods. Data Knowl. Eng. **25**, 161–198 (1998)
17. Ventura, J.: Análisis Estratégico de la Empresa. Paraninfo CENGAGE Learning (2008)

18. Wooldridge, M.: Multiagent Systems. A Modern Approach to Distributed Artificial Intelligence. The MIT Press, Cambridge (2000)
19. Wooldridge, M.: An Introduction to MultiAgent Systems, 2nd edn. Wiley, Chichester (2009)
20. Yongyong, S.: Design of Multi-agent Intelligent Decision Support System Based on Blackboard. Lecture Notes in Information Technology, vol. 12, pp. 236–241 (2012)

Reliable Interaction in Multiagent Systems

Dejan Mitrović, Mirjana Ivanović, Milan Vidaković and Zoran Budimac

Abstract The social ability is one of the defining characteristic of software agents. This paper presents an infrastructure for reliable agent interaction in dynamic environments. The term "reliable" is used to indicate that messages are always delivered to agents, regardless of hardware or software failures. Two concrete models of agent interaction are discussed: direct peer-to-peer messaging, and an action synchronization framework based on a general-purpose model for parallel computations.

1 Introduction

One of the key aspects of the agent technology is the *social ability* of agents [2]. That is, an agent rarely exists on its own, and is instead a member of a society of agents. More formally, a *multiagent system* represents a software system in which agents interact with each other and the environment in order to solve the problem [2].

Extensible Java EE-based Agent Framework (XJAF) is our multiagent middleware based on the modern enterprise technologies [5, 11]. One of the main characteristics of XJAF is that it operates on top of computer clusters. In this way, it provides the two main features: automatic load-balancing of deployed agents, and fault-tolerance of its internal components (including agents).

D. Mitrović (✉) · M. Ivanović · Z. Budimac
Department of Mathematics and Informatics, Faculty of Sciences,
University of Novi Sad, Novi Sad, Serbia
e-mail: dejan@dmi.uns.ac.rs

M. Ivanović
e-mail: mira@dmi.uns.ac.rs

Z. Budimac
e-mail: zjb@dmi.uns.ac.rs

M. Vidaković
Faculty of Technical Sciences, University of Novi Sad, Novi Sad, Serbia
e-mail: minja@uns.ac.rs

© Springer International Publishing Switzerland 2016
P. Novais et al. (eds.), *Intelligent Distributed Computing IX*,
Studies in Computational Intelligence 616,
DOI 10.1007/978-3-319-25017-5_9

93

The purpose of this paper is to present how XJAF achieves *reliable* agent communication and action coordination. Here, the term "reliable" is used to indicate that messages are always delivered to agents, regardless of hardware or software failures.

Besides the reliable peer-to-peer message exchange, in this paper we also propose a new action coordination framework for multiagent systems. The framework is based on the *Bulk-Synchronous-Parallel* (BSP) computational model [9]. BSP is a general-purpose model for parallel execution, which has generally not been used by the agent technology researchers or practitioners. However, as shown in this paper, it can be efficiently used as an action coordination framework.

The rest of this paper is organized as follows. Section 2 presents the related work. Section 3 discusses how the two interaction models are realized in XJAF, with the focus on various cluster-specific features and issues. Concrete demonstrations of the interaction architecture's reliability are given in Sect. 4, while the overall conclusions and future work are presented in Sect. 5.

2 Related Work

Over the years, the fault-tolerance in multiagent systems has been the focus of many researchers. However, the fault-tolerance of agents themselves has received significantly more interest than the reliability of the communication infrastructure [4]. Papers that deal with the communication reliability often do so in the context of mobile agents, i.e. how to efficiently deliver a message to an agent that moves across the network [1, 6, 12]. Instead of mobility, here we are concerned with the resilience to hardware and software failures.

The existing research that most closely matches the work presented in this paper is the architecture named *External Fault-Tolerant Layer* (EFTL) [7, 8]. ETFL relies on the *Java Message Service* (JMS) and its *publish-subscribe* model [3] for inter-agent communication.

XJAF also relies on the JMS for inter-agent communication, and has done so since its first version, published several years before ETFL [10]. Nonetheless, there are some important differences between the two approaches. Unlike ETFL, XJAF does not perform any manual inspection of the communication state. That is, the JMS realization used in XJAF is advanced enough so that our system does not include any separate components that need to check for faults.

The main difference, however, is in the communication model. XJAF uses the JMS *point-to-point* model, with *message-driven beans* acting as intermediaries (see Sect. 3). This is because in computer clusters the publish-subscribe model (used in ETFL) delivers the message once *per cluster node*. This means that an agent would simultaneously process the message on each cluster node.

3 Reliable Interaction in XJAF

As noted earlier, the purpose of this paper is to extend XJAF with two reliable communication models: the peer-to-peer message exchange, and the BSP-based action coordination. In this section, we present how the reliability is achieved, with the focus on technical and cluster-specific difficulties and their solutions.

3.1 Peer-to-peer Interaction

In XJAF, agents are represented by *Enterprise JavaBean* (EJB) components [3]. As discussed in [5, 11], this design approach simplifies the agent development process and brings numerous advantages. One of the advantages relevant here is the readily-available agent state replication and failover, which makes XJAF agents resilient to hardware and software failures. Now, we extend the fault-tolerance to the message exchange sub-system.

The exchange of messages in XJAF is achieved via the *Java Message Service* (JMS), one of one of the standard Java EE APIs for asynchronous communication of loosely-coupled components [3]. The general flow of messages in the JMS-based communication in XJAF is shown in Fig. 1. Messages are published to a *queue* and consumed by *message-driven beans* (MDBs) [3]. MDBs are organized in a pool, which can automatically grow (and shrink) according to the demand.

An MDB delivers the message to the target agent by invoking its method `onMessage`. Once the method is executed, the message is automatically acknowledged and removed from the queue. If, however, there is an error, the message is re-queued and the delivery is retried at a later time. After a number of unsuccessful deliveries, the message is stored in the so-called *dead-letter queue* and can be inspected and processed manually. All possible issues, such as message ordering, concurrent access, cluster-wide coordination, etc., are handled by JMS.

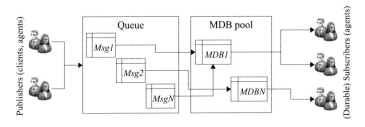

Fig. 1   Flow of messages in the JMS-based communication infrastructure realized in XJAF

3.2 The BSP Model

The standard BSP model incorporates a number of *components*, each with its own memory and processing abilities, and the capability of communicating with other components [9]. The overall computation is performed in so-called *supersteps*. Within a superstep, a component can receive messages sent to it during the previous superstep, in can perform a number of local calculations, and also send messages to other components. A superstep ends once all components reach a certain *barrier*.

Unlike the peer-to-peer model, the purpose of BSP within XJAF is for the agents to synchronize their action execution steps, rather than to exchange information. The BSP-based action synchronization process in XJAF revolves around the central component named *Barrier*. The Barrier keeps tracks of the supersteps and of the registered and active agents, it issues the appropriate executon signals, etc.

The overall execution flow is fairly simple and can be summarized as follows. Once started, the agent registers itself with the Barrier. At the beginning of a new superstep, the Barrier dispatches signals to all registered agents. It waits a certain amount of time for the agents to report back. If the time expires, the Barrier will try to determine if some of the pending agents have in the meantime become unavailable, and whether it should continue to wait.

In terms of reliability, the Barrier obviously represents the single point of failure. Therefore, it needs to be designed as a highly-available component. Given its functioning and operational requirements, the Barrier in XJAF is modeled as a regular agent, i.e. as an EJB component. This means that it achieves the state replication and failover "natively," and that all the communication between the Barrier and agents is done through the reliable peer-to-peer messaging system.

4 Examples

In order to show the reliability of agent interaction in practice, a pair of case studies was developed. The first case study demonstrates the fault-tolerance of the peer-to-peer messaging system in a dynamic environment. The second one shows how the Barrier's internal state is replicated as cluster nodes break down.

4.1 Reliability of the Peer-to-peer Messaging

To demonstrate the reliability of the peer-to-peer messaging, we developed an agent that follows a simple algorithm. Upon receiving a message, the agent prints the appropriate log information. It then simulates the message processing by sleeping for a number of seconds. Finally, the agent prints the total number of messages processed

Table 1 One execution flow of the first case study

Cluster node 1	Cluster node 2
	Processing a message... Total 1 msg
Processing a message... Total 2 msg	
	node abruptly terminated
connection failure detected	
Processing a message... Total 3 msg	
...	

Each column shows the log messages of the agent printed at the corresponding cluster node

so far. During the execution, however, cluster nodes are abruptly terminated, and the system observes if the agent eventually processes all messages.

One execution of the case study on a two-node cluster is shown in Table 1. Each column represents a cluster node and contains the agent's log messages printed at that node. The agent is automatically executed across the cluster in a round-robin fashion and its internal state is successfully replicated. While the agent was processing the third message, however, its host was abruptly terminated. As a result, its message was not acknowledged by the MDB and the delivery was repeated on the remaining node. Here, it is worth noting that the agent's internal state is replicated on the successful execution of its message processing method, as expected.

4.2 Reliability of the BSP Implementation

In order to test if the Barrier's internal state is replicated as expected, we developed a pair of BSP agents which simply log the superstep number once they receive a signal from the Barrier. The entire application (i.e. the Barrier and the agents) is first deployed and executed in a virtual cluster. Then, the cluster nodes are terminated at random. One possible execution in a two-node cluster is shown in Table 2.

The results show how the Barrier is executed on every cluster node in the round-robin fashion. Once in the middle of the superstep #3, the corresponding node is abruptly terminated. This change is detected by the enterprise server, which moves the agents and the Barrier itself to the remaining node. Since the agents continue to report the correct superstep numbers, it can be concluded that the Barrier's internal state is correctly replicated.

5 Conclusions and Future Work

In this paper, we have presented recent developments of the interaction infrastructure in our Extensible Java EE-based Agent Framework (XJAF). To focus is on reliability: the infrastructure's ability to deliver messages to target agents regardless of the underlying hardware or software failures.

Table 2 One execution flow of the second case study, involving a Barrier and a pair of BSP agents

Cluster node 1	Cluster node 2
	Agent1 executing superstep #1
	Agent2 executing superstep #1
Agent1 executing superstep #2	
Agent2 executing superstep #2	
	Agent1 executing superstep #3
	node abruptly terminated
connection failure detected	
Agent2 executing superstep #3	
...	

Two reliable interaction systems are introduced. The first one is the peer-to-peer messaging, which represents the standardized way for agents to exchange information. The second one represent a novel approach to distributed action coordination, and it's based on the *Bulk-Synchronous-Parallel* computational model.

The planned future work in this area will be focused on providing reliable communication at a higher level of abstraction. For example, we plan to realize many standardized interaction protocols. Although their realizations will be rooted in the present peer-to-peer messaging system, a special attention will be required in order to deal with unexpected failures of interaction participants.

Acknowledgments This work was partially supported by the Ministry of Education, Science and Technological Development of the Republic of Serbia, project no. OI174023: "Intelligent techniques and their integration into wide-spectrum decision support."

References

1. Ahn, J.: Atomic mobile agent group communication. In: Proceedings of the 7th IEEE Conference on Consumer Communications and Networking Conference. pp. 797–801. IEEE Press (2010)
2. Bădică, C., Budimac, Z., Burkhard, H.D., Ivanović, M.: Software agents: languages, tools, platforms. Comput. Sci. Inf. Syst. ComSIS **8**(2), 255–298 (2011)
3. Goncalves, A.: Beginning Java EE 6 platform with GlassFish 3, 2nd edn. Apress, New York (2010)
4. Isong, B.E., Bekele, E.: A systematic review of fault tolerance in mobile agents. Am. J. Softw. Eng. Appl. **2**(5), 111–124 (2013)
5. Mitrović, D., Ivanović, M., Budimac, Z., Vidaković, M.: Supporting heterogeneous agent mobility with ALAS. Comput. Sci. Inf. Syst. **9**(3), 1203–1229 (2012)
6. Murphy, A.L., Pietro Picco, G.: Reliable communication for highly mobile agents. Auton. Agent. Multi-Agent Syst. **5**(1), 81–100 (2002)

7. Tosic, M., Zaslavsky, A.: Reliable multi-agent systems with persistent publish/subscribe messaging. In: Ali, M., Esposito, F. (eds.) Innovations in Applied Artificial Intelligence, Lecture Notes in Computer Science, vol. 3533, pp. 165–174. Springer (2005)
8. Tosic, M., Zaslavsky, A.: Generic fault-tolerant layer supporting publish/subscribe messaging in the mobile agent systems. In: Chen, C.S., Filipe, J., Seruca, I., Cordeiro, J. (eds.) Enterprise Information Systems VII, pp. 207–214. Springer (2006)
9. Valiant, L.G.: A bridging model for parallel computation. Commun. ACM **33**(8), 103–111 (1990)
10. Vidaković, M., Konjović, Z.: EJB based intelligent agents framework. In: Proceedings of the 6th IASTED International Conference on Software Engineering and Applications. pp. 343–348 (2002)
11. Vidaković, M., Ivanović, M., Mitrović, D., Budimac, Z.: Extensible Java EE-based agent framework – past, present, future. In: Ganzha, M., Jain, L.C. (eds.) Multiagent Systems and Applications, Intelligent Systems Reference Library, vol. 45, pp. 55–88. Springer (2013)
12. Xu, W., Cao, J., Jin, B., Li, J., Zhang, L.: GCS-MA: a group communication system for mobile agents. J. Network Comput Appl. **30**, 1153–1172 (2007)

Fault Tolerant Automated Task Execution in a Multi-robot System

Stanislaw Ambroszkiewicz, Waldemar Bartyna, Kamil Skarzynski
and Marcin Stepniak

Abstract In multi-robot systems unexpected situations occur frequently, and cause
failures of robots performing tasks. Mechanisms for automation of failure handling
and recovery (if possible) are proposed. They are based on general protocols similar to the well known standard WS-TX for business transactions. The protocols and
mechanisms are implemented in the prototype Autero system as a software platform
for accomplishing complex tasks in open and heterogeneous multi-robot systems.

1 Introduction

Robots provide services that can be used to accomplish complex tasks in everyday
life, see for example [1].

The Autero system presented in this paper is a software platform for delegating
tasks (by human users) to be accomplished in a multi-robot system. The platform
is responsible for planning, distributing subtasks to the available robots, monitoring
and controlling the subtasks performance as well as handling recovery from failures.
All this is done in an automatic way on the basis of generic communication protocols, so that the architecture of Autero is appropriate for open distributed systems
consisting of heterogeneous robots providing services.

Since some ideas and methods are adopted form electronic business transactions,
realization of a task is called transaction. The transaction is successfully completed,
if the delegated task is accomplished. Special transaction mechanism to handling
failures is designed that has the following properties.

S. Ambroszkiewicz (✉) · W. Bartyna · K. Skarzynski · M. Stepniak
Institute of Computer Science, Polish Academy of Sciences, Jana Kazimierza 5,
01-248 Warsaw, Poland
e-mail: sambrosz@ipipan.waw.pl

S. Ambroszkiewicz · W. Bartyna · K. Skarzynski · M. Stepniak
Systems Research Institute, Polish Academy of Sciences, Newelska 6,
01-447 Warsaw, Poland

© Springer International Publishing Switzerland 2016
P. Novais et al. (eds.), *Intelligent Distributed Computing IX*,
Studies in Computational Intelligence 616,
DOI 10.1007/978-3-319-25017-5_10

1. Failed services may be replaced with other services during task realization.
2. General plan may be changed.
3. The transaction ends after successful completion of the task, or inability to complete the task, or cancellation of the task.

The classic meaning of the term *transaction* in Information Technology goes back to the ACID properties of modifying a database. Transactions are understood as a mechanism for guaranteeing a group of operations (as a whole) performed on database to be atomic (either all or nothing), to produce consistent results, to be isolated from other operations, and their result to be durably recorded.

Long-running transactions avoid locks on non-local resources, use compensation to handle failures, potentially aggregate smaller ACID transactions (also referred to as atomic transactions), and typically use a coordinator to complete or abort the transaction. In contrast to rollback in ACID transactions, compensation restores the original state, or an equivalent, and is domain-specific. For example, the compensating action for a failure when transporting a cargo by one robot, is arranging a second robot that can continue the transport to the destination, and charging (as a penalty) the owner of the first robot for the delay.

OASIS Web Services Transaction (WS-TX) [2] is the standard specification that describes coordination types that are used with the extensible coordination framework described in the WS-Coordination specification. It defines two coordination types: Atomic Transaction and Business Activity. These coordination type can be used for building applications that require consistent agreement (transaction) on the outcome of distributed activities (services).

Based on the WS-TX standard the transaction protocols have been designed for multi-robot systems, and implemented in the Autero. The system has been tested in a universal simulated environment implemented in Unity 3D. Tests performed in a real environment are always limited by the devices (robots) and their limited range and capabilities. From the point of view of the proposed information technology (the protocols) the fact that the environment is simulated is irrelevant.

Since services are provided by heterogeneous robots, there must be common and generic representation (ontology) of the environment where the robots operate. The ontology consists of concepts and relations between them, i.e. objects, object attributes, and the relations between objects. Each object is of a certain, pre-defined type. The object type is defined by its attributes, and by the internal (hierarchical) structure, i.e. object may consist of sub-objects and relations between these sub-objects. An elementary type has no internal structure, so it is defined only by attributes. A complex type has a hierarchical structure of subtypes. The ontology is defined as a hierarchical collection of types of objects (see [3]). Primitive attributes and relations are the key elements for constructing the types. The object itself, as an instance of its type, is defined by assigning specific values to its attributes and by specifying relations. Primitive attributes and primitive relations must be *measurable* and *recognizable* by the robots.

The ontology is common for all components of the system, so that the robots are context-aware (see [4]). The ontology is also used to specify tasks, and define types of services called *service interfaces* consisting of the following elements.

- Type of service, i.e. type of action that the service performs.
- Specification of the inputs and outputs of the service.
- The conditions required for input of the service (preconditions), and the effects of its execution (postconditions) specifying the output. These conditions are expressed as relations between objects in the environment (ontology).
- Service attributes.

Service attributes contain information about the static features of a service and are used during planning, for example, operation range for a transport service, and average realization time.

The procedure of reserving the services for a task realization (according to a fixed plan) is called *arrangement* and has a form of contract between client and service provider. A similar solution was applied in the ASyMTRe-D approach [5].

There are several multi-robot systems architectures, see for example [6], [7], and [8]. However there, the tasks are executed in a tightly coupled manner and dedicated to a restricted class of tasks. There are also architectures that coordinate task execution, e.g. ALLIANCE [9] and M+ [10] with mechanisms for handling the failures. However, they are based on direct low level control that requires dedicated algorithms dependent on the robot hardware. Viewing a robot function as a service (having common interface) allows to apply Service Oriented Architecture (SOA) paradigm to multi-robot systems.

2 The Architecture of Autero

Autero was designed according to the SOA paradigm [11]. The system components communicate to each other using generic protocols. Repository stores ontology and provides access to it by the other system components. It also has a graphical user interface (GUI) for developing the ontology, and its management.

Task Manager (TM for short) represents a client, and provides a GUI for the client to define tasks and monitor their realization. The Planner provides abstract plans for TM, that are used to construct a concrete plan on the basis of information of available services (provided by Service Registry) and by arranging these services (by the Arrangement Module (AM) in a business process. TM controls the plan realization by communicating with the services arranged in the plan.

Arrangement is performed by sending to services requests (in the form of intentions) and collecting answers as quotes (commitments). Service Registry stores information about services currently available in the system. Each service, in order to be available, must be registered in Service Registry via Service Manager (SM). It is a robot interface for providing its services for an external client, in this case, TM is the client. SM controls the execution of the subtask delegated by TM and reports success or failure to TM.

Task is defined as logical formula that describes the required final situation in the environment by using types, objects and relations from the ontology stored in Repository.

For a given task, Planner returns abstract plans, that when arranged and executed, may realize the final situation specified by the task in question. An abstract plan is represented as a directed graph where nodes are service types and edges correspond to causal relationship between the output of one service and the input of the second service. The relationships determine the order of arrangement and then the execution of a concrete plan that has also a form of a directed graph (called business process) however, its nodes are concrete arranged services. Sometimes it is not possible to arrange the whole business process before the start of execution phase. In this case, the business process includes the unassigned nodes for which the arrangement is to be done later on.

In a concrete plan its node may represent a composed service (as a subprocess) consisting of already arranged services. Plan may also include handlers responsible for a compensations and failure handling.

Task Manager initializes the service execution by sending the required input data to the Service Manager. The service is executed in accordance with the agreement made in the arrangement phase. After execution phase, Service Manager sends a response with the output data, being a confirmation of successful subtask completion, e.g. changing situation in the environment to the required one. Task Manager can also stop the service execution before its completion. This may be caused by the task cancellation by the client, a failure during concurrent execution of other services in the plan (that can not be replaced), or by changes in the environment making the plan infeasible.

Robot may not be able to successfully complete the task. In this case, its Service Manager notifies Task Manager by sending a detailed description of the problem. On this basis, TM can take appropriate actions. If Service Manager is not able to send such information, TM must invoke appropriate cognitive service (special patrolling robot, if available) to recognize the situation resulting from the failure.

3 Failure Handling

Failures during task accomplishing by robots may cause problems that must be handled. Mobile robotics is still a subject of extensive research, and at its current stage of development, failures occur frequently.

Task Manager is equipped with a failure handling mechanism based on the simple algorithm.

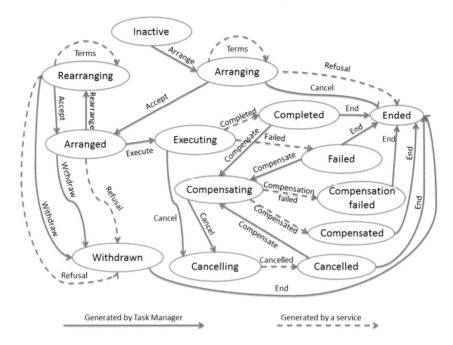

Fig. 1 Transaction protocol state transition diagram

All services are performed within a transaction that contains a dynamic set of participants. The transaction does not require all participants to successfully complete their tasks. The failure of a single service does not require the termination of the transaction. Termination is only necessary when it is not possible to continue the task.

Compensation is performed after a cancellation of a subtask execution by a service or the occurrence of a failure that interrupts the execution. It is designed to restore the original state of the environment before the execution. Since restoring that situation is sometimes impossible, the compensation may change the situation resulting from the failure to a situation from which the task realization can be continued. Note that even for such simple tasks (that seem to be trivial) a universal failure recovery mechanism and corresponding compensations are not easy to design and implement. A concrete plan should contain predefined procedures for failure handling and compensations.

Communication between Task Manager and Service Manager is done according to the transaction protocol which defines the states of the services and messages used to change them (see Fig. 1). It allows Task Manager to initialize particular phases of service invocation, monitor their progress, and perform additional actions, e.g. compensation. A service sends messages (according to the protocol) to notify Task Manager about the status of the delegated task performance.

All necessary data required for a task execution is a part of the transaction protocol message. This method allows to ensure the greater consistency of the system state. During the task execution messages are sent according to the specific sequences. They can create different combinations, but a set of possible messages in a given state of the service is strictly defined in the transaction protocol.

4 Conclusions

The presented work should be viewed as a preliminary study of the important and hard problem of designing mechanisms for handling failures and recoveries in open and heterogeneous multi-robot systems. The mechanisms must be based on generic protocols, so that problem is reduced to design such protocols.

The prototype system Autero verified that the proposed mechanisms of transactions are useful in the systems of heterogeneous robots in a simulated environment, and may improve their reliability. Task Manager can be configured to fully automate task accomplishment. Tasks can be delegated not only by human users but also by other system components or any software applications that can communicate according to the specified protocols.

Acknowledgments This work was carried out within the project RobREx—Autonomy for rescue and exploration robots, grant NRDC no PBS1/A3/8/2012. Marcin Stepniak was partially supported by the Foundation for Polish Science under International PhD Projects in Intelligent Computing.

References

1. Okada, K., Ogura, T., Haneda, A., Fujimoto, J., Gravot, F., Inaba, M.: Humanoid motion generation system on hrp2-jsk for daily life environment. In: 2005 IEEE International Conference on Mechatronics and Automation, vol. 4, pp. 1772–1777. doi:10.1109/ICMA.2005.1626828 (2005)
2. WS-TX 1.2 OASIS Standards.: https://www.oasis-open.org/committees/ws-tx/ (2009). Accessed 02 February 2009
3. Ambroszkiewicz, S., Bartyna, W., Faderewski, M., Terlikowski, G.: Multirobot system architecture: environment representation and protocols. Bull. Polish Acad. Sci. Tech. Sci. **58**(1), 3–13 (2010). doi:10.2478/v10175-010-0001-y
4. Go, Y.C., Sohn, J.C.: Context modeling for intelligent robot services using rule and ontology. In: The 7th International Conference on Advanced Communication Technology, ICACT 2005. vol. 2, pp. 813–816. IEEE (2005)
5. Tang, F., Parker, L.E.: A complete methodology for generating multi-robot task solutions using asymtre-d and market-based task allocation. In: 2007 IEEE International Conference on Robotics and Automation, pp. 3351–3358. IEEE (2007)
6. Long, M., Gage, A., Murphy, R., Valavanis, K.: Application of the distributed field robot architecture to a simulated demining task. In: Proceedings of the 2005 IEEE International Conference on Robotics and Automation, ICRA 2005. pp. 3193–3200. IEEE (2005)

7. Michael, N., Zavlanos, M.M., Kumar, V., Pappas, G.J.: Distributed multi-robot task assignment and formation control. In: IEEE International Conference on Robotics and Automation, ICRA 2008. pp. 128–133. IEEE (2008)
8. Pagello, E., Ferrari, C., d'Angelo, A., Montesello, F.: Intelligent multirobot systems performing cooperative tasks. In: 1999 IEEE International Conference on Systems, Man, and Cybernetics, IEEE SMC'99 Conference Proceedings. vol. 4, pp. 754–760. IEEE (1999)
9. Parker, L.E.: Alliance: an architecture for fault tolerant multirobot cooperation. IEEE Trans Robot Autom 14(2), 220–240 (1998)
10. Botelho, S.C., Alami, R.: M+: a scheme for multi-robot cooperation through negotiated task allocation and achievement. In: Proceedings of IEEE International Conference on Robotics and Automation, 1999. vol. 2, pp. 1234–1239. IEEE (1999)
11. Krafzig, D., Banke, K., Slama, D.: Enterprise SOA: service-oriented architecture best practices. Prentice Hall Professional, Upper Saddle River (2005)

Part III
Ambient Intelligence and Social Networks

Analysis of Mental Fatigue and Mood States in Workplaces

André Pimenta, Davide Carneiro, José Neves and Paulo Novais

Abstract Mental fatigue is a concern for a range of reasons, including its negative impact on productivity and quality of life in general. The maximal working capacity and performance of an individual, whether physical or mental, generally also decreases as the day progresses. The loss of these capabilities is associated with the emergence of fatigue, which is particularly visible in long and demanding tasks or repetitive jobs. However, good management of working time and of the effort invested in each task, as well as the effect of breaks at work can result in better performance and better mental health, delaying the effects of fatigue. In this paper a model and prototype are proposed to detect and monitor fatigue, based on behavioral biometrics (Keystroke Dynamics and Mouse Dynamics). Using this approach, the aim is to develop leisure and work context-aware environments that may improve quality of life and individual performance, as well as productivity in organizations.

1 Introduction

It is increasingly common for people to stretch their limits in order to have more time for work, for family and for leisure activities [7]. This extra time is usually obtained at the expense of rest and relaxation time. This, together with inadequate sleep patterns, is one of the leading causes of fatigue.

Fatigue and its effects may not be immediate or may not even be visible, but there are many consequences on health and safety at work, as well as on labor productivity.

A. Pimenta · D. Carneiro (✉) · J. Neves · P. Novais
Algoritmi Centre—University of Minho, Braga, Portugal
e-mail: apimenta@di.uminho.pt

D. Carneiro
e-mail: dcarneiro@di.uminho.pt

J. Neves
e-mail: jneves@di.uminho.pt

P. Novais
e-mail: pjon@di.uminho.pt

© Springer International Publishing Switzerland 2016
P. Novais et al. (eds.), *Intelligent Distributed Computing IX*,
Studies in Computational Intelligence 616,
DOI 10.1007/978-3-319-25017-5_11

111

This can be seen often in students preparing for exams, office workers, industrial workers, health care professionals, drivers, pilots or military personnel. Fatigue may even put people working in safety-sensitive jobs at risk, and any mistake on their part can lead to loss of lives [6, 18].

Mental fatigue can be seen as a state that involves a number of effects on a set of cognitive, emotional and motivational skills and usually results in overall discomfort, as well as the emergence of performance limitations. Some of these limitations imply that a tired person is often less willing to engage in tasks of effort, or perform the task of a conditioned form, well below their normal capacity [8]. Thus, mental fatigue may be characterized by a perception of a lack of mental energy. Persons who are affected by mental fatigue may feel like they have less energy than usual and are unusually tired and lethargic. Excessive activity and stimulation of the brain can cause a person to feel mentally exhausted, and the feeling is similar to what the body feels when a person is physically fatigued [19].

The effects of fatigue may occur at any moment and they may persist from only a few hours to several consecutive days. Depending on its duration and intensity, fatigue may make the carrying out of daily tasks increasingly difficult or even impossible. In severe or prolonged cases, it can cause illnesses such as depression or chronic fatigue syndrome [11, 18]. Thus, the detection and monitoring of fatigue are nowadays very important towards not only organizational performance but also worker well-being and health.

This paper describes an non-intrusive approach for fatigue monitoring and detection in real-time. It is especially suited for workplaces in which computers are used as it relies on the analysis of the workers interaction patterns with the computer. We conduct a case study to analyze how mental fatigue, mood and interaction performance interact. The main aim is to develop working and leisure environments that are aware of the user state and can react accordingly so as to improve the user's performance and well-being.

1.1 Related Work

There are many factors, both internal and external, that modulate the onset and presence of fatigue. These include sleep deprivation, naps, noise, heat, mood, motivation, time of day, and workload, and of course the individual profile [3].

The user's profile provides valuable information with respect to the potential level of fatigue. It can be seen as a predicted base level of fatigue in the sense that it establishes a baseline, according to the lifestyle of the individual [2]. These aspects have been thoroughly studied, mostly by psychologists, and encompass:

- **Age**—Defines the mental age of the individual. It is important to understand the expected cognitive abilities of the individual, which may have a tendency to degrade with old age.
- **Gender**—The mental states are different between men and women.

- **Professional occupation**—Many occupations are intrinsically more tiresome or exhausting than others.
- **Consumption of alcohol and drugs**—The use of certain substances for short or prolonged periods may cause dependencies and other effects that lead to a state of mental fatigue.

Mental fatigue is also affected by a number of other external factors. They may or may not be directly related to the individual's behavior. They include:

- **The mood** of the individual may influence decisively his or her mental state, with a particular effect on his or her motivation to work. Although tired, the individual may overcome (even if only temporarily) the effects of fatigue with a positive mood and motivation.
- **Stress** may be defined as the demands placed upon the individual's mind or body by external stimuli, requiring the individual to acclimatize to the dynamic requirements of the environment. However, these processes of acclimatization require an additional effort from the brain which, when prolonged over long or intense periods, will result in mental fatigue.
- **Mental Workload** as a result of the relationship between the amount of mental processing capacity and the amount required by the task.
- **Sleepiness** is often mistaken for mental fatigue or generalized as such. A difference exists and must be pointed out. However, the mistake is understandable since sleepiness is a symptom that is strongly connected to mental fatigue: it is one of the methods our brain uses to tell us that it is running out of resources. Sleepiness often results in a general loss of the individual vitality.

These factors can be assessed using validated tools such as the USAFSAM fatigue scale [15] for mental fatigue states, or as the NASA TLX [5] in the case of mental workload.

However, these instruments, based on the individual's subjective interpretation of the symptoms, do not fully take into account inter-individual differences. There are instruments that help to account for individual differences, such as the Profile of Mood States (POMS) [4] and the State-Trait Anxiety Inventory (STAI)[17]. However, their use in complex systems can prove to be complicated and confusing, due to the same problems that can be observed in subjective measures of fatigue detection [4, 10].

The POMS can also be used to estimate fatigue with the Vigor-Activity and Fatigue-Inertia factors. The POMS measures five aspects of affect or mood [9]. It consists of 65 adjectives describing feeling and mood to which the subject responds according to a five-point scale ranging from "Not at all" to "Extremely". Results are reported as six mood factors, namely:

- **Tension-Anxiety:** Heightened musculoskeletal tension including reports of somatic tension and observable psychomotor manifestation;
- **Depression-Defection:** Depression accompanied by a sense of personal inadequacy;

- **Anger-Hostility:** Anger and antipathy toward others;
- **Vigor-Activity:** Vigorousness, ebullience, and high energy;
- **Fatigue-Inertia:** Weariness, inertia and low energy level; and
- **Confusion-Bewilderment:** Bewilderment, muddleheadedness; appears to be an organized-disorganized dimension of emotion.

Because of its length, the POMS only results practical for occasional uses such as establishing a baseline and estimating the effects of excessive sleep deprivation or restriction.

Existing approaches on fatigue and mood analysis rely mostly on the described self-report mechanisms, which result unpractical for use in realistic scenarios. In fact, aspects such as fatigue, stress or mood change during the day as daily events occur. Following workers in real-time would imply to use these instruments multiple times during the day, with a significant impact on the workforce's routines. On the other hand, in the last years there have been emerging electronic performance monitoring tools. However, these rely on traditional productivity indicators that essentially quantify the work produced. This kind of approaches is often dreaded by the workforce, who feel spied. Moreover, the pressure that emerges from the need to produce in order to perform often has the opposite effect, i.e., decreased productivity [1, 16].

To address these issues we propose a non-invasive framework that can perform an analysis of performance that is not dependent on productivity indicators and should thus be more acceptable by the workforce. Moreover, it can be used continuously throughout the day without any impact on working routines. This framework is described in the following section.

2 Framework

In this paper we propose a non-invasive, non-intrusive, real-time approach to assess mental fatigue through the analysis of keyboard and mouse interaction patterns. The analysis can happen directly from the usage of an individual's computer as within the context of so-called desk-jobs. We build on the fact that computers are nowadays used as major work tools in many workplaces, to devise a non-invasive method for mental fatigue monitoring based on the observation of the worker's interaction with the computer, specifically the aspects related to the use of the mouse and the keyboard.

This system allows the detection of behaviors associated to fatigue in a non-invasive and non-intrusive way with the aim to develop leisure and work context-aware environments that may improve quality of life, mental health and individual performance, as well as productivity in organizations. To achieve this purpose we follow the guiding lines of Ambient Intelligence (AmI) [14], in which the technological aspects are hidden in the environment and the user is placed in the middle of the paradigm. There is also a focus on non-intrusiveness, with acquisition of information taking place without the need for explicit or conscious user interactions.

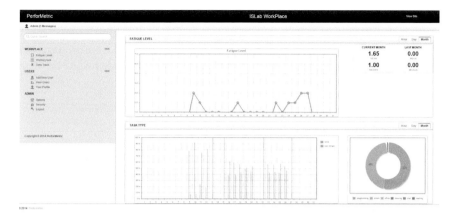

Fig. 1 Interface of the performance monitoring web service

Another major objective of the system is to support the decision-making processes of team managers or group coordinators. In this perspective, each element of a group/organization is seen as part of a whole which contributes to the general level of fatigue and the distraction of the group. Thus, the estimated fatigue level of the group is the average result of the fatigue level of its elements.

The developed system is also composed of a graphical web interface, which is available online and can be accessed remotely. It has different features according to the type of user. Specifically, a worker can analyze his data in real-time so as to know his current performance. He can also analyze personal historic data so as to know his evolution in the recent past (e.g. last day, last week). A higher-level login is also available in which similar features are provided but including data from all the users. The aim is that the group manager can, at a glance, have an estimation of the state of the workforce. This allows him to act both at a group level as well as at an individual level (e.g. advise someone to make a break when his individual performance drops significantly). The use of one or both types of login depends thus on the organization's policy, i.e., they may allow each worker to decide on working schedules and pauses at will (in which case both the worker and the manager have logins to the web interface) or they may decide that only the manager can take such decisions (in which case only the manager has access to the web interface) (Fig. 1).

3 Case Study

In order to test and validate the proposed system and approach, a case study was prepared using detection system models which were previously trained [12, 13]. To reach that goal we used two groups of volunteers who used the tool while they were providing feedback of their state of fatigue through USAFSAM Fatigue scale, as well as their mood through the POMS questionnaire.

Feedback values recorded during the monitoring period were used to validate the system and used to analyze the variation of moods.

The participants in the case study, fourteen in total (10 men, 4 women) were students from the University of Minho in the field of physics. Their age ranged between 18 and 25.

3.1 Methodology

The methodology followed to implement the study was devised to be as minimally intrusive as the approach it aims to support. Participants were provided with an application for logging the previously mentioned events of the mouse and keyboard. This application, which maintained the confidentiality of the keys used, needed only to be installed in the participant's computer and would run in the background, starting automatically with the Operating System. The only explicit interaction needed from the part of the user was the input of very basic information on the first run, including some personal identifying and profiling information.

As mentioned, two different groups of users were selected (7 participants in each group) who underwent the experience in different periods of the week. The first group was monitored on a Monday morning (starting at 9 am), while the second did so on a Thursday in the afternoon (starting at 14 pm). Each session had a duration of 3 h. During the session, in addition to the collection of interaction features, user feedback regarding their state was collected hourly.

3.2 Results

During the two sessions the mental fatigue monitoring system was used in real time. It not only allows recording the interaction patterns with the mouse and keyboard, but also to calculate the estimated individual level of mental fatigue for each user, and the estimated average level of each group. These values were compared with the feedback given by users.

Table 1 summarizes the results, showing that the first group evidences a less fatigued state than the second one. This can be confirmed through data collected during the session. It is also possible to check a RMSE (Root-mean-square Error) of 0.2 for group 1 and 0.4 for group 2. Taking into account the scale used, this is an acceptable error.

In addition to the subjective level of fatigue through the USAFSAM fatigue scale it was used the POMS factors in order to observe the influence of fatigue on the moods of the different groups, and therefore the emergence of fatigue and loss of vigor.

Through Table 2 and Fig. 2 it is visible that the average values are different between the two groups. In addition, the T-test was used to validate the differences

Table 1 Average level of fatigue of the different groups in USAFSAM fatigue scale of the estimated values and feedback values

Group	Estimated fatigue level (SD)	Fatigue from feedback (SD)	RMSE
Group 1	1.6 (0.7)	1.4 (0.8)	0.2
Group 2	3.3 (0.8)	3.2 (1.2)	0.4

Table 2 Average vales of Profile of mood states for group 1 and group 2, as well as the resulting *p*-value from the T-test

POMS factor	Group 1	Group 2	p-value
Tension-Anxiety	4.6	11.8	0.03
Depression-Dejection	4.4	9.1	0.02
Anger-Hostility	6.5	7.9	0.06
Vigor-Activity	19.4	7.6	0.04
Fatigue-Inertia	5.1	13.8	0.02
Confusion-Bewilderment	4.0	7.3	0.03

between the two groups, in order to determine if the two sets of data are significantly different from each other, a fact that was confirmed.

We can also observe that fatigue is higher in the second group and, on the contrary, vigor is lower in the second group. This is in accordance to the fatigue levels of each group.

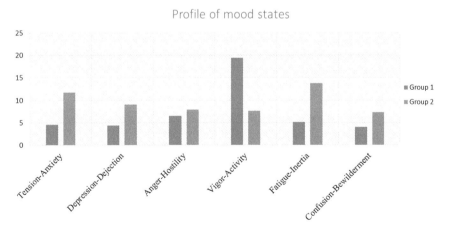

Fig. 2 Profile of mood states from groups 1 and 2: group 2 has more negative mood states

4 Conclusion

This paper described a prototype for monitoring mental fatigue in real-time of an individual or a group of individuals. The aim is to detect behaviors associated with mental fatigue within a group, so as to estimate the average level of mental fatigue in the group. Such information may be extremely useful in scenarios of critical tasks, which require that the individual performing the task is fully rested, as well as to assess the mental state of working groups as one.

The results of the experiment demonstrate that it is possible to evaluate the average degree of mental fatigue in a group using the prototype presented. The prototype used in the experiment uses a fatigue detection model trained in a previous study, with data of a different group of users but with a similar profile. This also shows that the changes induced by fatigue on different users can be generalized, i.e., we all react similarly.

Within the context of the CAMCoF project, which supports this work, the long-term goal is to develop environments that are autonomous and take measures concerning their self-management to minimize fatigue and increase the performance and well-being of a group of individuals.

Acknowledgments This work is part-funded by ERDF—European Regional Development Fund through the COMPETE Programme (operational programme for competitiveness) and by National Funds through the FCT—Fundaộo para a Ciência e a Tecnologia (Portuguese Foundation for Science and Technology) within projects FCOMP-01-0124-FEDER-028980 (PTDC/EEI-SII/1386/2012) and project Scope UID/CEC/00319/2013.

References

1. Aiello, J.R., Kolb, K.J.: Electronic performance monitoring and social context: impact on productivity and stress. J. Appl. Psychol. **80**(3), 339 (1995)
2. Akerstedt, T., Knutsson, A., Westerholm, P., Theorell, T., Alfredsson, L., Kecklund, G.: Mental fatigue, work and sleep. J. Psychosom. Res. **57**(5), 427–433 (2004). doi:10.1016/j.jpsychores.2003.12.001
3. Balkin, T.J., Wesensten, N.J.: Differentiation of sleepiness and mental fatigue effects. (2011)
4. Curran, S.: Short form of the profile of mood states (POMS-SF): psychometric information. Psychol. Assess. **7**(1), 80–83 (1995). doi:10.1037//1040-3590.7.1.80
5. Hart, S.G., Staveland, L.E.: Development of NASA-TLX (Task Load Index): results of empirical and theoretical research. Adv. Psychol. **52**, 139–183 (1988)
6. Jaber, M.Y., Neumann, W.P.: Modelling worker fatigue and recovery in dual-resource constrained systems. Comput. Ind. Eng. **59**(1), 75–84 (2010)
7. Kobayashi, H., Demura, S.: Relationships between chronic fatigue, subjective symptoms of fatigue, life stressors and lifestyle in Japanese high school students. Sch. Health **2**, 5 (2006)
8. Lorist, M.M., Klein, M., Nieuwenhuis, S., De Jong, R., Mulder, G., Meijman, T.F.: Mental fatigue and task control: planning and preparation. Psychophysiology **37**(5), 614–25 (2000)
9. McNair, D.M., Lorr, M., Droppleman, L.F.: Manual for the Profile of Mood States. Educational and Industrial Testing Service, San Diego CA (1971)
10. Newcombe, P.A., Boyle, G.J.: High school students' sports personalities: variations across participation level, gender, type of sport, and success. Int. J. Sport Psychol. **26**, 277 (1995)

11. Perelli, L.P.: Fatigue stressors in simulated long-duration flight. Subjective Fatigue, and Physiological Cost. Technical Report, DTIC Document, Effects on Performance, Information Processing (1980)
12. Pimenta, A., Carneiro, D., Novais, P., Neves, J.: Monitoring mental fatigue through the analysis of keyboard and mouse interaction patterns. In: Hybrid Artificial Intelligent Systems, pp. 222–231. Springer, Berlin (2013)
13. Pimenta, A., Carneiro, D., Novais, P., Neves, J.: Detection of distraction and fatigue in groups through the analysis of interaction patterns with computers. In: Intelligent Distributed Computing VIII, pp. 29–39. Springer, Berlin (2015)
14. Ramos, C., Augusto, J.C., Shapiro, D.: Ambient intelligence-the next step for artificial intelligence. IEEE Int. Syst. **23**(2), 15–18 (2008)
15. Samn, S.W., Perelli, L.P.: Estimating aircrew fatigue: a technique with application to airlift operations. Technical Report, DTIC Document (1982)
16. Smith, M.J., Carayon, P., Sanders, K.J., Lim, S.Y., LeGrande, D.: Employee stress and health complaints in jobs with and without electronic performance monitoring. Appl. Ergonomics **23**(1), 17–27 (1992)
17. Spielberger, C.D.: State-Trait Anxiety Inventory. John Wiley & Sons, Inc. (2010)
18. Tucker, P.: The impact of rest breaks upon accident risk, fatigue and performance: a review. Work Stress **17**(2), 123–137 (2003)
19. Williamson, R.J., Purcell, S., Sterne, A., Wessely, S., Hotopf, M., Farmer, A., Sham, P.C.: The relationship of fatigue to mental and physical health in a community sample. Soc. Psychiatry Psychiatr. Epidemiol. **40**(2), 126–132 (2005)

Agent-Based Simulation of Crowds in Indoor Scenarios

Rafael Pax and Juan Pavón

Abstract Crowd simulation models usually focus on performance issues related with the management of very large numbers of agents. This work presents an agent-based architecture where both performance and flexibility in the behaviour of the entities are sought. Some algorithms are applied for the management of the crowd of agents in order to cope with the performance in the processing of their movements and their representation, but at the same time some alternative reasoning mechanisms are provided in order to allow rich behaviours. This facilitates the specification of different types of agents, which represent the people, sensors and actuators. This is illustrated with a case study of the evacuation of the building of the Faculty of Computer Science, where different types of human behaviours are modelled for these situations. The result is the simulation of more realistic scenarios.

1 Introduction

There are several models for crowd simulation but in general these focus on scalability issues derived from the management of a large number of agents in real time, specially when considering their visualization or the way the agents find their way while avoiding obstacles and other agents [1]. Different techniques have been proposed to cope with these issues, by relying on specific assumptions of the problem under study. This has, however, an effect on limiting of the flexibility of agents' behaviour, which is quite homogeneous in most of the cases.

In this work the scope of the problem is the simulation in indoor scenarios, where the number of agents may be large (thousands) but not very large as in a city (mil-

R. pax · J. Pavón (✉)
Universidad Computense Madrid (Spain), Madrid, Spain
e-mail: rpax@ucm.es
Url:http://grasia.fdi.ucm.es

J. Pavón
e-mail: jpavon@ucm.es

© Springer International Publishing Switzerland 2016
P. Novais et al. (eds.), *Intelligent Distributed Computing IX*,
Studies in Computational Intelligence 616,
DOI 10.1007/978-3-319-25017-5_12

lions). This gives more room on performance constraints, which opens the possibility to modelling of individual agents with heterogeneous behaviours.

Several tools for simulation and design of how people behave in indoor scenarios exist [8, 13–15, 17], but they are still considering the agents more like a crowd that can be characterized by simple behaviours with a fixed number of parameters, instead of considering them as individuals. Other works, like [7, 9–12] have addressed the specification of richer agent behaviours, but the methodological aspects for a design process when developing them are not sufficiently exposed.

Taking this into account, this work proposes an agent-based model for indoor scenarios where both performance and flexibility in the behaviour of the entities are sought. Agents are specified and managed individually, but the effects of the crowd are taken into account by several methods that take advantage of characteristics of the indoor domain in order to cope with the efficiency and scalability issues in the processing of their movements and their visualization. At the same time, some alternative reasoning mechanisms are provided for each agent in order to allow modelling of rich and heterogeneous behaviours.

Section 2 introduces the MASSIS (Multi-agent System Simulation of InDoor Scenarios) architecture and components. They allow for the specification of the elements for indoor scenarios simulation. Special attention is given to the definition of the behaviour of humans under different situations, which includes the process for decision making of the agents. Other relevant aspects to model are interactions among agents and with their environment, the events on the environment, and the precise representation of the building.

The strategies that facilitate an efficient management of crowd simulation aspects are presented in Sect. 3. This is illustrated with a case study on the evacuation of the building of the Faculty of Computer Science in Sect. 4. The model facilitates the specification of different types of agents, which representing the people, and the sensors and actuators of the environment. Different views can be generated and manipulated during the simulation. Finally, Sect. 5 presents the conclusions.

2 Overview of MASSIS Architecture

MASSIS has a component-based architecture, where some part of the infrastructure is well-proven open source software. An agent-based framework above of the core infrastructure allows the specification of flexible agent behaviour types, as well as interactions among agents and the elements of the environment.

The environment initially is defined using SweetHome3D, a well known package that is used to model all components involved in an indoor environment, such as walls, doors, stairs, people, etc. In order to facilitate more flexibility of the characterization of the physical elements, it is possible to define a set of plugins, to specify the characteristics of the elements of the building, which will be represented by agents. For instance, in the case of sensors and actuators, they will be reactive agents, with simple behaviour and attributes. In the case of people, some physical characteristics can be defined such as weight, speed, but also some inherent attributes of the person (fear, courage, etc.) and a link to their behaviour. The types of behaviours are defined as components.

The simulation engine of MASSIS is MASON [5], a lightweight multi-purpose agent-based simulation library. Agents' behaviour is controlled with the Pogamut's POSH engine [4]. Some tests have been performed in order to check the performance of their integration in MASSIS. In the order of 10 thousand agents the experimentation has shown that execution times grow linearly with respect to the number of agents, so the results are satisfactory.

All the changes made in the environment are reflected in real time by 3D (Fig. 5) and 2D (Figs. 2, 3 and 8) displays, and can be logged in JSON format, as a single zipped file or in a SQLite database for further analysis. Although 3D display is more realistic and nicer for demonstration purposes, the 2D view is useful for analysis and debugging. Also, the 2D visualization API allows the creation of user-defined layers in order to filter the different elements involved in the simulation.

Once a simulation is performed, the exported data can be used to playback all events that have occurred during the execution of the simulation, i.e., the agents will behave in the same way they did during the simulation. This is interesting to allow the users to review the simulation when analysing what has happened.

Modelling human behaviour with agents have to consider two main aspects: those related with their perception and the interaction with the environment, and those dealing with the reasoning on the context and the decision making. They are implemented as low-level and high-level behaviour components, respectively. When the high level component decides *what to do next*, the action is executed by the low-level component, which performs all the necessary operations. The relationship between these components is illustrated in Fig. 1.

The low-level components deal with the perception of the environment and a set of basic behaviours for interacting with it. These behaviours are mostly a combination of *steering behaviours* [2], which are explained in Sect. 3.

The high-level behaviour components deal with decision making, learning and communications with other agents. Decisions are taken on the knowledge of the environment, which is provided by the low-level components.

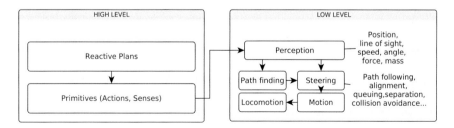

Fig. 1   MASSIS 's human behaviour agent model

Behaviour Oriented Design (BOD) [3] is applied to facilitate the design of agents. Developing a system of agents under BOD involves dividing their implementation into two different parts: A library of Behaviour modules and POSH Dynamic action selection scripts. Behaviour modules are a set of classes representing a set of modules for perception, action and learning (i.e. *primitives*). They can be called from the mechanism of action selection in order to determine *how* to do something. These senses and actions are created in the native language for the problem space (in the case of MASSIS, Java).

A POSH action selection script is a prioritized set of conditions and the related actions to be performed when certain conditions are satisfied. Figure 6 shows an example with the main elements: *drive collections*, *competences*, and *action patterns*. On the action selection step, the POSH engine executes the corresponding action pattern or competence from the drive collection with highest priority. Competences are similar to drive collections, but rules they do not interrupt each other. Finally, action patterns are simple, reusable sequences of actions.

The next sections present some of the components of the MASSIS framework that facilitate the efficient management of crowds of agents in a simulation.

3 Crowd Modelling and Simulation Issues

There are several aspects to take into account when modelling crowds of agents, with a trade-off between efficiency and flexibility of the specification of behaviours.

The building model, designed with SweetHome3D, is loaded and transformed into MASSIS internal representation to support the efficiency of algorithms implementation. This has an impact in the way agents interact with the environment, which is described below for different aspects: path finding, localization of elements and steering behaviours.

3.1 Path Finding

Pathfinding is one of the issues that has more impact when simulating crowd behaviours. Some models treat the crowd as a single entity or a group in order to simplify the number of calculations, such as in [6]. However, MASSIS, as it has been stated in the introduction, has as objective to support flexibility in agent behaviour, therefore the path finding model is implemented individually for each agent, but taking advantages of some assumptions from the problem domain in order to gain in efficiency.

When the action selection component (see Fig. 1) decides that the agent must go to a particular location, a path must be computed from the agent's location to the target. Its computation is done in MASSIS by an A* search over the polygons' visibility graph. The computational cost has been considerably improved by taking advantage of several characteristics of the environment:

1. The perimeter of rooms consists of walls or doors.
2. The path from a point A to a point B, being B in a different room than A, must pass necessarily through a door.
3. In an indoor scenario walls intersect each other *very often*.

With these assumptions, several aspects have been improved to gain in computational efficiency:

- **Faster visible edges computation.** In order to generate more realistic paths, the obstacle polygons are *inflated*, so the points of the path are slightly detached from the polygon edges. The advantage of this separation in the case of pathfinding, is that the edges of the obstacle polygons are now inside the rooms boundaries, and the number of possible reachable nodes from the agent's location decreases (only the nodes inside the room's boundaries are considered).
- **Obstacle polygons reduction.** The obstacles polygons that intersect can be merged, forming a bigger polygon. If most of the walls can be merged (as stated in assumption 3), the number of obstacles is drastically reduced (in the case of the Faculty of Computer Science, the number of wall obstacles are reduced from 2351 to 171, less than 8 % of the original) and consequently also the number of intersection tests.
- **The complete path is not needed at once.** The assumption 2 implies that it is not necessary to compute always the whole path between two points A and B. It is frequent, due to events in the environment, that the agent decides to change its current targets before it has reached the end of the path. To avoid unnecessary calculations, at the beginning of the simulation, a navigation graph based on door-room connections is created, and the doors are converted into waypoints. When the agent requests a path, the A* algorithm runs only in the current room.

3.2 Elements Localization

An intuitive way for storing the locations of the elements present during the simulation is using uniform grids. However, uniform grids are useful when the spatial data is distributed in an homogeneous way, which is rarely seen during simulation. The use of such grids can easily degenerate into situations where there are many redundant sparse cells. Furthermore, depending on the required accuracy, they can consume too many resources.

People, sensors and actuators need real time information about the elements that surround them. This implies that, during simulation, lots of query ranges must be performed. In order to minimize the impact of this calculation, MASSIS uses a QuadTree [18], a variable resolution data structure for retrieving agents' neighbours within a radius in an efficient manner (see Fig. 3). Although some CPU time is required in order to update the structure with the change of each agent, the gain obtained is worth it.

Fig. 2 Visibility graph and
merged walls layer

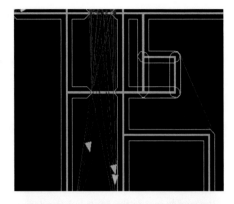

Fig. 3 Agents' Quad Tree
layer, showing the space
partitioning

3.3 Steering Behaviours

When the goals of the agent have been determined and the path to the target is known,
the agent can start moving in the environment, avoiding collisions with obstacles
(e.g., other agents, walls, etc.). These basic skills of the agent can be described with
steering behaviours [2], which need as parameters the agent's mass m, the agent's
location $\mathbf{L}$, a maximum force f_{max} and a maximum speed s_{max}.

Every step n, the computed steering forces are applied to the agent's location (limited by f_{max}), producing an acceleration whose magnitude is inversely proportional
to the vehicle's mass.

$$\mathbf{A}_n = \left(\frac{trunc(\mathbf{F}_n, f_{max})}{m} \right) \tag{1}$$

The velocity of the agent in every step n is approximated by the Euler integra-
tion. Adding the velocity at the previous step ($\mathbf{V}_{n-1}$) to the current acceleration($\mathbf{A}_n$),
produces a new velocity:

$$\mathbf{V}_n = trunc(\mathbf{V}_{n-1} + \mathbf{A}_n, s_{max}) \tag{2}$$

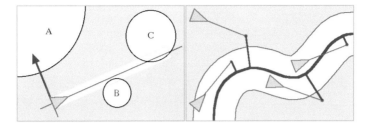

Fig. 4 Some of the steering behaviours implemented in MASSIS framework: seek and flee, obstacle avoidance and path following

Finally, the velocity is added to the agent's location.

$$\mathbf{L}_n = (\mathbf{L}_{n-1} + \mathbf{V}_n) \tag{3}$$

MASSIS provides a flexible implementation of several behaviours of this type (e.g., seek, arrival, separation, collision avoidance, wall containment, and path following, see Fig. 4), that can be grouped into more complex behaviours, like flocking or queuing.

4 Simulation of the Evacuation of a Building

It is common practice in public buildings to define some emergency protocols, which may involve, for instance, evacuation of the building. Planning and testing these protocols might be costly, but making simulations about these situations can help to this task (at least, as a first approach). This case study addresses this kind of situation for the building of the Faculty of Computer Science at UCM, which is represented in Fig. 5.

Three kinds of roles have been modelled based on the behaviours described by Proulx [16]:

- **Students** have some knowledge about the building. Their priority is the evacuation, but if they see someone needing help, they will try to assist. They also pay attention to the evacuation signals and indications displayed in the Faculty's CCTV. Their behaviour can be modelled as a POSH plan (see Fig. 6).
- **Well-trained staff members** are persons who work in the faculty (like a professor or administrative staff). They give instructions to the non-trained people, and try to assure that the evacuation is being done properly, following the established protocol.
- **Visitors** represents persons who have never been in the faculty, so the building is unknown to them. They interpret the fire alarm as something that is happening, so they will start searching for any person, expecting someone to tell them what

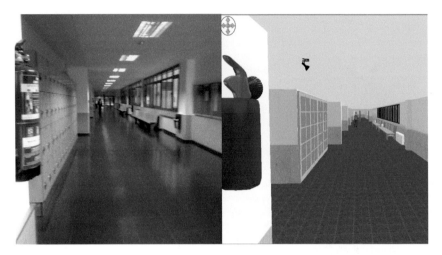

Fig. 5 MASSIS 3D representation versus a real photo

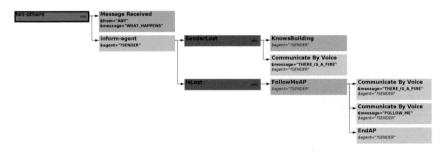

Fig. 6 Partial view of a student's POSH plan, having a drive element (*blue*), a competence(*cyan*), competence elements (*purple*), actions (*yellow*) and an action pattern (*pink*), which models the way in which the agent communicates with other people during the evacuation, reacting differently depending on the characteristics of the other agent. *Note* some elements were omitted for clarity

to do, if something serious is really happening. TV messages can help them to understand that they must evacuate the building, and also other people (an *Student* or a *Well-trained staff member*).

Elements of the environment (sensors and actuators) can be also modelled as agents, with their respective plans, which are usually simpler than those of agents representing people. For instance, Fig. 7 shows a fire detector's plan: the existence of a fire triggers its only action, which is the activation of the fire alarm.

These behaviour definitions suggest that the agents in this context must be capable of using their visual perception of the environment, what they hear and the ability of interacting with other agents, in order to accomplish higher-level goals. These abilities are modeled using low-level behaviours, managed by POSH primitives, which are the leaves of any reactive plan tree. Figure 6 illustrates parts of the reactive plans used in this case study for the people.

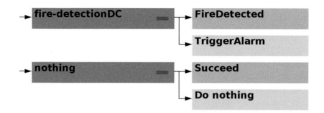

Fig. 7 Simpliest reactive plan: a fire detector

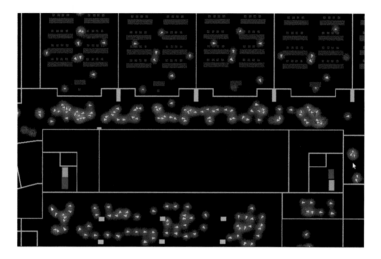

Fig. 8 Simulation 2D view, representing the different state of the agents by color

The simulation offers the option to work with 3D and 2D views. Usually 2D views are more practical for analysing what is happening in the simulation. For instance, Fig. 8 represents the state of the agents as colors. Other possibilities are 2D representations of crowd density, perception area, agent's IDs, paths, states, steering forces, etc. Other customized views can be easily integrated.

5 Conclusions

As it has been shown, MASSIS allows for the efficient simulation of indoor scenarios without losing the ability to specify rich and heterogeneous agent behaviours. This is achieved by simulating each agent individually, but with the support of several methods that take advantage of particularities of the indoor domain.

The extensibility of the MASSIS platform is well supported through its component-based architecture. For instance, different visualizations can be managed during the simulation, new algorithms and agent attributes can be supported and monitored. Also, simulation is logged in order to be able to replicate it and facilitate further data processing to analyse models in more detail.

Acknowledgments This work has been been supported by the Government of the Region of Madrid through the research programme MOSI-AGIL-CM (grant P2013/ICE-3019, co-funded by EU Structural Funds FSE and FEDER), and by the Spanish Ministry for Economy and Competitiveness, with the project Social Ambient Assisting Living—Methods (SociAAL) (grant TIN2011-28335-C02-01).

References

1. Schuerman, M., et al.: Situation agents: agent-based externalized steering logic. J. Vis. Comput. Anim. **21**(3–4), 267–276 (2010)
2. Reynolds, C.W.: Steering behaviours for autonomous characters. In: Proceeding of Game Developers Conference 1999, San Jose, California, pp. 763–782 (1999)
3. Bryson, J.: Intelligence by design. Ph.D. thesis, Department of Electrical Engineering and Computer Science, Massachusetts Institute of Technology (2001)
4. Gemrot, J., et al.: Pogamut 3 can assist developers in building AI (Not Only) for their videogame agents. Agents for Games and Simulations. LNCS, pp. 1–15. Springer, Heidelberg (2009)
5. Luke, S., et al.: Mason: a multiagent simulation environment. Simulation **81**(7), 517–527 (2005)
6. Treuille, A.C., et al.: Continuum crowds. In: ACM Transactions on Graphics Proceedings of SIGGRAPH vol. 25(3), pp. 1160–1168 (2006)
7. Massive Software Simulating Life.: http://www.massivesoftware.com/ (2002). Accessed March 2015
8. Serrano, E., Botia, J.: Validating ambient intelligence based ubiquitous computing systems by means of artificial societies. Inf. Sci. **222**, 3–24 (2013)
9. Algeria l.saifi et al.: Approaches to modeling the emotional aspects of a crowd. In: EUROSIM'13: Proceedings of the 2013 8th EUROSIM Congress on Modelling and Simulation, pp. 151–143 (2013)
10. Wu, S., Sun, Q.: Computer simulation of leadership, consensus decision making and collective behaviour in humans. PLoS ONE **9**(1), e80680 (2014). doi:10.1371/journal.pone.0080680
11. Tibor Bosse et al.: Modelling Collective Decision Making in Groups and Crowds: Integrating Social Contagion and Interacting Emotions, Beliefs and Intentions, vol. 6443. Springer, Berlin (2010)
12. Bicharra, A.C., et al.: Multi-agent simulations for emergency situations in an airport scenario. Adv. Distrib. Comput. Artif. Intell. J. **1**(3), 69–73 (2013)
13. Legion I Science in Motion.: http://www.legion.com (2015). Accessed March 2015
14. Galea, E., et al.: The EXODUS evacuation model applied to building evacuation scenarios. J. Fire Prot. Eng. **8**(2), 65–84 (1996)
15. PedGo—TraffGo HT.: http://www.traffgo-ht.com/ (2006). Accessed March 2015
16. Proulx, G.: Occupant behaviour and evacuation. In: Proceeding of 9th International Fire Protection Symposium. pp. 219–232 (2001)
17. Pathfinder—Thunderhead Engineering. http://www.thunderheadeng.com/pathfinder/ (2006). Accessed March 2015
18. Finkel, R.A., Bentley, J.L.: Quad trees : a data structure for retrieval on composite keys. Acta Informatica **4**, 1–9 (1974)

Forming Homogeneous Classes for e-Learning in a Social Network Scenario

Antonello Comi, Lidia Fotia, Fabrizio Messina, Giuseppe Pappalardo, Domenico Rosaci and Giuseppe M.L. Sarné

Abstract The use of network technology to provide online courses is the latest trend in the training and development industry and has been defined as the "e-Learning revolution". On the other hand, Online Social Networks (OSNs) represent today an effective possibility to have common and easy-to-use platforms for supporting e-Learning activities. However, as underlined by previous studies, many of the proposed e-Learning systems can result in confusion and decrease the learner's interest. In this paper, we introduce the possibility to form e-Learning *classes* in the context of OSNs. At the best of our knowledge, any of the approaches proposed in the past considers the evolution of on-line *classes* as a problem of matching the individual users' profiles with the profiles of the *classes*. In this paper, we propose an algorithm that exploits a multi-agent system to suitably distribute such a matching computation on all the user devices. The good effectiveness and the promising efficiency of our approach is shown by the experimental results obtained on simulated On-line Social Networks data.

A. Comi, L. Fotia · D. Rosaci
DIIES, University Mediterranea of Reggio Calabria, Reggio Calabria, Italy
e-mail: antonello.comi@unirc.it

L. Fotia
e-mail: lidia.fotia@unirc.it

D. Rosaci
e-mail: domenico.rosaci@unirc.it

G.M.L. Sarné
DICEAM, University Mediterranea of Reggio Calabria, Reggio Calabria, Italy
e-mail: sarne@unirc.it

F. Messina (✉) · G. Pappalardo
Department of Mathematics and Computer Science, University of Catania,
Viale A. Doria, 95125 Catania, Italy
e-mail: messina@dmi.unict.it

G. Pappalardo
e-mail: pappalardo@dmi.unict.it

© Springer International Publishing Switzerland 2016
P. Novais et al. (eds.), *Intelligent Distributed Computing IX*,
Studies in Computational Intelligence 616,
DOI 10.1007/978-3-319-25017-5_13

1 Introduction

In last years, e-Learning has emerged as a rising solution to lifelong learning and on-the-job work force training. E-learning can be defined as technology-based learning in which learning materials are delivered electronically to remote learners via a computer network. In contrast to the traditional classroom learning that centers on instructors who have control over contents and learning process, e-Learning provides a learner-centered, self-paced learning environment. Moreover, it allows time and location flexibility, low cost for learners, unlimited access to knowledge and archival capability for information reuse and sharing [7, 16, 22]. However, as underlined by previous studies [23, 24], the proposed e-Learning systems can result in confusion and decrease the learner's interest. To solve this problem, in this paper we introduce, in the context of On-line Social Networks (OSNs), the possibility to form e-Learning *classes*, that are groups of persons [13] with a particular interest who want to deepen their understanding of what it means to teach and learn. In this particular context, the e-Learning class is sub-network of users having similar interests or topics and sharing opinions and media contents [12].

In the past, some studies concerning the relationships existing between users and groups of persons in OSNs [4–6] have been proposed. For instance, in [8], a detailed analysis is presented about the influence of neighbours on the probability of a particular node to join with a group, on four popular OSNs. Also the studies presented in [10, 11] deal with the problem of giving recommendations to a group, instead of a single user. This is a key problem from the viewpoint of creating OSN groups that provide their users with a sufficient satisfaction. The problem is not simply to suggest to a user the best groups to join with, but also to suggest to a group the best candidates to be accepted as new members. If the existing research in OSN well covers the issue of computing individual recommendations, and the aforementioned issue begins to give attention to the issue of computing group satisfaction, however at the best of our knowledge any study has been proposed to consider the issue of managing the evolution of a OSN group of persons (i.e., a class) as a problem of matching the individual users' profiles with the profiles of the classes.

The notion itself of *class profile* is already unusual in OSN analysis, although the concept of *social profile* is not new in the research area of virtual communities. For example, in [9], following some theories originated in sociological research on communities, the authors present an approach to describe and manage the social environments of transactions that are provided in virtual communities. In particular, they present a model of a virtual community as a set of characteristics that the persons must have in common.

In this paper, we introduce the notion of *class profile* in the context of OSNs, giving to this notion a particular meaning coherent with the concept of OSN group. In our perspective, an OSN group is not simply a set of categories of interests, but also a set of common rules to respect, a preferred behaviour of its members, a communication style and a set of facilities for sharing media contents. Our definition of class profile is coherent with the definition of a *user profile*, that contains information comparable with those of a class profile. We exploit the above notion of class profile to provide each group of an OSN with a *class agent*, capable of creating, managing and continuously updating the class profile. Similarly, we associate a *user agent* with each user of the OSN. In keeping with the literature, we proposed in the past to build recommender systems for virtual communities [17, 19–21], introducing efficiency via the use of a distribute agent system, here we propose to exploit the agents above to automatically and dynamically computing a matching between user profiles and class profiles in a distributed fashion. The idea of associating an agent as a representative of a class is not completely new in a social network scenario, having been introduced also in [1–3]. However, in these past works this idea has been exploited to allow interoperability between different groups, without facing the problem of matching users and classes. We propose to provide the user agent with a matching algorithm able to determine the class profiles that best match with the user profile. This matching algorithm, named User-to-Classes (U2C), is based on the computation of a dissimilarity measure between user and class profiles. As a result of the U2C computation, the user agent will submit on behalf of its user some requests for joining with the best suitable classes. In particular, the agent of each class will execute the U2C algorithm to accept, among the users that requested to join with the class, only those users having profiles that sufficiently match with the class profile. This way, the dynamic evolution of the classes should reasonably lead to a more homogeneous intra-group cohesion.

The paper is organized as follows. In Sect. 2, we introduce the e-Learning social network scenario and describe how it is generated. In Sect. 3, we present the proposed U2C matching algorithm. Section 4 describes the experiments we have performed to evaluate our approach. The experiments are executed on a set of simulated users and classes confirm the above intuition, and also show the promising efficiency and the scalability of the proposed algorithm. Finally, in Sect. 5, we draw our conclusions.

2 The e-Learning Social Network Scenario

In our scenario, we deal with an e-Learning Social Network EN, represented by a pair $EN = \langle U, C \rangle$, where U is a set of *users*, C is a set of *classes*, where each class $c \in C$ is a subset of U (i.e., $c \subseteq U \; \forall c \in C$).

We assume that a single user u (resp., a single class c) is characterized by a profile $\langle T_u, E_u, A_u, F_u \rangle$ (resp., $\langle T_c, E_c, A_c, F_c \rangle$):

- $T_u(s) : S \rightarrow [0,1] \subset \mathbb{R}$ (resp., $T_c(s) : S \rightarrow [0,1] \subset \mathbb{R}$) is a mapping representing the the level of interest of the user u (resp., of the users of the class c) with respect to the categories of topics of interest (e.g., *mathematics, computer science, foreign languages*, etc.), which is denoted by S ($s \in S$). The values of this mapping are computed on the basis of the actual behavior of u (resp., of the users of c).

- E_u (resp. E_c) is a preference with respect to the level of expertise of the user (resp., the level of expertise of the classes). We assume that the level of expertise will take values in a finite set, e.g., {*base, medium, high*}.
- A_u (resp. A_c) represents the set of actions performed by u (resp. c) is actually performed or tolerated within the social network by u (resp. c). For example, the possible actions that a user could perform are: *(i)* "publishing more than 2 posts per hour", or *(ii)* "publishing posts longer than 500 characters". We suppose each possible action is represented by a boolean variable, that is equal to *true* if this action is adopted, *false* otherwise. We denote as A the set of possible actions, representing how u (resp., c) may behaves. For instance, if $A = \{b_1, b_2\}$, where b_1 represents the action "(i)" and b_2 represents the action of "(ii)" then a property $A_u = \{true, false\}$ characterizes a certain user.
- F_u (resp., F_c) is the set of friends of u (resp., the set of all the users that are friends of at least a member belonging to the class c).

We assume that a software agent a_u (resp., a_c) is associated with each user u (resp., class c), and automatically performs the following tasks:

- *Periodically updates $T_u(s)$ (resp. $T_c(s)$)*. Each time the user u publishes a post, or comments another post, dealing with a category of topic s, the new value $T_u(s)$ is updated as follows:

$$T_u(s) = \alpha \cdot T_u(s) + (1 - \alpha) \cdot \delta$$

that is a weighted mean between the previous interest value and the new value, where $\alpha, \delta \in [0, 1] \subset \mathbb{R}$, are arbitrarily assigned by the user itself. In detail, δ represents the increment to give to the u's interest in s consequently of the u's action, while α is the importance we want to assign to the past interest value with respect to the new contribution. Similarly, $T_c(s)$ is updated by the agent a_c each time $T_u(s)$ has changed for any $u \in c$ as:

$$T_c(s) = \sum_{u \in EN} T_u(s)$$

- *Updates A_u* each way the user u performs an action in the social network (e.g., publishing a post, a comment, etc.). Agent a_u analyzes the action and consequently sets the appropriate boolean values for all the variables contained in A_u (e.g., if A_u contains a variable representing the fact of publishing more than 2 posts per hour, then a_u checks if the action currently performed by u makes true or false this fact and, consequently, sets the variable). Furthermore, each agent a_c associated with a class c, updates the variables contained in A_c each time the administrator of c decides to change the correspondent rules (e.g., if A_c contains a variable representing the fact of publishing more than 2 posts per hour, and the administrator of c decides that this fact is not tolerated in the class, then a_c sets the correspondent variable to the value *false*).

- *Updates E_u (resp. E_c).* Whenever the user u (resp., the administrator of the class c) decides to change his/her preference with respect to the level of expertise, the agent a_u (resp., a_c) consequently updates E_u (resp., E_c).
- *Updates F_u (resp. F_c)* Whenever the user u (resp., a user of the class c) adds a new friend in his/her friends list, or deletes an assisting friend from his/her friends list, the agent a_u (resp., a_c) consequently updates F_u (resp., F_c). Note that the agent a_c computes the property F_c as the union of the sets F_u of all the users $u \in c$.

The core part of the presented model is represented by the *User-to-Class (U2C) Matching* algorithm, which is distributed among the network. of user agents and class agents. It is described in Sect. 3.

3 The U2C Matching Algorithm

The U2C matching is a global activity distributed on the user and class agents belonging to the agent network. At this regard, we assume that each agent can interact with each other, sending and receiving messages by means of any software agent communication facility. In particular, we assume that the *Directory Facilitator* agent (DF) is available in the network, in order to provide a service similar to Yellow Pages.

As the *U2C* matching algorithm is distributed over user agent and class agents, in the following we describe two different set of tasks which are sequentially executed bu user agents and class agents: the *User-side U2C algorithm* and the *Class-side U2C algorithm*.

In the following, we will call *epoch* the timestamp related to a complete execution of the U2C algorithm, and we denote as P the (constant) period between two consecutive epochs.

3.1 User-Side U2C Algorithm

Let X be the set of the classes the user u is joined with, with $||X|| = n \leq NMAX$, where $NMAX$ is the maximum number of classes which a user can join with. Let's suppose that a_u records into an internal cache the profiles of the classes $c \in X$ obtained in the past by the associated class agents and relative retrieval date, $date_c$. Let also m be the number of the class agents that at each epoch must be contacted by a_u and T the current epoch of execution. In such a situation, a_u performs the following actions:

- It randomly select, from the DF repository, m classes that are not present in the set X. Let Y be the set of these selected classes, and let $Z = X \bigcup Y$.
- For each class $c \in Y$, and for each class $c \in X$, such that $|T - date_c| \geq \psi$, it sends a message to the correspondent a_c, to request the profile p_c of the correspondent class.

- For each p_c, it computes a *dissimilarity measure* between the profile of the user u and the profile of the class c, defined as a weighted mean of four contributions c_T, c_E, c_A and c_F, associated with the components T, E, A and F of the profile (see Sect. 2). In particular:

  – c_T is computed as the average of the differences of the interests values of u and c for all the categories of topics present in the social network, that is:

  $$c_T = \frac{\sum_{s \in S} |T_u(s) - T_c(s)|}{|S|}$$

  – c_E is set equal to 0 (resp.,1) if E_u is equal (resp., not equal) to E_c.
  – c_A is computed as the average of all the differences between the boolean variables contained in A_u and A_c, respectively, where this difference is equal to 0 (resp., 1) if the two corresponding variables are equal (resp., different).
  – c_F is computed as the percentage of common friends of u and c, with respect to the total number of friends of u or c. That is:

  $$c_F = \frac{|F_u \bigcap F_c|}{|F_u \bigcup F_c|}$$

  Note that each contribution has been normalized in the interval $[0, 1] \subset \mathbb{R}$, in order to compare all the contributions. The dissimilarity d_{uc} of a class c with respect to the user u is then computed as:

  $$d_{uc} = \frac{w_T c_T + w_E c_E + w_A c_A + w_F c_F}{w_T + w_E + w_A + w_F}$$

- Let $\tau \in [0, 1] \subset \mathbb{R}$ a given threshold for the dissimilarity, then

  – each class $c \in Z$ is considered as a good candidate to join if (*i*) $d_{uc} < \tau$ and (*ii*) it is inserted by a_u in the set *GOOD*;
  – if there exist more than *NMAX* classes satisfying this condition, the *NMAX* classes having the smallest values of global difference are selected;
  – for each selected class $c \in GOOD$, when $c \notin X$, the agent a_u sends a join request to the agent a_c, that holds the profile p_u associated with u, otherwise, a_u leaves the class c.

3.2 Class-Side U2C Algorithm

Let K be the set of the users currently into the class c, where $|K| = k \leq n_{KMAX}$, being *KMAX* the maximum number of members allowed by the class administrator. We also suppose that a_c stores into an internal cache the profiles of the users $u \in K$ obtained in the past by the associated user agents, and the retrieval date $date_u$, and T

the current epoch of execution. Each time a_c receives a join request by a user agent r, that also contains the profile p_r associated with r, it behaves as follows:

- For each user $u \in K$ such that $|T - date_u| \geq \eta$, it sends a message to the user agent a_u, whose name has obtained by the DF, requesting it the profile p_u associated with the user.
- After the reception of the responses, it computes the *dissimilarity measure* d_{uc} between the profile of each user $u \in K \bigcup \{r\}$ and the profile of the class c.
- Now, let $\pi \in [0, 1] \subset \mathbb{R}$ a threshold for the dissimilarity, such that a user u is considered as acceptable to join if $d_{uc} < \pi$, then

    - agent a_c stores in a set *GOOD* those users $u \in K \bigcup \{r\}$ such that $d_{uc} < \pi$;
    - if there exist more than *KMAX* users satisfying this condition, the *KMAX* users having the smallest values of global difference are selected;
    - if $r \in GOOD$, a_c accepts its request to join with the class;
    - for any user $u \in K$, with $u \notin GOOD$, a_c deletes u from the class.

4 Experiments

We performed a set of experiments to evaluate the effectiveness of the U2C matching activity in making more homogeneous the classes of an OSN. To this purpose, we performed several simulations by means of ComplexSim [14, 15]. We simulated an OSN with 200.000 users and 100 classes along with user and class profiles, provided with the structure described in Sect. 2. Profiles p_u are generated as follows:

- $T_u(s)$ is randomly generated in its domain.
- E_u is assigned from its three possible values, i.e. *base*, *medium* and *high* basing on three different probability: P(*base*)=0.7, P(*medium*)=0.2, P(*high*)=0.1. This distribution values appear reasonable in a realistic OSN scenario.
- A_u contains six boolean variables $\{b_1 \ldots b_6\}$ whose value was uniformly distributed into the set $\{true, false\}$, representing the user's attitude to:

    - publish more than 1 post per day;
    - publish posts longer than 200 characters in most of the cases;
    - publish at least two comments per day to posts of other users;
    - respond to comments associated with its posts in most the cases;
    - leave at least 2 rates "I like it" per day;
    - respond to a message of another user in most the cases.

- F_u is randomly generated among different values with identical probability:

    1. users that have a dissimilarity with u smaller than 0.5 (the dissimilarity is computed in the same way of d_{uc} in Sect. 2);
    2. users randomly chosen from the set of the OSN users;
    3. 50 % of the users generated as in (1), the others as in (2).

We selected these distributions as they would represent three real typology of OSN users: (1) those that select their friends based on similar preferences and behavior, (2) those that randomly accept any friendship and (3) those that behaves in an intermediate fashion with respect to the first two attitudes.

Users are then randomly assigned to the available classes, in such a way that a user is joined at least with 2 classes and at most with 10 classes. The profile p_c of each class c is assigned by generating random values for properties E_c and A_c, while T_c and F_c are computed on the base of the corresponding members' values, following the formulas described in Sect. 2. The values of the parameters introduced in Sect. 3 are as follows: $\alpha = 0.5$, $\delta = 0.1$, $\psi = 10$, $\tau = 0.7$, $\eta = 10$, $\pi = 0.7$.

As a measure of the internal *cohesion* of a class, we use the concept of *average dissimilarity*, commonly exploited in Clustering Analysis [18], defined as the average of the dissimilarities between each pair of objects in a cluster, as in our scenario a class c is the equivalent of a cluster of users. Then we defined average dissimilarity (AD_c) and its global mean (*MAD*):

$$AD_c = \frac{\sum_{x,y \in c, x \neq y} d_{xy}}{|c|} \quad MAD = \frac{\sum_{c \in C} AD_c}{|C|}$$

Simulations started with $MAD = 0.479$, which indicates a population with a very low homogeneity, due to the completely random generation of the classes. Starting from this initial configuration, we have applied the U2C algorithm described in Sect. 3, simulating a number of 200 epochs of execution per each user, where each epoch simulated a time period of a day. The results of the simulation, in terms of *MAD* with respect to the epochs, are shown in Fig. 1. The results clearly show that the U2C algorithm introduces a significant increment of the cohesion in social network classes, that after a period of about 122 epochs achieves a stable configuration represented by $MAD = 0.176$. Moreover, we repeated the experiments above,

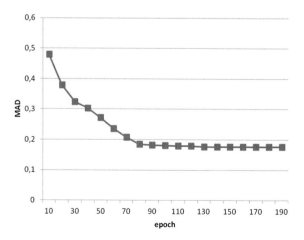

Fig. 1 The variation of MAD versus epochs

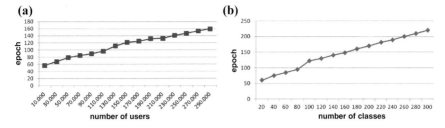

Fig. 2 **a** The stable MAD for different values of the users' number and **b** different values of the classes' number

changing the number of simulated users and classes. In particular, in Fig. 2a we have plotted the stable MAD for different values of the users' number, having fixed to 100 the number of the classes, while in Fig. 2b the stable MAD values are reported for different values of the classes' number, having fixed to 200.000 the number of the users. The results show that the number of the necessary epochs for achieving a stable configuration increases almost linearly with respect to the number of the classes, confirming a good scalability of the approach.

5 Conclusions

Differently from to the traditional classroom learning, e-Learning allows time and location flexibility, low cost for e-learners, unlimited access to knowledge and archival capability for information reuse and sharing. However, as underlined in the literature, the e-Learning systems can result in confusion and decrease the learner's interest. To solve this problem, the authors introduce the possibility to form e-Learning *classes* in OSNs.

The problem of making possible a suitable evolution in OSN groups of persons (i.e. *classes*), dynamically increasing the intra-group cohesion, is emerging as a key issue in the OSN research field. In this paper, we present a User-to-Class matching algorithm, that allows a set of software agents associated with the OSN users to autonomously manage the evolution of the classes. In particular, basing on a dissimilarity measure, it detects for each user the most proper classes to join with him. Moreover, the agents operate on behalf of the class administrators, such that a class agent accepts only those join requests that come from users profile compatible with the profile of the class. The experiments clearly show that the execution of the matching algorithm increases in time the internal cohesion of the classes.

Acknowledgments This work has been supported by project **PRISMA** PON04a2 A/F funded by the Italian Ministry of University and **NeCS** Laboratory of the Department DICEAM, University Mediterranea of Reggio Calabria.

References

1. De Meo, P., Nocera, A., Quattrone, G., Rosaci, D., Ursino, D.: Finding reliable users and social networks in a social internetworking system. In: Proceeding of the 2009 International Database Engineering and Applications Symposium, pp. 173–181. ACM, (2009)
2. De Meo, P., Nocera, A., Rosaci, D., Ursino, D.: Recommendation of reliable users, social networks and high-quality resources in a social internetworking system. AI Commun. **24**(1), 31–50 (2011)
3. De Meo, P., Quattrone, G., Rosaci, D., Ursino, D.: Dependable recommendations in social internetworking. In: Web Intelligence and Intelligent Agent Technologies, IAT, pp. 519–522 (2009)
4. De Meo, P., Messina, F., Rosaci, D., Sarné, G.M.L.: Improving the compactness in social network thematic groups by exploiting a multi-dimensional user-to-group matching algorithm. In: 2014 International Conference on Intelligent Networking and Collaborative Systems (INCoS), IEEE, pp. 57–64 (2014)
5. De Meo, P., Messina, F., Rosaci, D., Sarné, G.M.L.: Recommending users in social networks by integrating local and global reputation. In: Internet and Distributed Computing Systems, pp. 437–446. Springer International Publishing (2014)
6. De Meo, P., Messina, F., Rosaci, D., Sarné, G.M.L.: 2d-socialnetworks: away to virally distribute popular information avoiding spam. In: Intelligent Distributed Computing VIII, pp. 369–375. Springer International Publishing (2015)
7. Garruzzo, S., Rosaci, D., Sarné, G.M.L.: Isabel: A multi agent e-learning system that supports multiple devices. In: IEEE/WIC/ACM International Conference on Intelligent Agent Technology, IAT'07, pp. 485–488. IEEE (2007)
8. Hui, P., Buchegger, S.: Groupthink and peer pressure: social influence in online social network groups. In: ASONAM'09 International Conference on Advances in Social Network Analysis and Mining, pp. 53–59. IEEE, (2009)
9. Hummel, J., Lechner, U.: Social profiles of virtual communities. In: HICSS Proceeding of the 35th Annual Hawaii International Conference on System Sciences, pp. 2245–2254. IEEE, (2002)
10. Kasavana, M.L., Nusair, K., Teodosic, K.: Online social networking: redefining the human web. J. Hospitality Tour. Technol. **1**(1), 68–82 (2010)
11. Kim, J.K., Kim, H.K., Oh, H.Y., Ryu, Y.U.: A group recommendation system for online communities. Int. J. Inf. Manag. **30**(3), 212–219 (2010)
12. Messina, F., Pappalardo, G., Rosaci, D., Santoro, C., Sarné, G.M.L.: A distributed agent-based approach for supporting group formation in p2p e-learning. In: AI* IA 2013: Advances in Artificial Intelligence, pp. 312–323. Springer International Publishing (2013)
13. Messina, F., Pappalardo, G., Rosaci, D., Santoro, C., Sarné, G.M.L.: Hyson: A distributed agent-based protocol for group formation in online social networks. In: Multiagent System Technologies, pp. 320–333. Springer Berlin Heidelberg (2013)
14. Messina, F., Pappalardo, G., Santoro, C.: Complexsim: An smp-aware complex network simulation framework. In: 2012 Sixth International Conference on Complex, Intelligent and Software Intensive Systems (CISIS), pp. 861–866. IEEE (2012)
15. Messina, F., Pappalardo, G., Santoro, C.: Complexsim: a flexible simulation platform for complex systems. Int. J. Simul. Process Model. 6 **8**(4), 202–211 (2013)
16. Moore, J., Dickson-Deane, C., Galyen, K.: e-learning, online learning, and distance learning environments:are they the same? Internet High. Educ. **14**(2), 129–135 (2011)
17. Palopoli, L., Rosaci, D., Sarné, G.M.L.: A multi-tiered recommender system architecture for supporting e-commerce. In: Studies in Computational Intelligence 446, Intelligent Distributed Computing VI, pp. 71–81 (2013)
18. Pearson, R.K., Zylkin, T., Schwaber, J.S., Gonye, G.E.: Quantitative evaluation of clustering results using computational negative controls. In: Proceeding of 2004 SIAM International Conference on Data Mining, pp. 188–199 (2004)

19. Rosaci, D., Sarné, G.M.L.: Efficient personalization of e-learning activities using a multi-device decentralized recommender system. Comput. Intell. **26**(2), 121–141 (2010)
20. Rosaci, D., Sarné, G.M.L.: A multi-agent recommender system for supporting device adaptivity in e-Commerce. J. Intell. Inf. Syst. **38**(2), 393–418 (2012)
21. Rosaci, D., Sarné, G.M.L.: Recommending multimedia web services in a multi-device environment. Inf. Syst. **38**(2), 198–212 (2013)
22. Ruiz, J.G., Mintzer, M.J., Leipzig, R.M.: The impact of e-learning in medical education. Acad. Med. **81**(3), 207–212 (2006)
23. Welsh, E.T., Wanberg, C.R., Brown, K.G., Simmering, M.J.: e-learning: emerging uses, empirical results and future directions. Int. J. Train. Dev. **7**(4), 245–258 (2003)
24. Zhang, D., Zhao, J.L., Zhou, L., Nunamaker, J.F. Jr.: Can e-learning replace classroom learning? Commun. ACM **47**(5), 75–79 (2004)

Social-Based Arcs Weight Assignment in Trust Networks

Marco Buzzanca, Vincenza Carchiolo, Alessandro Longheu, Michele Malgeri and Giuseppe Mangioni

Abstract Virtual interaction with strangers often makes use of underlying trust networks. Usually, existing proposals address the evaluation of global (unique) trust for a given node within the network. In this paper we discuss about how to assess the local (direct) trust a node receives from each of his neighbors. Our proposal is social-based and takes into account both positive and negative experiences as well as the history of past feedbacks, ensuring good stability also when a node receives hundreds of positive feedbacks briefly followed by few negative feedbacks. In order to highlight the stability of this approach we performed several simulations with different networks.

1 Introduction

Trust networks have grown in popularity as a fundamental precautionary component that helps users in managing virtual interactions with (possibly total) strangers, either real people or virtual entities, in several contexts such as e-commerce, social networks, distributed on-line services and many others [5, 10, 12].

The research was supported by S.M.I.T. Sistema di Monitoraggio integrato per il Turismo—Linea di intervento 4.1.1.1 del POR FESR Sicilia 2007–2013.

M. Buzzanca · V. Carchiolo · A. Longheu · M. Malgeri (✉) · G. Mangioni
Dip. Ingegneria Elettrica Elettronica E Informatica (DIEEI),
Università Degli Studi di Catania, Catania, Italy
e-mail: michele.malgeri@dieei.unict.it

M. Buzzanca
e-mail: marco.buzzanca@dieei.unict.it

V. Carchiolo
e-mail: vincenza.carchiolo@dieei.unict.it

A. Longheu
e-mail: alessandro.longheu@dieei.unict.it

G. Mangioni
e-mail: giuseppe.mangioni@dieei.unict.it

© Springer International Publishing Switzerland 2016
P. Novais et al. (eds.), *Intelligent Distributed Computing IX*,
Studies in Computational Intelligence 616,
DOI 10.1007/978-3-319-25017-5_14

143

Generally, in trust models and frameworks developed in the past years [2, 8, 9, 11], the trust network is represented as a graph where nodes are agents and trusting relationships (arcs) are weighted against a measure of the *direct* trust value according to a given metric.

Existing literature mainly focus on the assessment of *global* trust, i.e. the unique value for a node that aims at mediating all local (direct) values that express different judgements the node received from others; in this work however, we consider in more detail how *direct* trust should be evaluated for each node *i*.

In particular, we take into account positive feedbacks that *i* receives after successful interactions, similarly to the well known proposals EigentTrust [6] and Gossip-Trust [13]. Moreover, in order to provide a weight assignment model based on real world social criteria, we define two additional factors, the *mistrust* as a measure of a lack of trust, and the *popularity* of the node as the total number of received feedbacks. We use mistrust to balance positive and negative feedbacks when selecting a node, and popularity to measure to what extent a certain trust or mistrust rating is relevant. This way, a node which has more positive feedbacks than negative feedbacks it is considered overall trustworthy, and a node with more feedback scores is considered more trustworthy than a node with less feedback scores, if the overall trust rating is the same.

We show through a set of simulations that this models exhibits greater stability compared to EigentTrust and GossipTrust, i.e. if a node changes its behavior, trust and mistrust ratings are not significantly affected, unless this behavior repeatedly occurs, as the proposed model also takes into account the node's history (if a node has received hundreds of positive feedbacks, it takes more than a few incoming negative feedbacks for it to be considered untrustworthy).

In Sect. 2 we first provide a few definitions together with existing approaches, then we introduce our social-based proposal and how we model direct trust accordingly, by using contributions coming from both a trust and a mistrust network. In Sect. 3 we elaborate on simulation results and on the properties our trust definition actually exhibits in different networks, discussing final remarks and further works in Sect. 4.

2 Weight Assignment

2.1 Notation and Existing Approaches

There are several possible intuitive approaches to assign weights to the arcs of a trust network. We will now analyze some of them and find possible shortcomings. Before this, we introduce a few definitions of the quantities involved for ease of notation:

Alongside these definitions, we introduce three new quantities:

$$p_i = \frac{r_i}{R}, \quad t_i = \frac{r_i^+}{r_i}, \quad m_i = \frac{r_i^-}{r_i} = 1 - t_i \tag{1}$$

Symbol	Description
R	Total number of interactions among all neighbors
r_i	Interactions with neighbor i
r_i^+	Positive interactions experienced with neighbor i
r_i^-	Negative interactions experienced with neighbor i

We name p_i the *popularity* of node i seen from the perspective of node n, t_i the *trust* that node n places in node i and m_i the *mistrust* that node n places in node i. Note that we use the term mistrust loosely and not as a synonym of *distrust*. Distrust is usually intended as a direct feedback about the unreliability of a certain node. Even though mistrust is still a measure of the unreliability of the node, it does not come from a direct feedback of distrust, but from a lack of trust. There are several works about the coexistence of trust and distrust in literature: in [4] propagation of trust and distrust is analyzed, whereas in [7] the authors manage to predict with acceptable accuracy whether a node is going to trust or distrust another node which is not connected to; finally, in [3], an extension of Pagerank which works on signed graphs named PageTrust is introduced.

Our approach attempts to take into account both trust and mistrust, incorporating them in a single weight assignment criterion. One of the most intuitive way to model the node's attitude in a trust network is to give a *Positive Feedback*. That is can be obtained by normalizing the positive interactions:

$$w_i^+ = \frac{r_i^+}{\sum_{k=1}^{N} r_k^+}$$

However, this solution does not make use of negative interactions. A node which experiences a negative interaction with a neighbor, should alter its attitude towards that neighbor accordingly. Intuitively, we want to avoid more those nodes we had negative interactions with, and this formula does not take this aspect into account.

Another intuitive way to model the node aptitude is to give *Net Feedback*, that tries to incorporate negative interactions into the weight formula, as done by EigenTrust. If we define $f_i^+ = max(0, r_i^+ - r_i^-)$ we have:

$$w_i^+ = \frac{f_i^+}{\sum_{k=1}^{N} f_k^+}$$

While this approach does include negative interactions in the arc weight, it has one important shortcoming: it reacts very poorly to feedback changes. Let us suppose we have a neighbour with 100 positive feedback ratings and 99 negative feedback ratings. With this formula, its f_i^+ would be 1. If our node completes another interaction with this neighbour positively, its f_i^+ will change to 2. This feedback gain essentially doubles the previous value, and this doesn't model well the mixed behavior of the node.

2.2 Social-Based Weight Assignment

Taking note of issues described so far, our idea is to provide a weight formula that models the attitude according to these features:

- Neighbors with more interactions should be preferred. If two neighbors have the same ratio $\frac{r_i^+}{r_i}$, the one with the largest r_i should be more likely to be contacted.
- The higher negative/total interactions ratio a neighbor has, the more it should be avoided.
- The higher positive/total interactions ratio a neighbor has, the more it should be contacted.
- It should take into account previous interactions.

The idea is that a node should weight both popularity and trust when making a decision about the trustworthiness of a certain node. A popular, trustworthy node is more likely to yield a successful interaction than a less popular node with the same level of trustworthiness. At the same time, we want to take into account the mistrust of other nodes. This is because the more untrustworthy a node is, the more we want to avoid having interactions with it. Our novel criterion to assign weights in a trust-based network, that we named *social weight assignment*, sports two different components: the trust towards node i and the average mistrust of all neighbors except i. Both components are linear dependent on the neighbor popularity. It is important to highlight that the novel weight assignment criterion defined in Sect. 2.3 can be effortlessly applied in conjunction with metrics that require that the sum of the node outlink weights is 1, as it guarantees this property by definition.

2.3 Definition

The social weight assignment criteria assigns the following weight to node neighbors:

$$w_i^+ = t_i p_i + \sum_{k=1, k \neq i}^{N} \frac{m_k p_k}{N-1} \tag{2}$$

where N is the number of neighbour nodes of node n. It is easy to prove that $\sum_{i=1}^{N} w_i^+ = 1$:

$$\sum_{i=1}^{N} w_i^+ = w_1^+ + w_2^+ + \cdots + w_N^+ = t_1 p_1 + \sum_{i=1, i \neq 1}^{N} \frac{m_i p_i}{N-1} + t_2 p_2 + \sum_{i=1, i \neq 2}^{N} \frac{m_i p_i}{N-1} +$$

$$+ \cdots + t_N p_N + \sum_{i=1, i \neq N}^{N} \frac{m_i p_i}{N-1} = \sum_{i=1}^{N} t_i p_i + \sum_{j=1}^{N} \sum_{i=1, i \neq j}^{N} \frac{m_i p_i}{N-1}$$

Note that in the double sum each $m_i p_i$ is repeated $N - 1$ times so we may reduce the notation as:

$$\sum_{i=1}^{N} t_i p_i + \sum_{i=1}^{N} m_i p_i = \sum_{i=1}^{N} (t_i + m_i) p_i = \sum_{i=1}^{N} \left(\frac{r_i^+}{r_i} + \frac{r_i^-}{r_i} \right) \frac{r_i}{R} = \sum_{i=1}^{N} \frac{r_i}{R} = 1$$

It is also possible to define a dual criterion with the same entities defined in (1), where t_i and m_i are swapped:

$$w_i = t_i m_i + \sum_{k=1, k \neq i}^{N} \frac{t_k p_k}{N - 1} \tag{3}$$

2.4 Meaning of Social Weight Assignment

The two criteria generate two different networks, a trust network for the social weight assignment, and a mistrust network for its dual. These two networks are not complementary, in-fact, while they share the same topology, the arc weights are different: weights of the trust network cannot be derived directly from the weights of the mistrust network, and vice-versa.

This implies that the node with highest weight in the trust network (the "best node") is not necessarily the node with the least weight in the mistrust network. Another consequence of the way the weights are assigned is the fact that the most reliable node isn't necessarily the best node, as its trust weight is multiplied by p_i: the popularity of the node (which essentially means the portion of the node interactions shared with neighbour i) impacts greatly the final weight. This is by design, as the more interactions the node has with a neighbour, the more reliable the trust (or mistrust) weight is: a node with 0.9 trust and 0.1 popularity has a lower weight than a node with 0.8 trust and 0.6 popularity in the trust network. This is in line with the behavior of human beings in a social context.

An aspect that is interesting to expand upon a bit is how a neighbor is judged in trust and mistrust networks. There are three possible situations:

- Trustworthy node: node has higher trust weight than mistrust weight. These nodes are obviously the most reliable, especially if they have high popularity.
- Mixed node: node has similar trust and mistrust weight. This can happen when the trust weight is near to the mistrust weight.
- Non-trustworthy node: node has higher mistrust weight than trust weight. These nodes should be avoided, regardless of popularity.

It is clear that a node should not react in a symmetric fashion to trust and mistrust: trust is easily shaken, but hard to build up. However, it is currently beyond the scope of this document to define a behavioral pattern for nodes using this weight assignment criteria. In conclusion, the two criteria provide the end-user with different infor-

Table 1 Network parameters

Random network parameters	
Number of vertices	2000
Min number of arcs for vertex	5
Max number of arcs for vertex	15
Scale free network parameters	
Number of vertices	2000
Number of line	No constraint
Average degree of vertices	10
Number of vertices in initial ER network	10
Initial probability of lines	0.2
α	0.25
β	0

mation, which can be used together to make informed decisions about which nodes to trust, and which nodes to avoid.

3 Simulation

In this section, we present a set of simulations showing that the social weight assignment is resilient to changes, and that it is sensible to nodes which behave in a different way from the others.

We defined a set of 5 simulation scenarios related to the weight assignment described by formula (2), which have a common set of rules. The networks in which weight assignment techniques are tested are either scale free or random. They are generated using the software Pajek [1] using the parameters in Table 1. After the networks are generated, the evolution of the network is simulated. for each simulation cycle, each node simulates a transfer with each of its neighbors. The outcome of a transfer is determined by a simple behavioral model: each node has a certain probability to reply to a transfer request unsuccessfully. The outcome for each neighbor is hence registered, and the weights for each neighbor are updated.

The behavioral pattern of a node is determined at the beginning of the simulation, and follows these rules:

- 95 % of the nodes have at most a 20 % chance of unsuccessfully replying to a transfer request.
- 5 % of the nodes have at least 80 % chance of unsuccessfully replying to a transfer request. These are the "bad nodes".
- The bad nodes are picked at random following a uniform distribution.

- Each node has a different chance based on a uniform random distribution, whether they are good or bad nodes.

Normally, a node never changes its behavior, and it always fair to all incoming transfer requests. An exception to this is the monitored node described below.

In each simulation, we exploit the pagerank compatibility of the social weight assignment to compute the pagerank for a specific node, and monitor its changes cycle after cycle. The node to be monitored is always the node which has the highest "topological pagerank", that is the pagerank calculated when all neighbors for each node have equal weight. The simulation graphs always show this pagerank for each cycle. As already mentioned, the monitored node does not follow the behavioral rules described above. Normally, these rules change from simulation to simulation.

The objective of Simulation 1 (Fig. 1) is to determine if the social weight assignment works as intended with a multitude of different networks. In order to do so, we made the monitored node always behave correctly for 500 cycles, and always misbehave for 500 cycles, with 10 different networks, 5 of them scale free and 5 random. No significant deviation from network to network is to be reported (except of course between scale free networks and random networks):

The objective of Simulation 2 (Fig. 2) is to compare the behavior of the Social Weight Assignment against the sample assignments described in Sect. 2, i.e. the Positive Feedback approach, and the Net Feedback approach.

In pictures, we can see that the net feedback approach behaves quite poorly in Simulation 2 (Fig. 2b), and drops much faster in Simulation 2 (Fig. 2a). The behavior of Social and Positive Feedback approaches is quite similar, but the Social approach is usually more stable (it reacts more slowly to changes). It is interesting to note that in Simulation 2 (Fig. 2)a the Social and Net Feedback lines cross. This is expected,

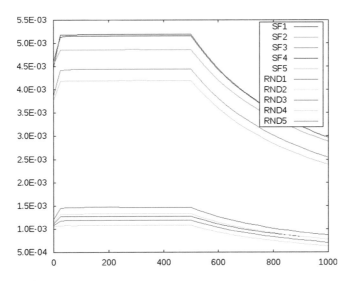

Fig. 1 Simulation 1

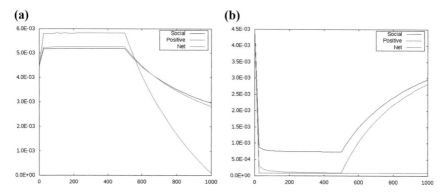

Fig. 2 Simulation 2. **a** Good to bad. **b** Bad to good

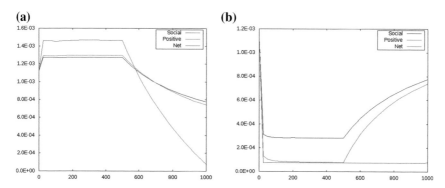

Fig. 3 Simulation 3. **a** Good to bad. **b** Bad to good

as the slope of the social graph is more gentle than the slope of the graph of Net
Feedback.

Simulation 3 (Fig. 3), like Simulation 2, but with a random network, aims at comparing Social Weight Assignment against the Positive Feedback approach, and the
Net Feedback.

The numbers are different, but the results pretty much follow what happened with
Simulation 2.

Simulation 4 (Fig. 4) has same setup of Simulation 2 but the error rates that has
been set to 0 (that means that all nodes always behave properly). The overall behavior
is similar to Simulation 2, but this time the slope of the social approach is more steep.
This is because the Social weight assignment tends to highlight neighbors which
behavior differs from the average.

Simulation 5 (Fig. 5), has same setup of Simulation 4 but with a random network
topology.

Again, the numbers are different, but the results pretty much follow what happened with Simulation 2.

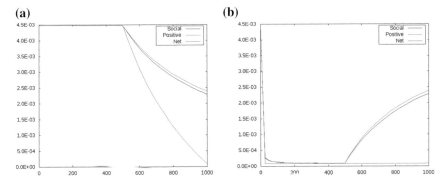

Fig. 4 Simulation 4. **a** Good to bad. **b** Bad to good

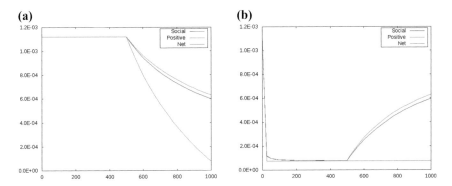

Fig. 5 Simulation 5. **a** Good to bad. **b** Bad to good

4 Conclusion

This work introduces a new, social-based proposal for weight assignment (local trust) in trust networks. Through simulations we show that such an approach provides greater stability compared to EigentTrust and GossipTrust; weights are not significantly affected if a node exhibits little changes to its behavior, since we're taking into account the node's past experiences.

Future works include:

- The introduction of the concept of aging for older interactions, because past experiences are usually less important than current ones,
- The definition of a behavioral pattern for the node using this weight assignment criteria,
- The definition of a global trust assessment mechanism expanding upon the ideas presented in this proposal.

References

1. Batagelj, V., Mrvar, A.: Pajek—program for large network analysis. Connections **21**(2), 47–57 (1998)
2. Corritore, C.L., Kracher, B., Wiedenbeck, S.: On-line trust: concepts, evolving themes, a model. Int. J. Hum. Comput. Stud. **58**(6), 737–758 (2003). Trust and Technology
3. De Kerchove, C., Van Dooren, P.: The PageTrust algorithm: how to rank web pages when negative links are allowed? In: Proceedings of the of the SIAM International Conference on Data Mining, pp. 346–352 (2008)
4. Guha, R., Kumar, R., Raghavan, P., Tomkins, A.: Propagation of trust and distrust. In: Proceedings of the 13th international conference on World Wide Web WWW '04, pp. 403–412. ACM, New York (2004)
5. Huang, J., Fox, M.S.: An ontology of trust: formal semantics and transitivity. In: Proceedings of the 8th international conference on Electronic commerce: The new e-commerce: innovations for conquering current barriers. obstacles and limitations to conducting successful business on the internet, ICEC '06, pp. 259–270. ACM, New York (2006)
6. Kamvar, S.D., Schlosser, M.T., Garcia-Molina, H.: The EigenTrust algorithm for reputation management in P2P networks. In: Proceedings of the Twelfth International World Wide Web Conference (2003)
7. Leskovec, J., Huttenlocher, D., Kleinberg, J.: Predicting positive and negative links in online social networks. In: Proceedings of the 19th international conference on World Wide Web, pp. 641–650 (2010)
8. Marsh, S.: Formalising trust as a computational concept. Ph.D. thesis (1994)
9. Riegelsberger, J., Sasse, M.A., McCarthy, J.D.: The mechanics of trust: a framework for research and design. Int. J. Hum. Comput. Stud. **62**(3), 381–422 (2005)
10. Thatcher, J.B., McKnight, D.H., Baker, E.W., Arsal, R.E., Roberts, N.H.: The role of trust in postadoption IT exploration an empirical examination of knowledge management systems. IEEE Trans. Eng. Manag. **58**(1), 56–70 (2011)
11. Walter, F.E., Battiston, S., Schweitzer, F.: A model of a trust-based recommendation system on a social network. J. Auton. Agents Multi-Agent Syst. **16**(1), 57 (2008)
12. Wang, Y.D., Emurian, H.H.: An overview of online trust: concepts, elements, and implications. Comput. Hum. Behav. **21**(1), 105–125 (2005)
13. Zhou, R., Hwang, K., Cai, M.: GossipTrust for fast reputation aggregation in peer-to-peer networks. IEEE Trans. Knowl. Data Eng. **20**(9), 1282–1295 (2008)

Reconfiguration of Access Schemes in Virtual Networks of the Internet of Things by Genetic Algorithms

Igor Saenko and Igor Kotenko

Abstract Virtual local area networks (VLAN) is a well-known technology of computer security in heterogeneous network infrastructures. It does not require significant computing resources. For this reason, it should find success in Internet of things. The VLAN access control scheme formation is divided into the tasks of initial configuration and reconfiguration. The paper presents an approach to the reconfiguration of VLAN access control schemes based on the improved class of genetic algorithms. Unlike the initial configuration, the reconfiguration additionally uses the previous access control scheme as input. Its search criterion is focused on minimizing the possible changes in the previous scheme. The paper shows that this problem is a special form of the Boolean matrix factorization. Main enhancements relate to generation of the initial population based on trivial solutions, using the columns of the connectivity matrix as the genes of chromosomes and applying in the fitness function the criterion of minimal cost to modify the access scheme. Experimental results demonstrate the proposed genetic algorithm has a high enough effectiveness.

1 Introduction

Virtual local area networks (VLAN) is a well-known technology of computer security in heterogeneous network infrastructures. Various means can be used to implement VLAN [1]. The main purpose of VLAN is counteraction to security threats that can occur from internal users (insiders). The essence of VLAN is to split the entire set of computers connected to a local network into logical chunks, known

I. Saenko (✉) · I. Kotenko
Laboratory of Computer Security Problems, St. Petersburg Institute for Informatics
and Automation (SPIIRAS), 39 14 Linija, St. Petersburg, Russia
e-mail: ibsaen@comsec.spb.ru

I. Kotenko
e-mail: ivkote@comsec.spb.ru

© Springer International Publishing Switzerland 2016
P. Novais et al. (eds.), *Intelligent Distributed Computing IX*,
Studies in Computational Intelligence 616,
DOI 10.1007/978-3-319-25017-5_15

155

as *virtual computer networks*, or, for simplicity, *virtual subnets*. VLAN tools ensure that if two computers do not belong to the same virtual subnet, then the exchange of information between them does not occur. VLAN technology is widely used in critical information infrastructures, where damage from loss of information due to possible malicious activity from insiders is great. In our view, VLAN technology should be also widely used in the networks of the Internet of things, which are characterized, on the one hand, by known limitations on the computing resources used for network security, and, on the other hand, by high heterogeneity of electronic devices, which are joined together by means of computer networks [2]. Moreover, the presence of electronic consumer devices ('things') is the cause of new types of attacks, such as the use of force to capture, disrupt and/or manipulate computing and information resources [3]. In such circumstances, a firm delineation of information flows between the computers, which is created by VLAN, is one of the most effective ways to ensure network security, which is cost-effective and opposed to new attacks.

The problem of VLAN design should be considered in two options: initial configuration and reconfiguration. At an *initial configuration* the access control scheme (or access scheme) for VLAN is formed on the basis of the given computer connectivity matrix. The VLAN access scheme shows the distribution of computers over virtual subnets. In [4], an approach to solve the initial configuration problem for VLAN access scheme on the basis of genetic optimization was offered. Criterion of optimization was the minimal quantity of virtual subnets. New contribution in comparison with the work [4] is a transition from an initial configuration problem for VLAN access scheme to a reconfiguration problem for it. The problem of *reconfiguration* also consists in formation of VLAN access scheme. However the existing access scheme and the given change of the connectivity matrix are additionally used as initial data. The solution of this problem is related to finding the change of the existing access scheme, which satisfies to a new connectivity matrix and is minimal.

The paper proposes to use genetic algorithms to solve the problem of reconfiguration of VLAN access schemes. This task, as well as the problem of the initial configuration, is a kind of Boolean Matrix Factorization (BMF) which we defined as Boolean Matrix Self-Factorization (BMSF) [4]. As BMF and BMSF methods are applied to solve NP-complete problems, it is necessary to consider that the problem of reconfiguration has the same computing complexity. The main theoretical contribution of this work consists in the following. First, the problem of reconfiguration of VLAN access schemes was not formulated earlier. Secondly, the paper shows that this problem is one of the BMF forms. At last, for solving the problem an advanced genetic algorithm is offered. The main improvements of the algorithm are formation of initial population on the basis of trivial decisions, use of columns of the connectivity matrix as genes of chromosomes, and application of the criterion of minimal costs to change the access scheme in the fitness function. The rest of the paper is structured as follows. Section 2 provides an overview of related work. Section 3 deals with mathematical foundations. The description of the genetic algorithm is provided in Sect. 4. Section 5 discusses the experimental results. Conclusions and directions of future research are presented in Sect. 6.

2   Related Work

Miettinen et al. [5–7] prove that BMF problems, which include initial configuration and reconfiguration of VLAN access schemes, are NP-complete, and some mathematical methods were proposed to find solutions of these problems. High complexity of BMF problems stipulates the search of heuristic methods of their decision.

Janecek and Tan [8] analyze the possibility to solve the BMF problems by algorithms based on populations, such as genetic algorithms, Particle Swarm Optimization, Differential Evolution, Fish School Search and Fireworks Algorithms. Snasel et al. [9, 10] consider a genetic algorithm for solving the BMF problem. Fitness-function is created on the basis of Euclidean distance between initial and resultant matrices. However this algorithm does not provide equality between matrices on which the initial matrix is decomposed. Therefore it cannot be applied to the BMSF problem.

Lu et al. [11, 12] show that some problems in information security, for example, a Role Mining Problem (RMP), can be reduced to BMF problems. In [13, 14] genetic algorithms are offered to solve RMP. These papers firstly offer to use columns of required matrices as genes of chromosomes. It provides the use of chromosomes of identical length in the algorithm and, respectively, the absence of invalid decisions after crossing and mutation operations. However, these innovations cannot be used for VLAN design, because the BMSF problem demands to search two the same matrixes instead of two different matrixes. Besides, as it will be shown below, the considered problem does not require chromosomes with different length.

There are a few works, devoted to application of data mining methods to design virtual subnets. Tai et al. [15] suggest to apply cluster analysis to form VLAN access schemes. This approach is focused on realization in mobile ad hoc networks. In [16] the genetic algorithm for optimization of virtual network schemes is considered. However this algorithm can find the computers connectivity matrix, but not distribution of computers on virtual subnets.

Thus, the analysis of relevant works shows that genetic algorithms should be considered as rather effective method to solve the considered problem. At the same time existing genetic algorithms cannot be directly used for this purpose.

3   Mathematical Foundations

Suppose that there are n computers in a network, and the scheme of the allowed information streams between these computers is determined by a Boolean matrix $\mathbf{A}$ $[n, n]$. If $a_{ij} = 1$ $(i, j = 1, …, n)$, the exchange between computers i and j is allowed. Otherwise this exchange is not possible.

Further suppose that we created k virtual subnets in the computer network. Each of these subnets unites two and more computers. We will set distribution of computers on subnets by means of a matrix $\mathbf{X}[n, k]$. If $x_{ij} = 1$ $(j = 1, …, k)$, the computer i belongs to subnet j. Otherwise the subnet j does not cover the computer i.

The matrix $\mathbf{X}$ is connected with the matrix $\mathbf{A}$ by the following expression:

$$\mathbf{A} = \mathbf{X} \oplus \mathbf{X}^\mathrm{T}, \tag{1}$$

where $\mathbf{X}^\mathrm{T}$ is a transposed matrix $\mathbf{X}$, symbol $\oplus$ stands for Boolean matrix multi-plication, which is a form of matrix multiplication based on the rules of Boolean algebra. Boolean matrix multiplication allows getting the elements of the matrix $\mathbf{A}$ by the following expression: $a_{ij} = \vee_{j=1}^{n}\left(x_{ij} \wedge x_{ji}\right)$.

To prove the expression (1) we consider Fig. 1 in which an example of mappings between computers $\mathbf{C} = \{C_n\}$ and virtual subnets $\mathbf{V} = \{V_k\}$ is presented. The strict proof of this expression for the general case is the direction of further research. From Fig. 1 it is visible that the subnet V_1 consists of computers C_1, C_2 and C_5, the subnet V_2—computers C_3 and C_4, and the subnet V_3—computers C_1, C_3 and C_5. Between computers and subnets there is a mapping $\mathbf{X}: \mathbf{C} \to \mathbf{V}$, between subnets and computers—the mapping $\mathbf{X}^\mathrm{T}: \mathbf{V} \to \mathbf{C}$. As a result, the mapping $\mathbf{A}$ from $\mathbf{C}$ to $\mathbf{C}$ is defined by multiplication of mappings $\mathbf{X}$ and $\mathbf{X}^\mathrm{T}$. It proves (1).

From the example given in Fig. 1, we have the following values $\mathbf{A}$ and $\mathbf{X}$:

$$\mathbf{A} = \begin{pmatrix} 1 & 1 & 0 & 1 & 1 \\ 1 & 1 & 1 & 1 & 1 \\ 0 & 1 & 1 & 0 & 1 \\ 1 & 1 & 0 & 1 & 1 \\ 1 & 1 & 1 & 1 & 1 \end{pmatrix}, \mathbf{X} = \begin{pmatrix} 1 & 0 & 1 \\ 1 & 0 & 0 \\ 0 & 1 & 1 \\ 0 & 1 & 0 \\ 1 & 0 & 1 \end{pmatrix}.$$

It is easy to see that the matrix $\mathbf{A}$ is always symmetric. In [4] we show that this task is a kind of the BMF problems. The BMF problem is reduced to finding Boolean matrixes $\mathbf{W}$ and $\mathbf{H}$ that are related with the given Boolean matrix $\mathbf{A}$ by the equation $\mathbf{A} = \mathbf{W} \oplus \mathbf{H}$, where $\mathbf{A} = \mathbf{A}[n, m]$, $\mathbf{W} = \mathbf{W}[n, k]$ and $\mathbf{H} = \mathbf{H}[k, m]$. Thus we consider that the problem (1) is more difficult than the BMF problem, as in expression (1) both matrices, which are a result of the matrix $\mathbf{A}$ decomposition, are related with each

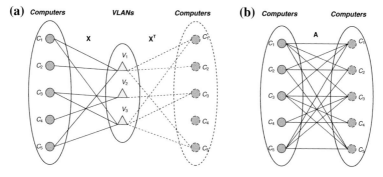

Fig. 1 An example of mappings between sets of computers and subnets. **a** the mappings between **C** и **V**. **b** the mapping between **C** и **C**

other, and in the BMF problem they are not related. For this reason we suggest to call the problem (1) as Boolean Matrix Self-Factorization (BMSF). The problem (1) is a problem of initial configuration of the VLAN access scheme, because the matrix $\mathbf{A}$ is an initial data, and it is required to find the matrix $\mathbf{X}$.

Now consider a problem statement for reconfiguration of the access scheme. Initial data are: initial matrix $\mathbf{A}_0[n, n]$; required matrix $\mathbf{A}_1[n, n]$; initial matrix $\mathbf{X}_0[n, k]$ that corresponds to the Eq. (1). It is required to find $\Delta\mathbf{X}[n, k]$ such that $\mathbf{X}_1 \oplus \mathbf{X}_1^T = \mathbf{A}_1$, where $\mathbf{X}_1 = \mathbf{X}_0 + \Delta\mathbf{X}$; and $Norm\,(\Delta\mathbf{X}) \to$ min. Operation '+' is considered as 'exclusive OR'.

The problem statement has the following sense. Let there is an access scheme $\mathbf{X}_0$, which satisfies to a connectivity matrix $\mathbf{A}_0$. Suppose that due to change of a security policy there is a need to change the matrix $\mathbf{A}_0$ with the value $\Delta\mathbf{A}$, thus a new connectivity matrix $\mathbf{A}_1 = \mathbf{A}_0 + \Delta\mathbf{A}$ can be defined. Therefore, it is necessary to create a new access scheme $\mathbf{X}_1$. The scheme $\mathbf{X}_1$ can be received, solving the initial configuration problem. However, as a result we can get the matrix, which is different from the initial matrix $\mathbf{X}_0$. This decision can demand from the administrator to perform a large number of actions to change the access scheme. Therefore as a criterion of decision search we choose a minimum of actions spent by the administrator for reconfiguration. This condition can be written as follows:

$$Norm(\Delta\mathbf{X}) = \sum_n \sum_k \Delta x_{ij} \to \text{min.} \tag{2}$$

Using the matrix $\Delta\mathbf{A}$ of connectivity changes, the problem of VLAN access scheme reconfiguration can be written as a problem of searching (at the given matrices $\mathbf{X}_0$, $\mathbf{A}_0$ and $\Delta\mathbf{A}$) the matrix $\Delta\mathbf{X}$ from the following equation:

$$(\mathbf{X}_0 + \Delta\mathbf{X}) \oplus (\mathbf{X}_0 + \Delta\mathbf{X})^T = (\mathbf{A}_0 + \Delta\mathbf{A}). \tag{3}$$

Comparing (1) and (3), it is visible that the problem of reconfiguration, as well as the problem of initial configuration, is a BMSF problem and, therefore, is NP-complete. However direct application of the genetic algorithm developed in [4] to solve the expression (1) is impossible. One of the reasons of it is that in (3) there are additional initial data which should be considered in the genetic algorithm. Besides, the criterion (2) differs from the criterion to minimize the quantity of columns in the matrix $\mathbf{X}$ used to solve the problem (1).

Let us consider an example. Let in the network presented in Fig. 1 it is necessary to pass to the logical connectivity matrix

$$\mathbf{A}_1 = \begin{pmatrix} 1 & 1 & 0 & 1 & 1 \\ 1 & 1 & 1 & 0 & 1 \\ 0 & 1 & 1 & 0 & 0 \\ 1 & 0 & 0 & 1 & 1 \\ 1 & 1 & 0 & 1 & 1 \end{pmatrix}.$$

If to solve this problem in the form (1), we get the decision $\mathbf{X}_2$ in which quantity of subnets $k = 3$. In the first subnet there are computers C_1, C_4 and C_5, in the second —C_1 and C_2, and in the third—C_2, C_4 and C_5. In this case the function *Norm* ($\mathbf{\Delta X}$), defined as quantity of '1'-elements in $\mathbf{\Delta X}$ is equal 7 that is confirmed as follows:

$$\mathbf{\Delta X} = \mathbf{X}_2 + \mathbf{X}_0 = \begin{pmatrix} 1 & 1 & 0 \\ 0 & 1 & 1 \\ 0 & 0 & 1 \\ 1 & 0 & 0 \\ 1 & 0 & 1 \end{pmatrix} + \begin{pmatrix} 1 & 0 & 1 \\ 1 & 0 & 0 \\ 0 & 1 & 1 \\ 0 & 1 & 0 \\ 1 & 0 & 1 \end{pmatrix} = \begin{pmatrix} 0 & 1 & 1 \\ 0 & 1 & 1 \\ 0 & 0 & 1 \\ 1 & 0 & 1 \\ 0 & 0 & 0 \end{pmatrix}.$$

When solving the problem (3), one of possible decisions is the matrix $\mathbf{X}_1$, corresponding to quantity of subnets plus 1, i.e. $k = 4$. In the first subnet there are computers C_1, C_2 and C_5, in the second—C_4 и C_5, in the third—C_1, C_2 and C_5, and in the fourth—C_2 и C_3. In this case the matrix $\mathbf{\Delta X}$ is as follows

$$\mathbf{\Delta X} = \mathbf{X}_1 + \mathbf{X}_0 = \begin{pmatrix} 1 & 0 & 1 & 0 \\ 1 & 0 & 1 & 1 \\ 0 & 0 & 0 & 1 \\ 0 & 1 & 0 & 0 \\ 1 & 1 & 1 & 0 \end{pmatrix} + \begin{pmatrix} 1 & 0 & 1 & 0 \\ 1 & 0 & 0 & 0 \\ 0 & 1 & 1 & 0 \\ 0 & 1 & 0 & 0 \\ 1 & 0 & 1 & 0 \end{pmatrix} = \begin{pmatrix} 0 & 0 & 0 & 0 \\ 0 & 0 & 1 & 1 \\ 0 & 1 & 1 & 1 \\ 0 & 0 & 0 & 0 \\ 0 & 1 & 0 & 0 \end{pmatrix}.$$

It is visible that in this case the function *Norm* ($\mathbf{\Delta X}$) is less, than in the previous case, and is equal 6. Therefore, the decision $\mathbf{X}_1$, in spite of the fact that there are more subnets than in previous case, is considered as better than $\mathbf{X}_2$ from the point of view of the expenses, which are carried out by the administrator.

4 Genetic Algorithm

Realization of the genetic algorithm is connected with the following key factors: formation of chromosomes and the initial population of decisions; definition of the fitness function; determination of crossing and mutation operations [17–19].

4.1 Formation of Chromosomes and the Initial Population

We use as a possible decision not a matrix $\mathbf{\Delta X}$, which is a variable in (3), but the matrix $\mathbf{X}_1$, which is defined as the sum (exclusive OR) of $\mathbf{\Delta X}$ and the matrix $\mathbf{X}_0$. Such choice allows to use for finding $\mathbf{X}_1$ the genetic algorithm proposed to solve the problem of initial configuration. In this case the vector $\mathbf{x1}_i = (x1_{i1}, x1_{i2}, \ldots, x1_{jn})$ is considered as a gene of chromosomes. The vector $\mathbf{x1}_i$ is a column of the matrix $\mathbf{X}_1$.

The quantity of genes in chromosomes is denoted by M, where M is a quantity of '1'-elements in the matrix $\mathbf{A}_1$, lying above the main diagonal. In [4] it is shown that M defines the quantity of columns in the trivial decision, in which in each subnet there are exactly two computers. It does not make sense to consider bigger number of columns of the matrix $\mathbf{X}_1$ as each additional column is a linear combination of columns of the trivial decision. For increasing a convergence of the genetic algorithm we make the assumption that concerns the formation of the initial population. We use a set of possible trivial decisions for its formation. We designate the population size through N and $P = 2^M$. If the condition $N \leq P$ is true, then the initial population consists completely of trivial decisions. Otherwise P individuals are trivial decisions and the others $(N \quad P)$ individuals are created randomly.

4.2 Fitness Function

For initial configuration of the VLAN access scheme the fitness function is as follows

$$F = (\alpha F_1 + \beta F_2)^{-1}, \tag{4}$$

where F_1—a function that needs to be minimized; F_2—a function which reflects full coincidence of the initial and resulting connectivity matrixes; α and β—the weight coefficients defining the direction of search of the decision. The ratio $\alpha \ll \beta$ between coefficients guarantees that firstly the decisions with $F_2 = 0$ are determined, and then—the decisions with smaller values F_1.

For reconfiguration of the VLAN access scheme the fitness function is based on (4) and is as follows

$$F = \left(\alpha \sum_i \sum_j \left(x1_{ij} - x0_{ij} \right)^2 + \beta \sqrt{\sum_i \sum_j \left(a1_{ij} - x1_{ij}x1_{ji} \right)^2} \right)^{-1}. \tag{5}$$

Comparing (2), (4) and (5), we see that $F_1 = Norm\ (\Delta\mathbf{X})$, and the function F_2 defines an extent of coincidence of the matrix $\mathbf{A}_1$ with the result of $\mathbf{X}_1 \oplus \mathbf{X}_1^T$.

4.3 Crossing and Mutation

Operation of crossing allows receiving two new descendant-individuals from a pair of parent-individuals by an exchange of parts of parental chromosomes. In traditional algorithm the exchange of genes of chromosomes is carried out. However such approach in our case is unacceptable as the individuals with zero columns can be never obtained and, respectively, we cannot minimize the value k. Therefore the

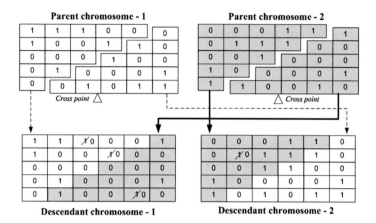

Fig. 2 Example of two-dimensional crossing

crossing operation should fulfill an exchange of parts of genes in addition. To provide such mode, we offer a two-dimensional mode of crossing, at which matrices of parental chromosomes before crossing are divided on two parts on a diagonal [20]. The essence of the two-dimensional crossing is visually illustrated in Fig. 2 for two parents, which are trivial decisions.

According to Fig. 2, the division of parental chromosomes is randomly carried out on the diagonal which middle passes through cross-point. The crossing operation can get as result two descendants, with some chromosomes represented by columns with a single '1'-element. It means that the corresponding virtual subnet contains only one computer. We cannot consider such subnets. Therefore we nullify these elements. In results, the descendant chromosome reflects the VLAN project with smaller number of subnets than the parental chromosomes. In particular, the first descendant chromosome reflects the project with three subnets, and the second —with five subnets.

We suggest carrying out an operation of mutation in two stages. At the first stage, according to traditional approach, with the given probability W_{gen} the genes of chromosome—columns of matrix $\mathbf{X}_1$—are selected. At the second stage with probability W_{el} the inversion of elements of the chosen columns is made.

5 Experimental Results

The experimental assessment of the proposed genetic algorithm for VLAN access scheme reconfiguration was carried out for a case study that imitates the Internet of things of 'the smart house'. The network includes 4 servers (supervisor, application server, database server and web-server), from 4 to 10 workstations, and from 4 to 36 user electronic devises ('things'). For imitation of things the Arduino Yún set is used. It consists of the ATmega 32u4 microcontroller and the Atheros AR9331 processor working under control of Linux [21]. In general, the dimension of the network was changed from 10 to 50 elements.

Experiments were made according to the following scheme. In the beginning the network size ($n = 10$; 25; 50) was chosen. Then the matrix $\mathbf{X}_0$ was randomly formed, and the matrix $\mathbf{A}_0$ was calculated by (1). Further the coefficient γ, determining the reconfiguration power, was selected. It is defined as Norm ($\Delta\mathbf{A}$)/B, where B is a quantity of elements in a matrix $\mathbf{A}_0$, lying above the main diagonal. Coefficient γ had a value 0.1, 0.25 and 0.5. At bigger γ reconfiguration loses meaning, and the initial configuration is necessary. Then, proceeding from the value γ, the matrix $\Delta\mathbf{A}$ was generated. At last, further by means of the proposed genetic algorithm the decision $\mathbf{X}_1$ and the matrix $\Delta\mathbf{X}$ were found, and the function Norm ($\Delta\mathbf{X}$) was calculated. The number T of iterations that are spent for search $\mathbf{X}_1$ was considered as the algorithm speed indicator.

The problem of initial configuration for $\mathbf{A}_0$ was also solved. For its decision $\mathbf{X}_2$ it was defined $\Delta\mathbf{X}_2 = \mathbf{X}_2 + \mathbf{X}_0$, which was compared to $\Delta\mathbf{X}$.

Results of experiments are given in Table 1. Values of indicators Norm ($\Delta\mathbf{X}$), Norm ($\Delta\mathbf{X}_2$) and T are received as averages on the random selection making 10 tests.

Analyzing the data in Table 1 it is possible to draw the following conclusions. At small dimension of the task ($n = 10$) and at different γ the values of functions Norm ($\Delta\mathbf{X}$) and Norm ($\Delta\mathbf{X}_2$) are approximately equal. It means that results of the solution of initial configuration and reconfiguration are almost identical. The same effect is observed at $n = 25$ or $n = 50$, but at small value $\gamma = 0.10$. At big γ (0.25 or 0.50) the result of the reconfiguration has advantage in comparison with the initial configuration, and the gain is larger, the larger n and γ.

The analysis of the proposed algorithm speed shows that with growth of n and γ the number of demanded iterations has the dependence that is close to linear. That is acceptable for a NP-complete task. Therefore, it is possible to draw a conclusion that the proposed genetic algorithm for access scheme reconfiguration is efficient.

The received experimental results allow making a number of recommendations for the administrator of the Internet of Things network. First of all, it should be

Table 1 Experimental results

n	B	γ	Norm ($\Delta\mathbf{A}$)	Norm ($\Delta\mathbf{X}$)	Norm ($\Delta\mathbf{X}_2$)	T
10	45	0.10	5	8.4	8.2	28.5
		0.25	11	18.2	17.9	37.3
		0.50	23	38.5	39.1	47.4
25	300	0.10	30	51.5	53.5	125
		0.25	75	119	137	210
		0.50	150	225	241	265
50	1225	0.10	123	175	186	565
		0.25	306	413	530	735
		0.50	613	512	633	850

noted that if the network is created for the first time then the administrator has to implement an initial configuration of VLAN which minimizes quantity of virtual subnets. Further actions for VLAN reconfiguration depend on dimension of the network. If the network has low dimension (it is possible to assume that VLAN is created on one router which has 8 or 16 ports), then at change of the given connectivity matrix **A** the initial configuration or reconfiguration has identical importance. We recommend using the first algorithm in interests of additional optimization of VLAN structure. If the Internet of things network has high dimension (such case takes place when VLAN is created on one router which has 24, 32 and more ports or on a set of routers connected in a cascade way) then at a change of the matrix **A** the administrator is obliged to use the reconfiguration which causes considerably smaller number of changes in the available VLAN scheme, than the initial configuration. It is necessary to notice that even if the matrix **A** changes for 10 percent, the number of necessary changes in the VLAN scheme is so big that the reconfiguration in the manual mode is almost impossible.

6   Conclusion

The paper proposes the approach to solve the problem of reconfiguration of VLAN access schemes based on the developed genetic algorithm. The analysis of the mathematical problem statement showed that the problem is the special kind of BMF, in which both matrices, on which the connectivity matrix is decomposed, are coincide. For this reason the use of known BMF methods is impossible. To solve the problem it is offered to use genetic algorithms. It is shown that known genetic algorithms cannot be directly used and demand improvement. The main improvements that are offered are formation of initial population on the basis of trivial decisions; using columns of the connectivity matrix as genes of chromosomes and implementing the criterion of the minimum costs of access scheme change. The experimental assessment of the proposed genetic algorithm showed its sufficiently high efficiency. Future research is related with usage of this approach to other areas of access scheme reconfiguration.

Acknowledgement This research is being supported by The Ministry of Education and Science of The Russian Federation (contract # 14.604.21.0033, unique contract identifier RFMEFI60414X0033).

References

1. Catalyst 2900 Series XL and Catalyst 3500 Series XL Software Configuration Guide.: Cisco IOS Release 12.0(5) WC(1). Cisco Systems, San Jose (2001)
2. Perera, Ch., Zaslavsky, A., Christen, P., Georgakopoulos, D.: Context aware computing for the internet of things: a survey. In: IEEE Commission Surveys and Tutorials, vol. 16(1) (2014)

3. Applegate, S.D.: The Principle of maneuver in cyber operations. In: 4th International Conference on Cyber Conflict, pp. 1–13 (2012)
4. Saenko, I., Kotenko, I.: A genetic approach for virtual computer network design. Stud. Comput. Intell. **570**, 95–105 (2014)
5. Miettinen, P., Vreeken, J.: Model order selection for boolean matrix factorization. In: 17th ACM SIGKDD Conference on Knowledge Discovery and Data Mining, pp. 51–59 (2011)
6. Miettinen, P.: Dynamic Boolean matrix factorizations. In: 2012 IEEE 12th International Conference on Data Mining, pp. 519–528 (2012)
7. Cergani, E., Miettinen, P.: Discovering relations using matrix factorization methods. In: ACM International Conference on Information and Knowledge Management, pp. 1549–1552 (2013)
8. Janecek, A., Tan, Y.: Using population based algorithms for initializing nonnegative matrix factorization. LNCS **6729**, 307–316 (2011)
9. Snasel, V., Platos, J., Kromer, P.: On genetic algorithms for boolean matrix factorization. In: Eighth International Conference on Intelligent Systems Design and Applications, vol. 2 (2008)
10. Snasel, V., Platos, J., Kromer, P., Husek, D., Neruda, R., Frolov, A.A.: Investigating boolean matrix factorization. In: Workshop on Data Mining using Matrices and Tensors (2008)
11. Lu, H., Vaidya, J., Atluri, V., Hong, Y.: Extended boolean matrix decomposition. In: Ninth IEEE International Conference on Data Mining, pp. 317–326 (2009)
12. Lu, H., Vaidya, J., Atluri, V.: Optimal boolean matrix decomposition: application to role engineering. In: 24th IEEE International Conference on Data Engineering, pp. 297–306 (2008)
13. Saenko, I., Kotenko, I.: Genetic algorithms for role mining problem. In: 19th International Conference on Parallel, Distributed and Network-based Processing, pp. 646–650 (2011)
14. Saenko, I., Kotenko, I.: Design and performance evaluation of improved genetic algorithm for role mining problem. In: 20th International Conference on Parallel, Distributed and Network-based Processing, pp. 269–274 (2011)
15. Tai, Ch.-F., Chiang, Tz.-Ch., Hou, T.-W.: A virtual subnet scheme on clustering algorithms for mobile Ad Hoc networks. Expert Syst. with Appl. 38(3):1269–2922 (2011)
16. Saenko, I., Kotenko, I.: Genetic optimization of access control schemes in virtual local area networks. LNCS **6258**, 209–216 (2010)
17. Goldberg, D.E.: Genetic Algorithms in Search, Optimization, and Machine Learning. Addison-Wesley Longman Publishing, Boston (1989)
18. Mitchell, M.: An Introduction to Genetic Algorithms. MIT Press, Massachusetts (1998)
19. Eiben, A.E., Smith, J.E.: Introduction to Evolutionary Computing. Springer, Berlin (2007)
20. Saenko, I., Kotenko, I.: Design of virtual local area network scheme based on genetic optimization and visual analysis. J. Wirel. Mob. Networks Ubiquitous Comput. Dependable Appl. (JoWUA) 5(4):86–102 (2014)
21. Schwartz, M.: Internet of Things with the Arduino Yún. Packt Publishing, Birmingham (2014)

GAMPP: Genetic Algorithm for UAV Mission Planning Problems

Gema Bello-Orgaz, Cristian Ramirez-Atencia, Jaime Fradera-Gil
and David Camacho

Abstract Due to the rapid development of the UAVs capabilities, these are being incorporated into many fields to perform increasingly complex tasks. Some of these tasks are becoming very important because they involve a high risk to the vehicle driver, such as detecting forest fires or rescue tasks, while using UAVs avoids risking human lives. Recent researches on artificial intelligence techniques applied to these systems provide a new degree of high-level autonomy of them. Mission planning for teams of UAVs can be defined as the planning process of locations to visit (way-points) and the vehicle actions to do (loading/dropping a load, taking videos/pictures, acquiring information), typically over a time period. Currently, UAVs are controlled remotely by human operators from ground control stations, or use rudimentary systems. This paper presents a new Genetic Algorithm for solving Mission Planning Problems (GAMPP) using a cooperative team of UAVs. The fitness function has been designed combining several measures to look for optimal solutions minimizing the fuel consumption and the mission time (or makespan). The algorithm has been experimentally tested through several missions where its complexity is incrementally modified to measure the scalability of the problem. Experimental results show that the new algorithm is able to obtain good solutions improving the runtime of a previous approach based on CSPs.

G. Bello-Orgaz (✉) · C. Ramirez-Atencia · J. Fradera-Gil · D. Camacho
Escuela Politecnica Superior, Universidad Autonoma de Madrid,
C/Francisco Tomas Y Valiente 11, 28049 Madrid, Spain
e-mail: gema.bello@uam.es

C. Ramirez-Atencia
e-mail: cristian.ramirez@inv.uam.es

D. Camacho
e-mail: david.camacho@uam.es

© Springer International Publishing Switzerland 2016
P. Novais et al. (eds.), *Intelligent Distributed Computing IX*,
Studies in Computational Intelligence 616,
DOI 10.1007/978-3-319-25017-5_16

1 Introduction

The potential applications of Unmanned Aerial Vehicles (UAVs) are varied, including surveillance [8], disaster and crisis management [11], and agriculture or forestry [7], among others. Therefore, over the past 20 years, a large number of research works related to this field have been carried out [5]. Due to the rapid development of the UAVs capabilities, these are being incorporated into many areas to perform increasingly complex tasks. Some of these tasks are becoming very important because they involve a high risk to the vehicle driver, such as detecting forest fire or rescue tasks, while using UAVs avoids risking human lives.

A mission can be described as a set of goals that are achieved by performing some tasks with a group of resources over a period of time. Specifically, mission planning for UAVs can be defined as the planning process of locations to visit (waypoints) and the vehicle actions to do (loading/dropping a load, taking videos/pictures, acquiring information), typically over a time period. There are some attempts to implement mission planning systems for UAVs in the literature. Doherty et al. [3] presented an architectural framework for mission planning and execution monitoring and its integration into a fully deployed unmanned helicopter. Then planning and monitoring modules use Temporal Action Logic (TAL) for reasoning about actions and changes, and the knowledge gathered from the sensors during plan execution is used in the process. Other novel approach formulates the mission planning problem as a Constraint Satisfaction Problem or CSP, where the tactic mission is modelled and solved using constraint satisfaction techniques [9].

These methods can be improved using stochastic search algorithms based on an objective function to be optimized, also known as Genetic Algorithms (GAs). There are many applications where GAs have been successful, from optimization [2] to Data Mining [1, 6]. GAs have demonstrated to be robust, able to find satisfactory solutions in highly multidimensional problems with complex relations between the variables. The Soliday et al. [10] approach developed a GA able to effectively solve UAV missions under complex constraints. The GA was constructed using a novel representation based on the nearest neighbour search, being each allele the N Nearest Neighbours. It uses a qualitative fitness function based on the number of mission objectives and the time allowed. Finally, other novel work has designed a graph based representation for mission planning of UAVs to carry out tasks in a flying space constrained with the presence of flight prohibited zones and radar sites [4].

This work aims to design and implement a new mission planning algorithm in order to improve the existing approaches using GAs. For this purpose a fitness function has been designed combining several measures to look for optimal solutions minimizing the fuel cost and the mission time (or makespan). These measures used in the fitness function and their weights can be changed in the algorithm settings. Then the algorithm is applied to real-world cases and a detailed analysis of the experimental results is carried out.

The rest of the paper is structured as follows. Section 2 describes the model designed for generating UAV missions. Section 3 presents the genetic algorithm, the encoding used and the fitness function implemented to solve UAV missions. Section 4 provides a description of the dataset used, the experimental setup of the algorithm and a complete experimental evaluation of it. Finally, in Sect. 5, the conclusions and some future research lines of the work are presented.

2 UAV Mission Plan Description

This section details the proposed structure to generate missions that the genetic algorithm receives as input, and also the output obtained. A UAV mission can be defined as a number n of *tasks* to accomplish for a set of *UAVs*. A task could be exploring a specific area or search for an object in a zone. One or more *sensors or payloads* belonging to a particular UAV may be required to perform a task. Each task must be performed in a specific geographic *zone*, at a specific time interval.

As can be seen in Fig. 1, the GA receives as input a list of tasks to be performed in specific zones ($[T_i, z_j]$), and a set of UAVs that can be used to perform these tasks. After the execution of the GA for an input mission, the output will be the possibles assignments of UAVs to tasks ($[T_i, z_j, UAV_k]$ where T_i is a task, z_j is the zone where the task is performed and UAV_k UAV is a vehicle from the set of available UAVs).

To perform a mission, there are m available UAVs, each one with specific characteristics such as fuel consumption rate, list of available payloads, an attribute indicating if the UAV is able to fly within restricted zones, and a position (geographical coordinates). A UAV can be equipped with one or more payloads that allows to perform different types of tasks:

- **Camera EO (Electro Optical)**: to take photos of large amplitude and long distance.
- **Camera IR (Infra-red)**: to take photos and videos at night or in conditions of very low luminosity. This sensor is also capable of performing thermal photos, especially used to forest areas analysis and fire detection and prevention.
- **Radar SAR (Synthetic Aperture Radar)**: to take images of an object allowing its representation in 2D and 3D. It can be used to track a zone.

Fig. 1 Input/Output example of the genetic algorithm for mission planning

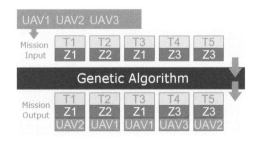

3 GA for Mission Planning Problems

The Genetic Algorithm for Mission Planning Problems (GAMPP) is a genetic algorithm to solve mission planning problems using a team of UAVs. This section describes this algorithm, including the encoding, the fitness function, and the genetic operators applied. The algorithm is implemented according to the structure of a simple genetic algorithm as can be seen in Algorithm 1.

3.1 Encoding

A possible solution for a mission planning problem consists of the assignment of each task to a specific UAV. If the mission contains a number N of tasks, the genotypes (chromosomes) will be represented as a integer vector of size N. Each allele represents a UAV assigned to a task. Therefore, if M is the number of UAVs to solve the mission, the value of each allele is between 0 and $M - 1$ (see Fig. 2). In this example, two UAVs (UAV0 and UAV1) perform most of the tasks, meanwhile the rest of tasks have been assigned to other UAVs, therefore they can be performed in parallel.

3.2 Genetic Algorithm

In the new approach used to solve mission plans, the population evolves using a standard GA as it is shown in Algorithm 1. The algorithm performs an Elitism selection method, where the n-best chromosomes of the population are copied to the new population (line 8 in Algorithm 1). This prevents losing the n-best found solutions. Finally the genetic operators work as follows:

- **Crossover**: One-point crossover operator is applied. Firstly, two individual are selected by tournament as parents, and a randomly point from the genome is chosen (see lines 11 and 12 in Algorithm 1). Then the information of both parents is swapped from this point to create two new offspring.
- **Mutation**: Uniform mutation is applied. A value of the genome is randomly chosen, and this value (with a predefined mutation probability) changes from 1 to M, where M is the number of available UAVs. See lines 13 and 14 in Algorithm 1.

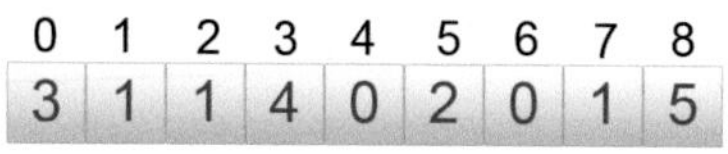

0	1	2	3	4	5	6	7	8
3	1	1	4	0	2	0	1	5

Fig. 2 Chromosome representing a solution for a mission planning problem. Each allele represents a particular assignment of a UAV to a task of the mission. This example is the solution for a mission that contains 9 tasks and 6 UAVs for accomplishing these tasks

Input: A mission $M = (T, U)$ where T is a set of tasks to perform denoted by $\{t_1, \ldots, t_n\}$ and U is a set of UAVs denoted by $\{u_1, \ldots, u_m\}$ representing the available vehicles to perform the tasks. And positive numbers *generations*, *population*, μ, λ and *mutprobability*
Output: The chromosome $S_i = \{a_1, a_2, \ldots, a_n\}$ such that *Fitness*(S_i) is maximized
$S \leftarrow$ randomly generated set of *population* of p chromosomes of size n, and the value of each allele is from 1 to m
$i \leftarrow 1$
convergence $\leftarrow 0$
while $i \leq$ *generations* $\wedge$ *convergence* $= 0$ **do**
    $F \leftarrow \emptyset$
    for $j \leftarrow 1$ *to* p **do**
        $F \leftarrow$ *Fitness*(S_j)
    end
    Sbest $\leftarrow$ *SelectNBest*(λ, F)
    $S \leftarrow$ *Sbest*
    for $j \leftarrow 1$ *to* λ **do**
        $p1, p2 \leftarrow$ *TournamentSelection*(*Sbest*)
        $i1, i2 \leftarrow$ *OnePointCrossover*$(p1, p2)$
        $i1 \leftarrow$ *Mutation*$(i1, mutprobability)$
        $i2 \leftarrow$ *Mutation*$(i2, mutprobability)$
        $S \leftarrow I \cup \{i1, i2\}$
    end
    $i \leftarrow i + 1$
    convergence $\leftarrow$ *CheckConvergence*(*Sbest*)
end
return *SelectBest*(S,F)
Algorithm 1: Genetic Algorithm for Mission Planning Problems (GAMPP)

3.3 Fitness Function

The fitness function implemented consists of two distinct phases to evaluate the generated individuals. Firstly, the criteria ensuring that the mission can be resolved successfully are evaluated. Afterwards, the quality of the mission is measured. For this purpose, the fitness function has been designed combining various measures to find an assignment of UAVs to the mission tasks minimizing the *fuel cost* and the *makespan*. To look for optimal solutions, a weighted function based on these criteria is used. The weights can be changed in the algorithm settings, and the fitness function is calculated as follows:

$$F(i) = (Mak(i) \cdot w_{mak} + Fuel(i) \cdot w_{fuel}) \cdot \alpha \qquad (1)$$

where $w_{dur} \in [0, 1]$, $w_{fuel} \in [0, 1]$, $w_{dur} + w_{fuel} = 1$ and α is defined as:

$$\alpha = \prod_{i=0}^{n} checkPayloads(i) \cdot checkDur(i) \cdot isResZone(i) \qquad (2)$$

3.3.1 Validation Criteria

These criteria ensure that the mission can be solved successfully, avoiding invalid
solutions. Invalid solutions are discarded giving them the lowest value of the fitness
function, which is 0. To validate the solutions three types of constraints are checked:

- **Payload Constraint**: checks whether each UAV carries the corresponding payload to perform the task assigned to it.
- **Temporal Constraint**: ensures that each UAV does not perform tasks at the same
  time.
- **Restricted Zone**: checks whether only UAVs with permissions to fly within
  restricted zones perform tasks developed in these restricted zones.

3.3.2 Optimization Criteria

Secondly, the quality of the solution is measured for valid individuals. A mission performed with a lower duration and fuel consumed, is usually better. For this purpose,
the fitness function combines two different criteria:

- **Makespan**: Time required to perform the complete mission. The different UAVs
  can perform tasks simultaneously. Therefore, the mission duration is given by the
  time interval from the start time of the first task to the end time of the last task.
- **Fuel cost**: Sum of the fuel consumed by each UAV at performing its assigned
  tasks. The fuel cost for a UAV k performing a task i is $fuel_i = fuelConsume *
  distance_{k \to z}$, being $fuelConsume$ the fuel consumption rate per distance unit of
  a UAV, and $distance_{k \to z}$ the distance from position k to position z given in geographic coordinates (latitude, longitude and altitude). This distance is calculated
  using the positions of the UAVs and the zones where tasks are performed.

4 Experimental Results

4.1 Dataset Description

In this work, 15 missions have been designed, each one composed by an increasing
number of tasks from 1 to 15. The first mission is composed of one task; the second,
two tasks; and so on up to 15, which is the most complex mission to solve. In order
to solve these missions, there are a set of UAVs with specific equipments. Each task
needs a particular payload to be performed, and each UAV has different available
types of payloads.

Table 1 Experimental setup for the genetic algorithm

Mutation probability	0.1
Generations	300
Population size	1000
Selection criteria ($\mu + \lambda$)	100 + 1000
Fitness function (w_{fuel})	0.7
Fitness function (w_{dur})	0.3

4.2 Experimental Setup

The GA parameters and the weights of the fitness function were obtained experimentally by performing several tests with different range of values. Table 1 shows the parameters used throughout the experimental phase where $\mu + \lambda$ is the selection criteria used, being λ the number of offspring (population size), and μ the number of the best parents that survive from the current generation to the next.

4.3 Results

Firstly, an analysis of the optimal solutions found is carried out. This can be seen in Table 2. Considering the values obtained to the fitness function, they begin close to 1 (very close to the best possible value). However, these values decrease as the num-

Table 2 Optimal solutions found for missions with 1–15 tasks

Task Number	F_{val}	Duration (min)	Fuel Consumed (l)
1	0.964	20	10.341
2	0.967	20	16.568
3	0.971	30	18.791
4	0.978	50	25.463
5	0.985	65	36.583
6	0.981	65	49.036
7	0.983	60	55.708
8	0.978	50	76.724
9	0.957	120	127.095
10	0.907	135	181.803
11	0.901	135	211.715
12	0.899	135	227.060
13	0.899	200	249.076
14	0.858	200	253.859
15	0.858	200	255.859

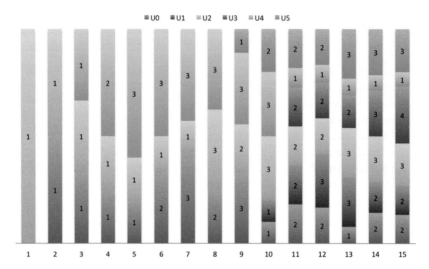

Fig. 3 Task assignments between the different available UAVs for solving the missions

ber of tasks increase. Analysing the task assignments, as can bee seen in Fig. 3, the algorithm carries out an equitable distribution of tasks between different available UAVs. There are several tasks performing in parallel, and therefore the mission duration is lower. It means that as the complexity of the missions increases, the quality

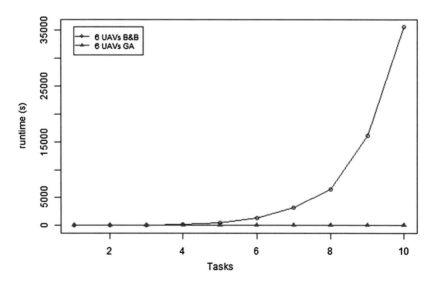

Fig. 4 Comparative assessment of runtime using other approach for the mission planning problem

of the optimal solution found decreases. But the algorithm is able to find solutions performing the complete mission with enough quality.

Finally, to study the computational performance of the algorithm, the runtime spent is compared with other approach based on a CSP model using Branch & Bound (B&B) [9]. The same dataset and optimization function ($0.7 \cdot Fuel(i) + 0.3 \cdot Mak(i)$) is used in both approaches. The results obtained are represented in Fig. 4. The time difference observed is very high, being the time needed to assign 10 tasks in the order of seconds (2, 3 s) to the genetic algorithm, whereas the CSP based approach is in a order of minutes (4 min). It can be appreciated that the runtime of the GA has a linear growth directly proportional to the number of tasks, whereas the runtime spent in the approach based on CSPs grows exponentially.

5 Conclusions and Future Work

In this work, a new mission planning algorithm for UAVs using GAs has been designed and implemented. For this purpose, a model to generate UAV missions is designed. Using this model, a mission is defined as a set of tasks to be performed in specific zones by several UAVs with some capabilities. In order to guide the search of possible solutions, a fitness function has been designed to look for optimal solutions minimizing the fuel cost and the makespan. The new algorithm has been tested using several UAV missions, and the experimental results obtained are analysed. Regarding the quality of the solutions, the algorithm performs a task assignment where the mission tasks are equitably distributed between the different UAVs available. Additionally, a comparative assessment of runtime to solve the mission planning problem is carried out. The experimental results obtained show that the new approach reaches a better runtime than a previous approach based on CSPs.

Finally, some improvements can be made to the algorithm. It is important to remark that the results obtained are highly dependant on the mission designed and on the topology of the zones the missions are developed in. Therefore, further works should consider different mission scenarios and topologies. In addition, the tactical scenarios for the missions are on real-time and dynamic. Many changes can affect the pre-loaded planning during its execution (UAVs failures, weather conditions, new tasks, etc.). Therefore an on-line distributed variant of the algorithm could be very useful.

Acknowledgments This work is supported by Comunidad Autónoma de Madrid under project CIBERDINE S2013/ICE-3095, Spanish Ministry of Science and Education under Project Code TIN2014-56494-C4-4-P and Savier Project (Airbus Defence & Space, FUAM-076915). The authors would like to acknowledge the support obtained from Airbus Defence & Space, specially from Savier Open Innovation project members: José Insenser, César Castro and Gemma Blasco.

References

1. Bello-Orgaz, G., Camacho. D.: Evolutionary clustering algorithm for community detection using graph-based information. In: IEEE Congress on Evolutionary Computation (CEC), 2014, IEEE, pp. 930–937 (2014)
2. Bin, X., Min, W., Yanming, L., Yu, F.: Improved genetic algorithm research for route optimization of logistic distribution. In: Proceedings of the 2010 International Conference on Computational and Information Sciences, ICCIS '10, IEEE Computer Society, pp. 1087–1090, Washington (2010)
3. Doherty, P., Kvarnström, J., Heintz, F.: A temporal logic-based planning and execution monitoring framework for unmanned aircraft systems. Auton. Agents Multi-Agent Syst. **19**(3), 332–377 (2009)
4. Geng, L., Zhang, Y.F., Wang, J.J., Fuh, J.Y.H., Teo, S.H.: Cooperative task planning for multiple autonomous uavs with graph representation and genetic algorithm. In: 10th IEEE International Conference on Control and Automation (ICCA), IEEE, pp. 394–399 (2013)
5. Kendoul, F.: Survey of advances in guidance, navigation, and control of unmanned rotorcraft systems. J. Field Robot. **29**(2), 315–378 (2012)
6. Menendez, H.D., Barrero, D.F., Camacho, D.: A co-evolutionary multi-objective approach for a K-adaptive graph-based clustering algorithm. In: IEEE Congress on Evolutionary Computation (CEC), IEEE, pp. 2724–2731 (2014)
7. Merino, L., Caballero, F., Martínez-de Dios, J.R., Ferruz, J., Ollero, A.: A cooperative perception system for multiple uavs: Application to automatic detection of forest fires. J. Field Robot. **23**(3–4), 165–184 (2006)
8. Pereira, E., Bencatel, R., Correia, J., Félix, L., Gonçalves, G., Morgado, J., Sousa, J.: Unmanned air vehicles for coastal and environmental research. J. Coast. Res. pp. 1557–1561 (2009)
9. Ramírez-Atencia, C., Bello-Orgaz, G., R-Moreno, M.D., Camacho, D.: Branching to find feasible solutions in unmanned air vehicle mission planning. Intelligent Data Engineering and Automated Learning–IDEAL 2014, pp. 286–294. Springer, Switzerland (2014)
10. Soliday, S.W., et al.: A genetic algorithm model for mission planning and dynamic resource allocation of airborne sensors. In Proceedings of the1999 IRIS National Symposium on Sensor and Data Fusion. Citeseer (1999)
11. Wu, J., Zhou, G.: High-resolution planimetric mapping from uav video for quick-response to natural disaster. In IEEE International Conference on Geoscience and Remote Sensing Symposium, IGARSS, IEEE, pp. 3333–3336 (2006)

FSP Modeling of a Generic Distributed Swarm Computing Framework

Amelia Bădică, Costin Bădică and Marius Brezovan

Abstract Swarm computing emerged as a computing paradigm for solving complex optimization problems using a nature-inspired approach. A swarm of particles populates a virtual space that mimics the physical environment. Virtual particles modeled as computational objects are behaving in the virtual space according to the laws of nature, seeking to solve a mathematical optimization problem. In this paper we propose a formal model of a generic distributed framework for swarm computing based on Finite State Process algebra. The model is simple, clear and technology-independent, and it can serve as a basis for concurrent or distributed implementation using available software technologies.

1 Introduction

Swarm computing emerged as a new computing paradigm for solving complex optimization problems using a nature-inspired approach. A swarm of particles populates a virtual space that mimics the physical environment. Virtual particles modeled as computational objects are behaving in the virtual space according to the laws of nature, seeking to solve a mathematical optimization problem.

Swarm computing has been proposed as an approach for solving difficult computational problems using methods of collective intelligence, decentralization and self-organization. The entities of a swarm can interact by exchanging information either directly or more often, indirectly, via the environment. They usually emulate the behavior of various types of natural entities like: insects, molecules, particles, birds or animals.

A. Bădică · C. Bădică (✉) · M. Brezovan
University of Craiova, Craiova, Romania
e-mail: cbadica@software.ucv.ro

A. Bădică
e-mail: ameliabd@yahoo.com

M. Brezovan
e-mail: mbrezovan@software.ucv.ro

P. Novais et al. (eds.), *Intelligent Distributed Computing IX*,
Studies in Computational Intelligence 616,
DOI 10.1007/978-3-319-25017-5_17

177

While the general paradigm of swarm computing is still not well-understood, currently, a large variety of nature-inspired algorithms is included under its umbrella.
Well-known examples are Ant Colony Optimization [5] and Particle Swarm Optimization [7].

The virtual environment acts as a container of entities, as well as an information
repository, where entities can deposit and retrieve information. Several approaches
have proposed the use of parallel and distributed computing for implementing the
entities' environment [6, 10]. The main idea is to partition the state space of the
entities and to allocate each set of the partition to a unique processor or computational
node.

In this paper we propose a formal model based on Finite State Process algebra
(FSP hereafter), of a generic distributed framework for swarm computing inspired
by the approach introduced in [6] for Ant Colony Optimization. Nevertheless, we
think that this approach is quite general and it can be applied to obtain distributed
variants of other swarm computing models.

We think that the main contribution of this paper is methodological. It presents a
simple, clear and technology-independent model of a distributed swarm computing
framework. The model can serve as a basis for concurrent or distributed implementation using available software technologies.

The paper is structured as follows. After this introductory section, we follow in
Sect. 2 with a brief discussion of related works. Section 3 introduces FSP, as well
as our proposed framework of distributed swarm computing. In Sect. 4 we describe
in detail the FSP model of our application and its implementation using Java and
Jason. Finally, in Sect. 5 we present our conclusions and point to possible future
developments.

2 Related Work

The paper [10] introduces a method for parallel state-space search that is based
on partitioning the state space and assigning each partition to a separate process.
The mapping of states to partitions is done using a hash function. The method is
applied to model checking algorithms similar to those employed by the SPIN model
checker [1]. This approach has some similarities to our framework, as it employs the
same technique of partitioning the state space. On the other hand, swarm algorithms
can also be described as state space search algorithms based on iterative improvement.

The paper [6] introduces a framework for distributed Ant Colony Optimization.
The framework was developed to run on a cluster of workstations and was applied
to solving path optimization problems in graphs, including the Traveling Salesman
Problem. The set of graph vertices was evenly partitioned and each partition was
assigned to a software agent. Then, the system was deployed using agent middleware
on a cluster of workstations, such that each computer ran a single agent. Basically,

the formal model that we propose here is an abstraction of the distributed system introduced in [6].

FSP is presented in detail in textbook [8]. This formal modeling language can be used to model multi-threaded Java programs in concurrent programming. At least in principle, the model developed in our paper can use the same methodology for its implementation using a multi-threaded approach. Nevertheless, other distributed computing technologies can be used for implementing the model, including multi-agent middleware, inter-process communication, as well as distributed object systems [4].

FSP was used for formal modeling in different areas including business processes, service composition and interaction protocols in multi-agent systems. In paper [2] the authors introduced a formal model of a service-oriented auction server, while in paper [3] the authors used FSP for modeling interaction protocols involving requester, provider and middle agents. The resulting models can be checked for correctness using a variant of Linear Temporal Logic, adapted for the action-based approach of FSP, known as Fluent Linear Temporal Logic—FLTL. This logic is also introduced in [8]. We can apply the same logic language and methodology for checking liveness and safety properties of our model.

Swarm computing attracted researchers working in high-performance computing. Consequently, there is a quite rich research literature covering parallel, distributed and high-computing approaches applied to swarm algorithms. For example, an overview and classification of parallel computing approaches to ACO was reported by the authors of [9]. This work introduces an interesting and novel classification scheme for parallel ACO algorithms. According to this classification, the cellular model of the ant colony that is structured as a set of small and overlapping neighborhoods is the closest to our approach. Nevertheless, none of the models included in this classification was analyzed at the level of providing a formal definition, as we propose here.

3 Background

3.1 Finite State Process Algebra

Finite State Process algebra is an algebraic specification technique of concurrent and cooperating computational processes as finite state labeled transition systems (LTS hereafter). FSP allows a more compact and easy to manage description of an LTS, rather than by directly describing it as a list of states and transitions between states.

FSP is an action-based language, rather than a state-based language, as for example SPIN [1]. Basically this means that the main modeling elements are abstract actions, rather than states and their transformations. The advantage is that an FSP model has a more compact representation. However, details regarding states are difficult to model using the FSP approach. Nevertheless, we found this modeling paradigm useful for our purpose, as our modeling is emphasizing the movement (i.e. actions) of swarm particles, rather than the details of their state transformation.

An FSP model contains a finite set of sequential and/or composite process definitions. The definition of a sequential process contains a sequence of one or more definitions of local processes. A local process definition $PN = PT$ consists of a process name PN associated to a process term PT.

FSP uses a rich set of operators for constructing process terms (see [8] for details). There are sequential, as well as composite process terms.

Sequential processes are defined by sequential process terms using the following constructs:

(i) Action prefix $(a \rightarrow P)$. It denotes a sequential process that executes action a and then behaves like sequential process P.
(ii) Nondeterministic choice $(P|Q)$. It denotes a sequential process that nondeterministically chooses to behave either like P or like Q.

Composite processes are defined by composite process terms using the following constructs:

(i) Parallel composition $(P||Q)$. It denotes a composite process that represents the parallel composition of composite or sequential processes P and Q using action interleaving semantics. Additionally, P and Q are constrained to proceed together on actions from their common alphabet (this constrained behavior is called synchronization).
(ii) Re-labeling $(P/\{new_1/old_1, \ldots, new_k/old_k\})$, where old_i are action labels and new_i are either action or sets of action labels.

FSP has an operational semantics given by the translation of process terms into an LTS. The mapping of an FSP term to an LTS is described in detail in [8]. The operational semantics is compositional, following the intuitive meaning of each FSP operator.

3.2 Framework of Distributed Swarm Computing

Our framework proposes a system composed of a set computational nodes, called just nodes in what follows. Each node acts as a container of swarm entities (just entities henceforth). Each node of the system is capable to receive entities from other nodes, to process (i.e. to change the state of) entities, as well as to send entities to other nodes. The block diagram of our proposed architecture of a distributed swarm computing system, instantiated for a set of 3 nodes, is shown in Fig. 1.

The system is using a mapping function that maps each entity e based on the contents of its state, to a node. So, if $\mathscr{E}$ is the set of entities and $\mathscr{N}$ is the set of nodes, the mapping function is defined as $\mu : \mathscr{E} \rightarrow \mathscr{N}$. For each entity $e \in \mathscr{E}$, node $\mu(e)$ is responsible with processing entity e. The framework is generic, i.e. it allows

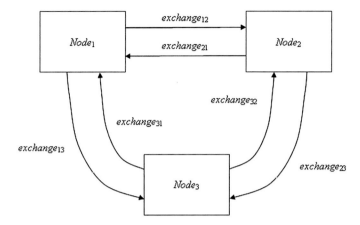

Fig. 1 Architecture of a distributed swarm computing system

the definition of static as well as dynamic mappings, for example inspired by parallel graph searching methods [10].

Additionally, each node has a local memory where it is able to store locally the information resulting after processing of each incoming entity. Note that this information depends on the particular type of swarm computing algorithm. For example, it can be pheromone information for an Ant Colony Optimization algorithm. Moreover, updates of state entities can take into account the local information of a node. Using this simple mechanism entities are able to exchange information using the "environment" of nodes, in a similar way insects or particles can indirectly exchange information or energy via the physical environment.

Nodes are able to exchange entities via communication channels. Whenever the state of an entity is updated, the mapping function determines if the updated entity should remain on the same node or it should be transferred to another node. Using this mechanism the computation load taken by the entities' state update operations is distributed among the nodes, at the communication cost for transferring the entities between nodes. Moreover, entities are able to indirectly exchange information via the local memory of nodes. An entity can deposit information in the local memory of a node, while the other entities that visit the node can read this information and update their states accordingly.

A node is composed of:

(i) *An interface for receiving entities.* This component is responsible with receiving entities from other nodes.
(ii) *A queue of entities.* Each received entity is stored into the queue. Then, entities are extracted from the queue and processed one by one, according to the state update rules of the swarm algorithm. Each updated entity is re-mapped to a node. If the resulting node is the same as the current node, then the entity

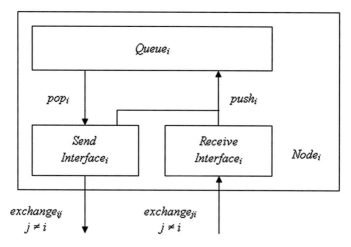

Fig. 2 Architecture of a swarm computing node

goes back into the queue. Otherwise, the entity is serialized and sent to the destination node.

(iii) *An interface for sending entities.* This component is responsible with sending entities to other nodes.

The block diagram of the node architecture presenting node components and their interaction is shown in Fig. 2.

4 Modeling Distributed Swarm Computing Using FSP

4.1 Model Presentation

Using abstraction as a basic principle of modeling, we abstract away from the actual state of an entity, as well as from the details of processing entity state information.

Let us assume that in our system there are P entities and N nodes. To simplify the presentation we assume that P is a multiple of N. Each node is initialized to hold $K = P/N$ entities. At each moment in time a node $I \in \{1, 2, \ldots, N\}$ contains in the corresponding queue X_I entities. Obviously we have $0 \le X_I \le P$ and $\sum_{I=1}^{N} X_I = P$.

For each $I \in \{1, 2, \ldots, N\}$ we use $Node(I)$ to refer to the Ith node, and we refer to its respective components as $Queue(I)$, $SendInterface(I)$, and $ReceiveInterface(I)$.

We start with the modeling of the queue component of $Node(I)$. As we abstract away from state entities, we are only interested in the number of entities held by each queue in the system, rather then their actual values. With this assumption, the model of $Queue(I)$ is shown in Fig. 3.

```
// Number of computing nodes
const N = 3
// Number of entities
const P = 12
// Initially there are K = P/N entities in each queue
const K = P/N
// Entities' queue
// x = number of entities in the queue
// I = index of the node to which the queue belongs
// Always X1 + ... + XN = P, i.e. there are P entities in the system
Queue(I=1) = Queue[K],  // Queue initialized with K entities
Queue[x:0..P] = (
  when x > 0 pop[I] -> Queue[x-1] |
  when x < P push[I] -> Queue[x+1]).
```

Fig. 3 FSP model of the queue of entities *Queue(I)*

```
ReceiveInterface(I=1) = (
  when I>1 exchange[j:1..I-1][I] -> push[I] -> ReceiveInterface |
  when I<N exchange[j:I+1..N][I] -> push[I] -> ReceiveInterface).
```

Fig. 4 FSP model of the receiving interface *ReceiveInterface(I)*

The second component is the *ReceiveInterface(I)*. It recognizes the messages of type *exchange[j][I]* transporting entities sent by *Node(j)* to *Node(I)* with $j \neq I$. The model of *ReceiveInterface(I)* is shown in Fig. 4.

The third component is the *SendInterface(I)*. It extracts an entity from the queue and then it processes the entity. As we abstract away from the processing details (including also the mapping functions that assigns an entity to a node), we model this situation with the nondeterministic choice of one action from the set of possible actions *process[I][i]* where $i \in \{1, 2, \ldots, n\}$. If $i = I$ then the updated entity is pushed into *Queue(I)*. Otherwise, if $i \neq I$ the updated entity is sent out to *Node(i)*. The model of *SendInterface(I)* is shown in Fig. 5.

The *Node(I)* component is described as a parallel composition of *Queue(I)*, *SendInterface(I)* and *ReceiveInterface(I)* components. Then, the system is defined as the parallel composition of all *Node(I)* components for $I \in \{1, 2, \ldots, N\}$. *Systems* and *Node(I)* processes are described in Fig. 6.

Note that both processes *SendInterface(I)* and *ReceiveInterface(I)* are independently pushing entities onto *Queue(I)*. Each of these push operations represent different actions of process *Node(I)*. This aspect is captured by renaming the *push* operations of *SendInterface(I)* and *ReceiveInterface(I)* with *push1* and *push2* respectively. Consequently, the component *Queue(I)* of process *Node* must allow

```
SendInterface(I=1) = (
  exchange[I] -> process[I][i:1..N] -> ContinueSendInterface[i]),
ContinueSendInterface[i:1..N] = (
  when (i == I) push[I] -> SendInterface | // entity i is processed
                                           // by the current node
  when (i != I) exchange[I][i] -> SendInterface). // send from node I
                                                  // to node i
```

Fig. 5 FSP model of the receiving interface *SendInterface(I)*

```
||Node(I=1) = (
      Queue(I)/{{push1,push2}/push} ||
      SendInterface(I)/{push1/push} || ReceiveInterface(I)/{push2/push}
).

||System = forall[i:1..N] Node(i).
```

Fig. 6 FSP model of the *Systems* and *Node(I)* components

both *push*1 and *push*2 operations to proceed independently. This aspect is modeled by renaming the *push* action of *Queue(I)* as either *push*1 or *push*2, so relabeling with the set {*push*1, *push*2} is used (see Fig. 6).

Summarizing, two FSP processes are accessing the *Queue(I)* to push entities:

 (i) *ReceiveInterface(I)* receives entities from other nodes and those entities must be pushed onto the *Queue(I)*. This is achieved using action *push*2 in Fig. 6.
(ii) *SendInterface(I)* extracts an entity from *Queue(I)*, processes the entity and if the updated entity is mapped back to the same node then it must be pushed onto the *Queue(I)*. This is achieved using action *push*1 in Fig. 6.

In order to assure the correct synchronization on the right "push" operation we have used the action renaming in *Queue(I)/{{push1, push2}/push}* before the composition of the queue process with the send and receive interfaces, in the definition of *Node(I)* from Fig. 6.

4.2 Experiment and Discussions

The FSP model introduced in this paper was developed using the Labeled Transition System Analyzer tool—LTSA, version 3.0.[1]

LTSA allows the preparation, visualization and analysis of an FSP model. The model can be edited and translated into its LTS representation. The resulting LTS model can be visualized and navigated through using different graph layouts, with the help of the Defroge extension of the LTSA tool.[2]

We used LTSA to analyze our model for the following values of the parameters: $N = 3$ nodes, $P = 12$ entities and each node is initialized with $K = 4$ entities. We recorded the number of states and transitions of each component in Table 1.

Analyzing the values from Table 1, it can be easily observed that the size of the resulting LTS of a complex system is increasing exponentially with respect with the number of its components. For example, the number of states of component *Node* is the product of the number of states of each component, i.e. $195 = 13 \times 3 \times 5$. This can make infeasible the analysis of large systems using LTSA.

---

[1]http://www.doc.ic.ac.uk/ltsa/. The model can be downloaded from: http://software.ucv.ro/~cbadica/fsp/idc2015_models.zip.
[2]http://lvl.info.ucl.ac.be/Tools/LTSADelforge.

Table 1 An example of a Table

Component	# states	# transitions
Queue	13	24
ReceiveInterface	3	4
SendInterface	5	7
Node	195	517
System	141255	728514

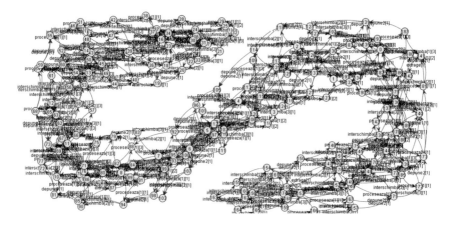

Fig. 7 LTS of *Node* process

As an example Fig. 7 shows the graphical representation of the LTS of a *Node* process, generated using LTSA Defroge, using the "Space-filling ISOM" layout. It can be noticed that, although interesting, even for a simple system with less than 200 of states, this type of visualization is difficult to follow.

As part of our analysis we observed that our *System* is free of deadlocks. This is a basic safety correctness property of a system. A safety property is satisfied if and only if for each execution scenario of the system it is true in each system state. Deadlock describes a state of a distributed system other than normal termination when none of its processes can make any further legal action. In our example, deadlock means that the system reached a state where none of its nodes can receive, process or sent swarm entities. For the definition and analysis of other more interesting properties of the system we can use the FLTL framework [8].

Moreover, from the mathematical point of view, the use of this model checking approach for verifying the deadlock property is of limited value, as it is valid only for the specific values of the parameters considered in the analysis. A rigorous analysis still needs a mathematical proof of the property—deadlock in this particular case. This proof is left as future work.

5 Conclusion

In this paper we proposed a simple FSP formal model of a generic framework for distributed swarm computing. The model can serve as a basis for development and implementation of various swarm computing algorithms using the available distributed computing technologies. As future work we would like: (i) to expand the initial formal analysis of this model by defining correctness properties and give the corresponding mathematical proofs; (ii) to investigate various computational methods of implementing systems by refining our proposed formal model, for example using multi-threading technology.

References

1. Ben-Ari, M.: Principles of the Spin Model Checker. Springer, Berlin (2008)
2. Bădică, A., Bădică, C.: Specification and verification of an agent-based auction service. In: Papadopoulos, G.A., Wojtkowski, W., Wojtkowski, G., Wrycza, S., Zupancic, J. (eds.) Information Systems Development, pp. 239–248. Springer, New York (2010)
3. Bădică, A., Bădică, C.: FSP and FLTL framework for specification and verification of middle-agents. Appl. Math. Comput. Sci. **21**(1), 9–25 (2011)
4. Coulouris, G., Dollimore, J., Kindberg, T., Blair, G.: Distributed Systems. Concepts and Design, 5th edn. Addison Wesley, London (2011)
5. Dorigo, M., Stützle, T.: Ant Colony Optimization. MIT Press, Cambridge (2004)
6. Ilie, S., Bădică, C.: Multi-agent approach to distributed ant colony optimization. Sci. Comput. Program. **78**(6), 762–774 (2013)
7. Kennedy, J., Eberhart, R.: Particle swarm optimization. In: Proceedings of the IEEE International Conference on Neural Networks, ICNN'1995, vol. 4, pp. 1942–1948. IEEE (1995)
8. Magee, J., Kramer, J.: Concurrency. State Models and Java Programs. World Wide Series in Computer Science, 2nd edn. Wiley, New York (2006)
9. Pedemonte, M., Nesmachnow, S., Cancela, H.: A survey on parallel ant colony optimization. Appl. Soft. Comput. **11**(8), 5181–5197 (2011)
10. Petcu, D.: Parallel explicit state reachability analysis and state space construction. In: Proceedings of the Second International Symposium on Parallel and Distributed Computing, ISPDC'2003, Ljubljana, Slovenia, pp. 207–214. IEEE (2002)

Part V
Distributed Computing

Heuristic-Based Job Flow Allocation in Distributed Computing

Victor Toporkov, Anna Toporkova, Alexey Tselishchev, Dmitry Yemelyanov and Petr Potekhin

Abstract In this paper, we propose a meta-data based approach for a deliberate job flow distribution in computing environments, such as utility Grids. Under conditions of a heterogeneous job flow composition and a variety of resource domains, we examine how different job and resource characteristics affect the efficiency of the scheduling process. Based on the most significant job flow and resource domain characteristics a heuristic distribution quality indicator is introduced. Additional simulation study is performed to verify the indicator in different distribution strategies and to compare them with a random job flow allocation.

V. Toporkov (✉) · D. Yemelyanov · P. Potekhin
National Research University "MPEI", Ul. Krasnokazarmennaya,
14, Moscow 111250, Russia
e-mail: ToporkovVV@mpei.ru

D. Yemelyanov
e-mail: YemelyanovDM@mpei.ru

P. Potekhin
e-mail: PotekhinPA@mpei.ru

A. Toporkova
National Research University Higher School of Economics, Ul. Myasnitskaya,
20, Moscow 101000, Russia
e-mail: AToporkova@hse.ru

A. Tselishchev
European Organization for Nuclear Research (CERN), Geneva 23 1211, Switzerland
e-mail: Alexey.Tselishchev@cern.ch

© Springer International Publishing Switzerland 2016
P. Novais et al. (eds.), *Intelligent Distributed Computing IX*,
Studies in Computational Intelligence 616,
DOI 10.1007/978-3-319-25017-5_18

1 Introduction

The fact that resources of utility Grids are non-dedicated makes challenges in the scheduling problem solution. In distributed computing with a lot of different participants and contradicting requirements the most efficient approaches are based on economic principles [3, 7, 11, 16]. A matter of the utmost importance for virtual organizations (VO) is to efficiently manage available computational resources with such quality of service indicators as an average job execution time and a number of required scheduling cycles while fulfilling requirements of all VO stakeholders: users, resource owners and VO administrators. The complexity of resource management and scheduling in distributed computing is determined by geographical distribution, resource dynamism and inhomogeneity of jobs and execution requirements defined by users of VOs [4, 8].

Diverse approaches to job scheduling can be classified based on job-flows allocation methods. In decentralized job-dispatching process, schedulers usually reside and work on the client side and fulfill end-user requirements (AppLeS [2], PAUA [5]). Centralized job-dispatching implies that a meta-scheduler ensures the efficient usage of all the resources. The meta-scheduler works with meta-jobs accompanied by resource requests, that contain resource characteristics required for the job execution. Such a hierarchical model is used in X-Com [18], GrADS [6] and other systems [4]. It is possible to evaluate job resource requirements statistically or by using expert systems [10]. Generally, the job-flow scheduling problem is solved using standard methods or algorithms, which include First-Come-First-Served, backfilling, user ranking mechanisms and resource sharing [1, 9, 17]. Within these approaches it is important to maintain the queue order and user priorities when executing these jobs. Even more"honest" queue forming is based on economic principles [11], which take into account single job features and their impact on the queue. Cycle job-flow scheduling [14] allows fulfilling VO requirements to a greater extent. Such scheduling is based on the set of dynamically updated information about the load of available resources. Three problems are being solved within each scheduling cycle: (1) job selection from a global flow; (2) forming jobs framework; (3) jobs scheduling and allocation based on the selected VO policy. During the job batch execution the VO policy, as a rule, has higher priority than single batch jobs preferences. This allows optimizing overall job batch execution parameters. For example, in a similar solution [19], it is described how a problem of minimizing the total energy consumption is solved during the job batch execution. However, at the same time queue order can be affected. There are two main steps in the cycle scheduling scheme (CSS) [14] for a single job batch: firstly, several execution alternatives are found for each job for a given scheduling interval and, secondly, the set of alternatives (one alternative for each job) is chosen following the VO policy [13]. Several execution alternatives allow optimizing the schedule for a batch of independent jobs. In order to fulfill VO user requirements the job batch is populated with the jobs with the highest priority (e.g. those in the beginning of a standard queue). Execution alternatives allocation is also performed sequentially for each job, which, in its turn, guarantees, that the priorities are followed. When additionally, user optimization criteria are used, one can guarantee "fair" scheduling of the whole job batch [12, 13]. However, it is worth noting, that job selection using simple user priorities can negatively impact the scheduling efficiency of the whole job batch. In other words, in order to increase the whole job batch scheduling efficiency according to the VO requirements one should evaluate different methods of job framework ranking.

In our previous works we considered job scheduling problems on an application level [15] and a job flow level in a single resource domain (including researches of an overall scheduling efficiency [13] and fair share scheduling mechanisms [14]).

In this paper, we study problems of a job flow distribution for the cycle scheduling in multi-domain environments. However resulting principles may be implemented in a wide range of scheduling approaches. Based on a job-flow scheduling simulation, a heuristic job and resource domain "compatibility" indicator is proposed. This indicator can be used to increase the job-flow scheduling efficiency in terms of resources utilization or user jobs execution characteristics. The rest of this paper is organized as follows. Sections 2 and 3 present basic job flow and distributed environment parameters, and simulation results necessary to determine the most significant pre-scheduling factors. A heuristic distribution quality indicator is presented in Sects. 4 and 5 contains comparative job flow scheduling simulation results. Conclusion and next steps are defined in the summary section.

2 Job Flow Distribution Problem

One of the key problems of a job flow hierarchical distribution structure is a deliberate user jobs allocation between the different job flows or available resource domains. To formalize the job flow distribution process, which can be considered as a pre-scheduling step, it is reasonable to rely on a meta information from a user job resource request. The resource request represents user requirements and expectations for the job service quality. The CSS model considered in this paper has the following basic resource request requirements to computational nodes.

1. Number n of computing nodes to be simultaneously allocated for the job execution.
2. Minimal computing nodes' relative performance indicator p required for the job execution.
3. Resource reservation time T required for the job execution on computing nodes with relative performance p.
4. Maximal job execution (hence a resource reservation) budget S allocated by user. Thus, a maximal price allowed per a computing node time unit is $c = \frac{S}{T*n}$.

To describe an individual computing node the following parameters are considered.

1. Computing node relative performance indicator p_0.

2. Price c_0 per a time unit specified by the resource owner.
3. Computing node local schedule during the scheduling interval.

Using the relative resource performance indicator p_0 allows to estimate job execution time on resources with different performance. For example, job execution time may decrease according to the performance level of the nodes the job is allocated to.

For the effective job flow distribution between the available resource domains it is important to determine the factors which contribute most to a job and a resource domain "compatibility". In this context a term "compatibility" means some indicator of how resource domain average parameters satisfy job resource requirements. Thus, the higher value of the compatibility indicator means the higher possibility of a job successful execution during the considered scheduling interval.

3 Significant Job Flow Allocation Parameters

In order to correlate resource request parameters with an individual resource domain one needs to formalize the domain's behaviour and to determine the most significant characteristics in terms of a job allocation. The following resource domain characteristics averaged over a single scheduling interval were considered for this purpose:

1. n_0—a total number of computing nodes in a resource domain; n_0^*—a total number of computing nodes satisfying the resource request requirements with performance indicator value $p_i > p$;
2. p_0—an average resource domain computing nodes performance;
3. N_l—a total number of slots available in the resource domain during the scheduling interval;
4. l_0—an average slot time length in the resource domain;
5. $V_0 = N_l * l_0$—a total amount of processor time available in the resource domain (sum of all available slots);
6. c_0—an average computing node utilization price per time unit;
7. Q_0—an average value of a price/cost factor in the resource domain; for an individual computing node $Q_i = \frac{c_i}{p_i}$ is a price of a unit of performance for a unit of time; the corresponding user job request price/quality indicator can be calculated as the following: $Q = \frac{c}{p} = \frac{S}{t*n*p}$ (the ratio of a maximal price per a time unit to a required minimal node performance).

The scheduling of each job flow related to a resource domain generally could be performed with any suitable scheduling approach. However for the current research we propose to use the cycle scheduling scheme [14] as it provides the following two important features. First, the job scheduling results are strongly dependent on a job queue order, hence the higher job's position in the queue means the higher probability of successful scheduling on a considered time interval. In that sense CSS has queue compliance properties similar to a traditional widely used backfilling algorithm. Second, CSS feature is that for each job there is an attempt to allocate as much execution alternatives as possible. Analyzing numbers of execution alternatives found for each job can provide an additional insight into which meta parameters affect the scheduling process and how.

A series of job flow scheduling experiments were carried out to determine the most significant job allocation and resource domain characteristics with a simulator [15]. Overall more than 5000 scheduling cycles with 40 randomly generated jobs per cycle were simulated using the CSS. A resource domain composition and a local schedule were generated once and remained unchanged during the whole series. At the same time parameters for a resource domain and 20,000 different resource requests with the final distribution results were stored in a file for a more thorough analysis.

A module for a visual scheduling results presentation in a given coordinates was developed to help determine characteristics affecting the scheduling results most. Every individual job is presented as a filled circle and the more execution alternatives were allocated for a job the bigger the circle's radius is. For example, Fig. 1 shows jobs for which no suitable allocations were found: all circles have the same default radius.

Figure 1 shows each job in $Q - p$ coordinates taken from the resource requests. A vertical line represents an average price/quality coefficient Q_o for a resource domain. A horizontal line represents average performance p_0 of the resource domain computing nodes. As can be seen from Fig. 1, the successful distribution probability increases when job's Q and p values are exceed the corresponding resource domain's average values. The number of alternatives found also increases for jobs with a relatively high Q value and undemanding to resources. It can be seen from a relatively small number of circles right to a vertical line and down from a horizontal line.

The job flow distribution results were also considered in coordinates of other resource request and resource domain's parameters. For example, Fig. 2 shows job

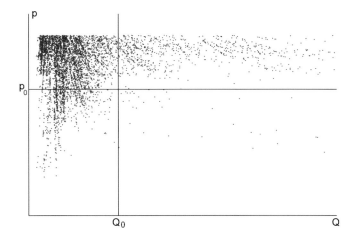

Fig. 1   Jobs with no suitable allocation found in $Q - p$ coordinates

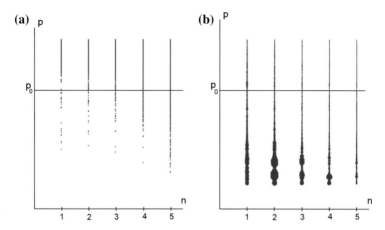

Fig. 2 Job flow allocation results in $n - p$ coordinates: (**a**)—no suitable allocation found; (**b**)—allocation found

flow distribution results in $n - p$ coordinates and the horizontal line represents an average resource domain nodes performance level p_0. Additionally Fig. 2b shows how a number of the possible execution alternatives depends on job p and n requirements.

As a result of the conducted job flow scheduling simulation and the following analysis the resource domain average characteristics which affect the scheduling result were determined. These parameters include Q_0, p_0, l_0, V_0. When values of these characteristics are compared to the corresponding resource request's parameters, one is able predict the job's probable scheduling outcome. Obviously when the user is ready to pay for high performance resources (and provides the high Q value) or does not request much processing power, chances for successful distribution increase. At the same time when a resource domain has the relatively high average performance p_0, it is possible to execute more user jobs at the same scheduling period.

In order to allocate resources for jobs with high demands (for example for jobs requesting a large number of simultaneously available nodes n), it is essential to provide enough processing time during the each scheduling cycle(related) l_0 and V_0 resource domain characteristics.

4 Distribution Quality Coefficient

As a compatibility measure of an individual job and a resource domain we propose an empirical D_q coefficient. D_q describes chances for a job to be scheduled and executed successfully during the present resource domain scheduling interval. D_q can have positive (high chance to be executed) or negative (low chance to be executed) values.

This coefficient consists of several summands, corresponding to the various characteristics of a resource domain and a user job. For each of the summands adjusting parameters are introduced: K_r—a weight coefficient determining the importance of the summand; C_r—a threshold value, approximately determining the value at which at least one alternative for the job is likely to be found. The values of the adjusting parameters can be formed based on statistics of the previous scheduling cycles or expert estimates. Thus, each summand can be presented in the following form:

$$D_{qr} = \frac{K_r}{C_r} * (C_r - \frac{r}{r_0}),$$ (1)

where r is a job characteristic, r_0 is a corresponding resource domain parameter. Additionally, r_0 may be modified to account for jobs already pending in the considered resource domain job queue.

Thus, for example, in accordance to (1) a term:

$$D_{qv} = \frac{K_v}{C_v} * (C_v - \frac{T * p * n}{V_0})$$ (2)

characterizes the ratio of slot utilization time required to execute the job and total processor time available during the considered scheduling cycle. In (2), K_v is a weight coefficient and C_v is a threshold value.

Using D_q coefficient it is possible to assign jobs to resource domains in different ways. We assume, that apart from an obvious strategy to assign jobs to the domains providing the highest D_q indicator value, it is reasonable to consider domains with the minimal positive D_q value—"threshold" domains. This threshold policy allows balancing of job flow execution during many cycles and providing the most efficient resource utilization. Otherwise, high performance resource domains could be overloaded with relatively small jobs, while resource demanding jobs will stuck in the queue.

5 Job Flow Allocation Simulation Study

The goal of the study is to estimate the properties of D_q indicator and it's particular qualities for a job flow distribution. It should be noted that one of D_q core benefits is that it does not require a preliminary job scheduling [7] for selecting a suitable resource domain which implies a relatively high job flow distribution processing speed.

The following is the input data for the experiment: the computational environment consists of 8 resource domains with different randomly generated characteristics (the resource domains' computational nodes number varies from 10 to 25 units; the scheduling interval length is defined as 600 time units of Grid simulator [15]; the job queues for each domain already contain 19 randomly generated high prior-

Table 1 Job distribution results

Strategy	Successful, %	Alternatives	Time	Cost
Random	33,1	2,44	40,14	1070,48
Threshold	57,9	2,39	39,13	1101,15
Best	70	2,86	36,89	943,65

ity user jobs); the next single user job is taken (generated) from the global job flow
for the assignment to one of the available domains. The goal of the experiment is to
select the best allocation. Three different distribution strategies are considered in the
experiment:

1. *Random*—the job is assigned to a random resource domain;
2. *Best*—the job is assigned to a resource domain providing the maximum D_q value;
3. *Threshold*—the job is assigned to a resource domain providing the minimum positive D_q value.

The *Random* strategy represents a general job queue First-Come-First-Served distribution policy with no use of job meta-parameters. The *Best* strategy assigns the
job to a resource domain with the highest chances of successful execution according
to D_q value and better suited for individual jobs execution. The *Threshold* strategy is
intended to maximize resource domains utilization level, though it requires a fine tuning of the D_q threshold and weight coefficients. The simulation results for different
distribution strategies: percentage of jobs executed successfully, an average number
of execution alternatives, job execution time and cost are presented in Table 1.

The results show that the *Best* strategy provides the best values for all considered
job distribution and scheduling efficiency parameters: the maximum successful job
execution percentage, the minimum job execution time and cost. According to D_q
value the most suitable, sometimes even too advanced domains were selected for
the job and hence the strategy provides the best assignment for the particular job.
From another hand, the *Threshold* strategy shows a decent successful job execution
percentage result especially compared to *Random*. Other considered parameters (the
strategy provides the maximum job execution cost and time) demonstrate that the
Threshold distribution allows to execute jobs in the domains roughly satisfying the
jobs resource requests thereby keeping domains with higher performance capacity
in reserve. The *Random* strategy shows inferior results both in terms of a single
job distribution optimization (compared to *Best*) and in terms of resource domains
utilization level maximization (compared to *Threshold*).

Thus, D_q indicator can be used for job flow distribution process depending on the
certain distribution strategy and optimizing execution of some particular jobs or the
whole computational environment utilization and throughput.

6 Conclusions and Future Work

In this work, the problem of a deliberate job ow distribution in such heterogeneous distributed computing environments as utility Grids is considered. In order to increase the distribution efficiency it is reasonable to rely on a job and a chosen resource domain meta information. Based on simulation studies we define significant factors affecting the job allocation and propose the general job and resource domain "compatibility" indicator D_q. We consider two different job flow distribution strategies: *Best* maximizes D_q values and hence the successful scheduling probability; *Threshold* is intended to maximize resource domains utilization level by allocating jobs to domains with minimum positive D_q values. The simulation study shows advantage of both strategies over the *Random* First-Come-First-Served job flow distribution approach. These results confirm advantages of the proposed heuristic D_q indicator for effective job flow distribution in distributed computing environments. Further research is aimed at developing methods for job allocation between several resource domains and forming a job framework while fulfilling requirements of all VO participants.

Acknowledgments This work was partially supported by the Council on Grants of the President of the Russian Federation for State Support of Young Scientists and Leading Scientific Schools (grants YPhD-4148.2015.9 and SS-362.2014.9), RFBR (grants 15-07-02259 and 15-07-03401), the Ministry on Education and Science of the Russian Federation, task no. 2014/123 (project no. 2268), and by the Russian Science Foundation (project no. 15-11-10010).

References

1. The Moab adaptive computing suite. http://www.adaptivecomputing.com/products/moab-adaptive-computing-suite.php
2. Berman, F., Wolski, R., Casanova, H.: Adaptive computing on the Grid using AppLeS. Trans. Parallel Distrib. Syst. **14**(4), 369–382 (2003)
3. Buyya, R., Abramson, D., Giddy, J.: Economic models for resource management and scheduling in Grid computing. J. Concurr. Comput. **14**(5), 1507–1542 (2002)
4. Cafaro, M., Mirto, M., Aloisio, G.: Preference-based matchmaking of Grid resources with CP-Nets. J. Grid Comput. **11**(2), 211–237 (2013)
5. Cirne, W., Brasileiro, F., Costa, L., Paranhos, D., Santos-neto, E., Andrade, N., Grande, C.: Scheduling in bag-of-task grids: the PAUA case. In: Proceedings of the 16th Symposium on Computer Architecture and High Performance Computing, pp. 124–131. IEEE Computer Society Press (2004)
6. Dail, H., Sievert, O., Berman, F., Casanova, H., Yarkhan, A., Vadhiyar S., Dongarra, J., Liu, C., Yang, L., Angulo, D., Foster, I.: Scheduling in the grid application development software project. In: Nabrzyski, J., Schopf, J.M., Weglarz, J. (eds.) Grid Resource Management. State of the Art and Future Trends, pp. 73–98. Kluwer Academic Publisher (2003)
7. Ernemann, C., Hamscher, V., Yahyapour, R.: Economic scheduling in Grid computing. In: Feitelson, D., Rudolph, L., Schwiegelshohn, U. (eds.) JSSPP, vol. 18, pp. 128–152. Springer, Heidelberg (2002)
8. Garg, S.K., Konugurthi, P., Buyya, R.: A linear programming-driven genetic algorithm for meta-scheduling on utility Grids. J. Par. Emergent Distr. Syst. **26**, 493–517 (2011)

 9. Kannan, S., Roberts, M., Mayes, P.: Workload management with LoadLeveler (2001)
10. Kurowski, K., Oleksiak, A., Nabrzyski, J.: Multi-criteria grid resource management using performance prediction techniques. In: Gorlatch, S., Danelutto, M. (eds.) Integrated Research in GRID Computing, pp. 215–225. Springer, Berlin (2007)
11. Mutz, A., Wolski, R., Brevik, J.: Eliciting honest value information in a batch-queue environment. In: *8th IEEE/ACM International Conference on Grid Computing*, pp. 291–297, New York. ACM (2007)
12. Soner, S., Ozturan, C.: Integer programming based heterogeneous CPU-GPU cluster scheduler for SLURM resource manager. In: 14th IEEE International Conference on High Performance Computing and Communication and 9th IEEE International Conference on Embedded Software and Systems, pp. 418–424, Liverpool. IEEE (2012)
13. Toporkov, V., Toporkova, A., Tselishchev, A., Yemelyanov, D.: Slot selection algorithms in distributed computing. J. Supercomput. **69**(1), 53–60 (2014)
14. Toporkov, V., Toporkova, A., Tselishchev, A., Yemelyanov, D., Potekhin, P.: Preference-based fair resource sharing and scheduling optimization in Grid VOs. Procedia Comput. Sci. **29**, 831–843 (2014)
15. Toporkov, V., Tselishchev, A., Yemelyanov, D., Bobchenkov, A.: Composite scheduling strategies in distributed computing with non-dedicated resources. Procedia Comput. Sci. **9**, 176–185 (2012)
16. Toporkov, V.V., Yemelyanov, D.M.: Economic model of scheduling and fair resource sharing in distributed computations. Program. Comput. Softw. **40**(1), 35–42 (2014)
17. Tsafrir, D., Etsion, Y., Feitelson, D.: Backfilling using system-generated predictions rather than user runtime estimates. In: Transactions on Parallel and Distributed Systems, pp. 789–803. IEEE (2007)
18. Voevodin, V.: The solution of large problems in distributed computational media. Autom. Remote Control **68**(5), 773–786 (2007)
19. Zhou, Z., Lan, Z., Tang, W., Desai, N.: Reducing energy costs for IBM Blue Gene/P via power-aware job scheduling. In: Seventeenth Workshop on Job Scheduling Strategies for Parallel Processing, pp. 96–115, Massachusetts (2013)

A Data Processing Framework for Distributed Embedded Systems

Ichiro Satoh

Abstract A MapReduce-based framework for processing data at nodes on the Internet of Things (IoT) is presented in this paper. Although MapReduce processing and its clones have been designed for high-performance server clusters, the processing itself is simple and generalized, so it should be used in non-high-performance computing environments, e.g., IoT and sensor networks. The proposed framework is unique among the other MapReduce-based processing approaches, because it can locally process the data maintained in nodes on the IoT rather than within high-performance server clusters and data centers. It deploys programs for data processing at the nodes that contain the target data as a map step and executes the programs with the local data. Finally, it aggregates the results of the programs to certain nodes as a reduce step. The architecture of the framework, its basic performance, and its application are also described here.

1 Introduction

The Internet of Things (IoT) connects devices such as everyday consumer objects and industrial equipment onto the network, enabling for information gathering and management of these devices via software to increase efficiency, enable for new services, or achieve other health, safety, or environmental benefits. IoT generate large quantities of data that need to be processed and analyzed in real time. IoT generally transfers massive amounts of small message sensor data to data centers or cloud computing environments for processing, because the computational resources of IoT devices are assumed to be limited. Most existing approaches assume to process a large amount of data at data centers. MapReduce is one of the most typical and modern computing models among them for processing large data sets in distributed systems. It was originally studied by Google [2] and inspired by the *map* and *reduce*

I. Satoh (✉)
National Institute of Informatics, 2-1-2 Hitotsubashi,
Chiyoda-ku, Tokyo 101-8430, Japan
e-mail: ichiro@nii.ac.jp

© Springer International Publishing Switzerland 2016
P. Novais et al. (eds.), *Intelligent Distributed Computing IX*,
Studies in Computational Intelligence 616,
DOI 10.1007/978-3-319-25017-5_19

199

functions commonly used in parallel list processing (LISP) or functional programming paradigms. *Hadoop*, is one of the most popular implementations of MapReduce and was developed and named by Yahoo!.

The processing of large quantities of data generated from IoT devices in real time will increase as a proportion of inbound traffics and workloads in networks from IoT to data centers. However, bandwidth of networks between IoT and these data centers tend to be slow and unreliable. IoT threatens to generate massive amounts of input data from sources that are globally distributed. Transferring the entirety of that data to a single location for processing is not technically and economically viable.

Modern IoT devices tend to have certain amounts of computational resources. For example, a Raspberry Pi computer, which has been one of the most popular embedded computers, has a 32 bit processor (700 MHz), a 512 MB memory, an Ethernet port, and USB ports. Therefore, such IoT devices have potential capabilities to execute a small amount of data processing. In fact, we have already installed and evaluated Hadoop on Raspberry Pi computers running Linux, but the performance is impractical even when the size of the target data is small, e.g., less than 10 MB.

Hadoop has been essentially designed to be executed on high performance servers and it is complicated so that it is almost impossible to redesign it for IoT devices, e.g., embedded computers. Therefore, a MapReduce framework is proposed here that is available on limited computers and network in IoT, e.g., Raspberry Pi computers, independently of Hadoop. The framework has three key ideas to save computational resources at a node. The first is to deploy and execute programs for data processing at IoT nodes that include the target data. The second is to introduce management functions into programs for data processing. The third is to provide Key-Value Store (KVS) for MapReduce processing available with limited memory.

The author previously proposed another data processing framework based on mobile agent technology [7]. The framework proposed in this paper is constructed independently of the previous one except for the notion of the deployment of programs for data processing.

2 Related Work

The tremendous number of opportunities to gain new and exciting value from big data are compelling for most organizations, but the challenge of managing and transforming it into insights requires new approaches, such as MapReduce processing. It originally supported *map* and *reduce* processes [2]. The first is invoked by dividing large scale data into smaller sub-problems and assigning them to worker nodes. Each worker node processed the smaller sub-problems. The second involves collecting the answers to all the sub-problems and aggregating them as the answer to the original problem it was trying to solve. There have been many attempts to improve Hadoop, which is an implementation of MapReduce by Yahoo! in academic or commercial projects. However, there have been few attempts to implement MapReduce itself except for Hadoop. For example, the Phoenix [9] and MATE systems [5] supported multiple core processors with shared memory. Also, several researchers have focused on iteratively and efficiently executing MapReduce, e.g., Twister [3], Haloop [1], and MRAP [8]. These implementations, except for SSS, assume the data in progress is stored in temporal files rather than key-value stores in data nodes. They assume the data will be stored on high-performance servers for MapReduce processing, instead of at the edges.

Google's MapReduce, Hadoop, and other existing MapReduce implementations have assumed their own distributed file systems, e.g., the Google file system (GFS) and Hadoop file system (HDFS), or shared memory between processors. For example, Hadoop needs to move target data from the external storage systems to HDFS via networks before processing them.

Our MapReduce system does not move data between nodes. Instead, it deploys program codes for defining the processing tasks to the nodes that include data by using the deployment of the components corresponding to the tasks and it executes the codes with their current local data. Hadoop and its extensions are unsuitable for use in sensor networks and embedded computers, because its file system, HDFS, tends to become a serious bottleneck in the operation of Hadoop and it often requires wide band networks, which may not be available at sensor nodes or embedded computers. In the literature on sensor networks, IoT, and machine-to-machine (M2M) communications, several academic or commercial projects have attempted to support data at the edge, e.g., at sensor nodes and embedded computers. For example, Cisco's *Flog Computing* and EMC's computing intend to integrate cloud computing over the Internet and peripheral computers. However, most of them do not support the aggregation of data generated and processed at the edge.

3 Requirements

Let us discuss requirements, before explaining our system.

- Unlike existing data processing frameworks, e.g., Hadoop, our framework should be available in IoT, because it generates a large amount of data from sensor nodes.
- Networks in IoT tend to be wireless or low-band wired, like industry-use networks. They have non-neglectable communication latency and are not robust to congestion. The transmission of such data from nodes at the edge to server nodes seriously affects performance when analyzing data and this results in congestion in networks.
- Most nodes in IoT, have non-powerful 32-bit processors with small amounts of memory, like Raspberry Pi computers. We assume that our framework is available on a distributed system consisting of Raspberry Pi computers.
- In IoT, a lot of data are generated from sensors. Nodes at IoT locally save their data inside their storage, e.g., flash memory.
- Every node may be able to support the management and/or data processing tasks, but may not initially have any codes for its tasks.

- Unlike other existing MapReduce implementations, including Hadoop, our framework should not assume any special underlying systems.[1] There is no centralized management system in IoT. Our framework should be available without such a system.

Our approach assumes data can be processed without exchanging data between nodes. In fact, in IoT data that each node contains is generated from the node's sensors so that the data in different nodes are independent of one another. Therefore, this assumption is reasonable. One of the most popular extensions of MapReduce, including the extensions of Hadoop, is to improve the performance of the iterative processing of the same data. However, our framework does not aim at such a iterative processing, just because most of the data at sensor nodes or in embedded computers in IoT are processed only once or a few times. Consider the analysis of logs in network equipment. Only the updated log data are collected and analyzed every hour or day instead of the data that were already analyzed.

4 Approach

We propose the following design principles to solve the requirements discussed in the previous section.

- *Dynamically deployable component*   Our framework enables us to define the data processing tasks as dynamically deployable components. To save network traffic, the task should be deployable on computers that have the target data. In fact, the sizes of programs for defining tasks tend to be smaller than the sizes of the data, so the deployment of tasks rather than data can reduce the network traffic.
- *Data processing-dependent networking*   Communication between nodes in MapReduce processing tends to depend on application-specific data processing. Each node, including master and data nodes in Hadoop, must have general-purpose runtime systems to support a variety of data processing rather than peculiar purpose one. However, such runtime systems tend to consume more memory. Our framework enables networking for MapReduce processing to be defined as the deployment of programs for defining the data processing.
- *MapReduce's KVS for limited memory*   In general, MapReduce processing tends to consume a lot of memory in its reduce phase, because the phase combines two data entries via KVS. The KVS that our approach introduces should be designed to save memory. Reducing data entries in the number of the KVSs, which are located on different computers, tends to increase the amount of traffic. Our framework transmits data between the nodes in a desynchronized way to avoid congestion.

The framework introduces the deployment of software components as a *Map* phase in MapReduce processing like Hadoop. However, the components are autonomous

---

[1]Hadoop has not been available in Windows because it requires a permission mechanism that is peculiar to Unix and its families.

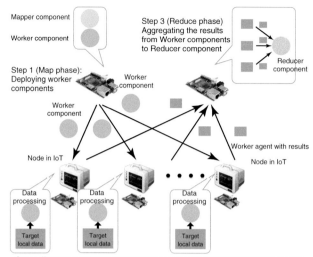

Fig. 1 Basic approach

in the sense that each component can control its destinations and itineraries under its own control. The framework allows developers to define their MapReduce processing from three parts, which are the map, reduce, and data processing, as Java classes, which can satisfy the specified interfaces. The map and reduce classes have similar methods as these used in the `Mapper` and `Reducer` classes in Hadoop. The data processing parts are responsible for the data processing at the edges. They consist of three methods corresponding to the following three functions: reading data locally from the nodes at the edge, the data processing of the data, and storing their results in a key-value store format. Figure 1 outlines the basic mechanism for processing.

- *Map phase*   A *Mapper* component makes copies of the *Worker* component and dispatches the copies to the nodes that locally include the target data.
- *Data processing phase*   Each of the *Worker* components executes its processing at its current data node. After executing its processing, it stores its results at the KVS of its current node.
- *Reduce phase*   The KVS of each of the nodes returns only the updated data to the computer that the *Reducer* component is running on according to their networking. The *Reducer* component collects the results from the *Worker* components via its KVS.

Each of the *Worker* components assumes to be executed independently of the others. The *Mapper* and *Reducer* components can be running on the same node.

5 Design and Implementation

Our framework consists of two layers: runtime systems and components (Fig. 2). Each runtime system was implemented with the Java language and operated on a Java virtual machine. The latter is defined with built-in Java classes by users as Jar-components corresponding to *map* and *reduce* processing and data processing tasks, where they are defined as the *Mapper*, *Reducer*, and *Worker* components. We call a set of *Mapper*, *Reducer*, and *Worker* components a *session*.

5.1 Runtime System

Each runtime system runs on a computer and is responsible for executing the *Mapper*, *Worker*, and *Reducer* components. It also establishes at most one TCP connection with each of its neighboring systems in a peer-to-peer manner without needing any centralized management server and exchanges control messages, components, and KVS-formatted data through the connection. Each runtime system is light so that it can be executed on embedded computers, including the JVMs for embedded computers. Each runtime system tries to maintain a minimal information about the other existing runtime systems, e.g., their network addresses, in a peer-to-peer manner through UDP multicasting.

5.2 Key-Value-Store

MapReduce processing should be executed on each of the data nodes independently as much as possible to reduce the number and amount of data transmissions through networks. However, data may not be divided into independent pieces. To solve this problem, Hadoop enables data nodes to exchange data with one another via HDFS, because HDFS is a distributed file system shared by all data nodes in a Hadoop

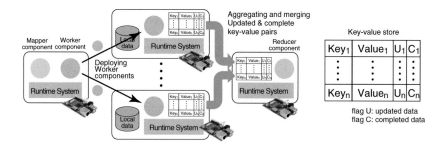

Fig. 2 System structure

cluster, like the GFS. Our framework provides a tree structure key value store (KVS) instead, where each KVS maps an arbitrary string value and arbitrary byte array data and is maintained inside its component, and the directory servers for the KVSs. Each session consists of a tree structure KVS, where its root tree is managed by a *Reducer* component and its subtrees are located at nodes that the session's *Worker* runs at. To support the *reduce* processing, the root of the KVS merge its subtrees into itself. To reduce the amount and congestion of the data transmission when merging subtrees, each of the keys' entries in a subtree has two flags.

- The *update flag* specifies whether the value of the key needs to be merged into the root tree. Only when the flag is positive does the runtime system transmit a pair of its key and value to the node that the *Reducer* component runs. The flag is useful to reduce the amount of data transmission.
- The *complete flag* specifies whether or not the value of the key will be changed. After the flag becomes positive, the runtime system starts to transmit a pair of its key and value to the node that the *Reducer* component runs, but not after the *Worker* component completes its process. The flag is useful to avoid congestion in the network between the *Worker*'s and *Reducer*'s nodes.

The runtime systems for executing *Worker* components transmit only the results on the KVS whose update and complete flags are positive before the components finish so that the data transmission from the *Worker* to *Reducer* components can be minimized and temporally distributed. Therefore, we can relax the limitations on the networks in IoT.

5.3 Component

Our framework supports the data processing on nodes, e.g., sensor nodes and embedded computers, which may be connected through non-wideband and unstable networks, whereas the already existing MapReduce implementations aim at data processing on high-performance servers connected through wideband networks. Therefore, we cannot directly inherit a programming model for the already existing MapReduce processing. In comparison with other MapReduce processing, including Hadoop, our framework explicitly divides the *map* operations into two parts in addition to the part corresponding to the *reduce* operation in MapReduce.

- *Duplication and deployment of tasks at data nodes*   Developers specify a set of the addresses of the target data nodes that their data processing are executed on or the network domains that contain the nodes. If they still want to define a more complicated MapReduce processing, our framework is open to extending the *Mapper* and *Reducer* components.
- *Application-specific data processing*   They define the following three functions: reading data locally from the nodes at the edge, the data processing of the data, and storing their results in a key-value store format. These functions can be isolated so

that the developers can define only one or two of the functions according to their data processing requirements.

- *Reducing data processing results*   They define how to add up the answers for the data processing stored in a key-value store.

Although the first is constructed in the *Mapper* and *Worker* components, the second in only the *Worker* components, and the third in both the *Worker* and *Reducer* components, the developers focus on the above three parts independently of their runtime systems. Our framework enables us to easily define the application-specific *Mapper*, *Reducer*, and *Worker* components as subclasses of the three template classes that correspond to the *Mapper*, *Reducer*, and *Worker* components, respectively, using several libraries for the KVS. When *Mapper* component gives one or more *Worker* components no information, we can directly define the component from the template class for *Mapper*. It can create specified application-specific *Worker* components according to the number of one or more specified data and deploy them at the nodes. When the *Reducer* components support basic calculations, e.g., adding up, averaging, and discovering the maximum or minimum values received from one or more *Worker* components through the KVSs based on the keys, we can directly define them as our built-in classes. In the current implementation, the locations of the *Worker* components are specified as Unix's environment variables, because the external management systems often notice which nodes have the target data. As a result, the *Mapper* does not need to know the network addresses of the computers that will execute the *Worker* component.

5.4 Fault-Tolerance

The runtime system that executes the *Mapper* component is responsible for detecting and handling any failures in data nodes. It periodically sends inquiry messages to all its data nodes. When it receive no answer from the node within a specified time, it treats the node as crashed. It leaves out the crashed node from the list of the target data nodes and informs about the crashed node to the *Reducer*. Even when the crashed node can be restarted or continues to work, the *Reducer* component does not wait for any results from the data node and then adds the node to the list, because data nodes periodically informs about their to the *Reducer* component. To mask failures in the nodes that runs the *Mapper* or *Reducer* components, the current implementations provide slave nodes that can run both of these components.

5.5 Security

The current implementation is a prototype system for dynamically deploying the components presented in this paper. Nevertheless, it has several security mechanisms. For example, it can encrypt components before migrating them over the network and can then decrypt them after they arrive at their destinations. Moreover, since each component is simply a programmable entity, it can explicitly encrypt its individual fields and migrate itself using them and its own cryptographic procedure. The JVM can explicitly restrict components so that they could only access specified resources to protect computers from malicious components. Although the current implementation cannot protect components from malicious computers, the runtime system supports authentication mechanisms for migrating components so that all runtime systems can only send components to and only receive them from trusted runtime systems.

6 Performance Evaluation

Although the current implementation was not constructed for performance, we evaluated that of several basic operations in a distributed system consisting of five networked embedded computers used as data nodes connected through Fast (100 Mbps) Ethernet via an Ethernet switch. Each embedded computer was a Raspberry Pi, where its processor was a Broadcom BCM2835 (ARMv6-architecture core with floating point) running at 700 MHz and it had a 512 MB memory, a Fast Ethernet port, and SD card storage (16 GB SDHC), with Raspbian, which was a Linux optimized to Raspberry Pi, and OpenJDK 6. The Java heap size was limited to 384 MB. We compared the basic performances of our framework and Hadoop. Among the five computers, one executed our *Mapper* and *Reducer* components or the master node in Hadoop. The others were data nodes in our framework and Hadoop. The Reducer component added up the numbers of each of the words received from the four *Worker* components for the word counting obtained from their nodes via the KVS. We compared our system to the Hadoop system. Figure 3 shows the costs of counting words using our framework and Hadoop. The former was faster than the latter, because the former was optimized to be executed in IoT.

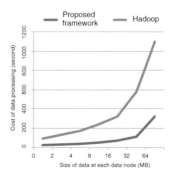

Fig. 3 Performance evaluation of propose framework and Hadoop

The readers may think that the application is unrealistic. However, we evaluated our approach for an abnormal detection from the data measured using sensors. It detected anomalous data, which were beyond the range of specified maximum and minimum values. This evaluation assumed each data node would have 0.01 % of abnormal data in its stream data generated from its sensor every 0.1 s and each data entry was 16 bytes. We detected abnormal values from the data volume corresponding to the data stream for one year at each of the eight data nodes. The total amount of the whole data came to about 5.04 GB and the amount of abnormal data was 504 KB in each node. When we used Hadoop, we need to copy about 40 GB data, i.e., multiply 5.04 GB by 4, from the data nodes to the HDFS.

7 Conclusion

We presented a distributed processing framework that was based on MapReduce processing. It was designed for analyzing the data at the edges of networks, IoT. It could distribute programs for data processing to nodes at the edges as a *map* operation, execute the programs with their local data, and then gather the results according to the user-defining *reduce* operation at a node. As previously mentioned, our framework is useful for thinning out unnecessary or redundant data from the large amounts of data stored at nodes in IoT, e.g., sensor nodes and embedded computers, connected through low-bandwidth networks. It enables developers to focus on defining the application-specific data processing at the edges without any knowledge on the target distributed systems.

References

1. Bu, Y., Howe, B., Balazinska, M., Ernst, M.D.: HaLoop: efficient iterative data processing on large clusters. In: Proceedings of the VLDB Endowment, vol. 3, p. 1 (2010)
2. Dean, J., Ghemawat, S.: MapReduce: simplified data processing on large clusters. In: Proceedings of the 6th Conference on Symposium on Opearting Systems Design and Implementation (OSDI'04) (2004)
3. Ekanayake, J., Li, H., Zhang, B., Gunarathne, T., Bae, S.H., Qiu, J., Fox, G.: Twister: a runtime for iterative MapReduce. In: Proceedings of the 19th ACM International Symposium on High Performance Distributed Computing (HPDC'10). ACM (2010)
4. Grossman, R., Gu, Y.: Data mining using high performance data clouds: experimental studies using sector and sphere. In: Proceedings of the 14th ACM SIGKDD International Conference on Knowledge Discovery and Data Mining (KDD'08), pp.920–927. ACM (2008)
5. Jiang, W., Ravi, V.T., Agrawal, G.: A map-reduce system with an alternate API for multi-core environments. In: Proceedings of 10th IEEE/ACM International Symposium on Cluster, Cloud, and Grid Computing (2010)
6. Satoh, I.: Mobile agents. In: Handbook of Ambient Intelligence and Smart Environments, pp. 771–791. Springer, Berlin (2010)
7. Satoh, I.: A framework for data processing at the edges of networks. In: Proceedings on 24th International Conference on Database and Expert Systems Applications (DEXA'2013), LNCS, vol. 8056, pp. 304–318. Springer, Berlin (2013)

8. Sehrish, S., Mackey, G., Wang, J., Bent, J.: MRAP: a novel MapReduce-based framework to support HPC analytics applications with access patterns. In: Proceedings of High Performance Distribute Computing (HPDC 2010) (2010)
9. Talbot, J., Yoo, R.M., Kozyrakis, C.: Phoenix++: modular MapReduce for shared-memory systems. In: Proceedings of 2nd International Workshop on MapReduce and Its Applications (MapReduce'11). ACM Press (2011)

A Distributed Reputation-Based Framework to Support Communication Resources Sharing

Antonello Comi, Lidia Fotia, Fabrizio Messina, G. Pappalardo, Domenico
Rosaci and Giuseppe M. L. Sarné

Abstract Improvements in computational and network capabilities of mobile
devices along with the wide spread of wireless networks for smart cities, allow
users to make accessible temporarily unused communication resources for free or
for a fee. Such interesting scenario introduces, however, some critical issues due to
reliability of users sharing/consuming resources which are a further concern with
respect to those strictly referred to the security. To tackle such problems, in this
paper we propose a distributed multi-agent framework, exploiting reputation information, which considers both the price paid by mobile users for resources and some
countermeasures to detect malicious users, as confirmed by experimental results.

1 Introduction

Improvements in wireless and mobile technologies draw new scenarios where mobile
users, in place of the Internet access offered by traditional mobile providers, might
take advantage from the same service made available via wireless by residential users, as recently proposed by Vodafone Italy [27]. In such a way, residential

A. Comi · L. Fotia · D. Rosaci · G.M.L. Sarné
DIIES, University "Mediterranea" of Reggio Calabria,
Loc. Feo di Vito, 89122 Reggio Calabria, Italy
e-mail: antonelo.comi@unirc.it

L. Fotia
e-mail: lidia.fotia@unirc.it

D. Rosaci
e-mail: domenico.rosaci@unirc.it

G.M.L. Sarné
e-mail: sarne@unirc.it

F. Messina (✉) · G. Pappalardo
Department of Mathematics and Computer Science, University of Catania,
V.le A. Doria 6, 95125 Catania, Italy
e-mail: messina@dmi.unict.it

© Springer International Publishing Switzerland 2016   211
P. Novais et al. (eds.), *Intelligent Distributed Computing IX*,
Studies in Computational Intelligence 616,
DOI 10.1007/978-3-319-25017-5_20

users might exploit their unused communication resources to realise hot-spot access points, based on their own private Wi-Fi networks [7], and mobile users might have the opportunity of choosing between the Internet access available via their mobile providers or via residential providers.

A similar scenario is currently applied to provide users with the access to the Web over areas uncovered by traditional mobile providers, for instance into rural districts [8], or to solve specific, often marginal, necessities [29]. This scenario will apply also into an urban area, i.e. a highly dynamic context where residential users could offer for free or for pay, via Wi-Fi, communication services.

However, in such a scenario some critical issues need to be solved, as the nature of relationships occurring among peers could encourage anomalous and/or also fraudulent behavior [25]. Therefore, each involved actor should be supported in order to have a certain level of confidence on the reliability of his/her counterpart, especially if the involved relationships include fee payments. Note that issues strictly related to authentication mechanisms are orthogonal with respect to the focus of this proposal, therefore, they are not dealt in this paper.

To tackle all the aforementioned questions, a distributed reputation-based agent framework called *Federated Wireless Community* (FWC) is conceived. FWC is managed by a *Framework* agent which provides some services to all the FWC affiliated. Furthermore, each peer is supported by a pair of associated agents, respectively called *Personal* and *Hosted* agents, running on its own device. A personal agent acts on behalf of its owner by monitoring user activities and to negotiate resources with other peers. The Hosted agent is tamper-proof, it manages and disseminates the reputation of its own host by interacting with its own peers in a safe manner [11].

Therefore, reputation is spread within the framework [9, 20, 21] whenever agents have to interact for services and each Hosted agent will maintain its own reputation by itself. To verify the performances of the FWC framework we ran a set of simulations which provided satisfactory and promising results.

The plan of the paper is the following. In Sect. 2 we introduce the Federated Wireless Community, i.e. the architecture of the proposed approach, while the reputation model is provided in Sect. 3. The results of the experiments are reported in Sect. 4, while Sect. 5 includes a comparison with related work. Finally, in Sect. 6 we draw our conclusions.

2 The Federated Wireless Community

In this section we describe the Federated Wireless Community (FWC) agent framework, with reference to Fig. 1. For convenience, we refer to a basic scenario including mobile users, denoted by u, and residential providers, denoted by r. In particular, a residential provider holds network resources which can be shared for free or for pay with mobile users by means of a private Wi-Fi access point.

The structure of the framework is based on three type of agents. The *Personal* agent, denoted by d_x^p, and the *Hosted* agent, denoted by h_x^p, are designed to live on

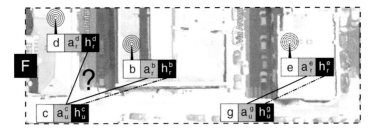

Fig. 1 The FWC scenario and the different relationships occurring among the agents

the peers. We denote with p the peer and with $x = \{u, r\}$ his/her mobile (i.e. u) or residential (i.e. r) role. A third agent, the *Framework* agent, denoted by F, manages some basic services into the FWC framework.

Affiliation. We assume that a generic peer p, in order to join with a FWC, has to register its own device on the associated Framework agent F, whose identity is certified by a *Certification Authority* $\mathcal{A}$ (some approaches, as in [12], exploit the SIM so that identity changes necessarily require another SIM). Once the identity of F is verified, the peer can register its device (i.e. Personal agent a_x^p). Therefore it *will receive an Hosted agents* h_x^p which will store an initial reputation $\mathcal{R}$ (see Sect. 3).

Hosted agent activities. The main activities of device agents depend on the residential/mobile role of its peer owner (host). Each *Hosted agent* will perform the following activities: (i) interacting with the associated Personal agent to mutually coordinate the activities related to the interactions with other peers; (ii) sending information to its associated Personal agent about identities and reputation of counterparts; (iii) managing peer's reputation, as described in the next Section, by interacting with the other Hosted agents.

Personal agent activities are divided into *Meta-data Synchronization* and *Service Provisioning*. More in detail, *Meta-data Synchronization* will consist of the following two stages:

- *Presentation.* Whenever a mobile user looks for a service in a specific area, all the Hosted agents in the same area mutually exchange their own credentials (storing identity and reputation) in order to verify the truthfulness of the information about identity and reputation received by the the mobile peer.
- *Service offering.* A tuple $OS = \langle SD(s), \mathcal{R}^{req}(s), C(s) \rangle$, is sent by each Personal agent a_r to the mobile peer agent a_u which has initiated the lookup for services. $SD(s)$ is the descriptor of the service, $\mathcal{R}^{req}(s)$ is the reputation score required by the agent for accessing to s and $C(s)$ its cost.

The *Service Provisioning* task is divided into two phases:

- *Interest for the service.* In order to allow a mobile user to exploit the service s offered by a residential user, it has to be $R_u > \mathcal{R}(s)^{req}$, i.e. the mobile user's reputation (i.e. R_u) required for the service has to be over the threshold (i.e. $R(s)^{req}$) specified by the residential user offering that resource. In this particular approach, the threshold $R(s)^{req}$ is said to be a *hazard threshold* [6, 22, 26], i.e. it represents the probability of "failure" that the provider (residential user) is willing to accept.

- *Service provisioning.* Once the requirement on the hazard threshold have been verified, the Hosted agent h_u signs the agreed *OS* and sends it to the Hosted agent peer h_r that, in turn, authorizes the associated Personal agent to make accessible the service. Once the service has been consumed, Personal agents compute their feedbacks, then Hosted agents h_r and h_u work together to exchange such feedbacks to update their own reputation, as explained in Sect. 3.

3 The FWC Reputation Model

The reputation model associated with the scenario discussed in the previous section is based on the following definition of reputation [1]: *an expectation about the user's behavior based on information about the observations of his past behavior.* Basing on the definition above, in this approach, the user's history—i.e. the behavior, in terms of feedbacks of the user—is included in the computation of reputation.

We assume that, once service provisioning phase is concluded, users (i.e. Personal and Hosted agents) have to compute and exchange *feedbacks* and update their own reputations. Feedbacks reflect both users' behavior and take into account some more aspects which may differ between mobile and residential users. For instance, when giving a feedback on the residential side, the mobile user's behavior over Internet might be considered while, vice versa, on the mobile side the quality of the provided service with respect to that agreed on the OS should be evaluated.

Let be $f_{j,i}^s \in [0, 1] \subset \mathbb{R}$ the feedback coming from the user u_j about the user u_i for the service s (where 0/1 means the minimum/maximum appreciation). Remember that feedbacks are managed by the Hosted agents of u_j (i.e. h^j) and u_i (i.e. h^i) in a safe manner. The latter will compute a further value $f_{j,i}^{*s}$ and, finally, the new value of $\mathcal{R}_i^{new}$ as follows:

$$
R_i^{new} = \begin{cases} (1-\alpha) \cdot R_i^{old} + \alpha \cdot f_{i,j}^{* \, s} & f_{i,j}^* > 0 \vee R_{old}^i \geq 0.5 \\ R_i^{old} & \text{otherwise} \end{cases} \tag{1}
$$

where

$$
f_{j,i}^{* \, s} = H_j \cdot \left(\frac{\pi_{j,i} + \beta_{j,i}}{2} \right) \cdot f_{j,i}^s
$$

Once h^i has computed the new value of reputation for its host, it also provides to update the u_i's credential which will include the new value of reputation. In the expression of R^{new}, parameter $\alpha \in [0, 1] \subset \mathbb{R}$ has the function to take into account the old reputation R_i^{old} (i.e. the past behavior of the node), which is, in fact, combined with the recent feedback $f_{i,j}^s$ provided by a_j. Parameter α affects the reputation system behavior (see Sect. 4), since the higher is the value of α, the lower will be the sensitivity of $\mathcal{R}$, and vice versa. Finally, parameters $H_{j,i}$, $\beta_{j,i}$ and $\pi_{j,i}$, depend from the reputation of agent j, the cost of the service and the past interactions between i and j, as discussed below. Please note that, reputation is not always updated through parameters α and f^* (basing on the logical condition $(f^*_{i,j} > 0 \vee R^i_{old} \geq 0.5)$). As we discuss later, this is done in order to be compliant with countermeasures discussed later and quantified by parameters H, π and β.

Computation of $H_{j,i}$, $\pi_{j,i}$ and $\beta_{j,i}$. In this approach we assume that the reputation of an agent is also a measure of its reliability in providing a honest feedback [23]. Therefore, values $H_{j,i}$ are computed by means of the step function described in the left part of Eq. 2. It is tuned by the parameter $\Gamma^* \in [0, 1] \subset \mathbb{R}$.

$$H_{j,i} = \begin{cases} 1 & R_j \geq \Gamma^* \\ 0 & Otherwise \end{cases} \qquad \pi_{j,i}(r) = \begin{cases} 1 & C = 0 \\ \dfrac{C(s)}{C_{Max}(s)} & Otherwise \end{cases} \qquad (2)$$

Parameter $(\pi_{j,i})$ takes account of the monetary cost $C(s)$ payed for the service s (right part of Eq. 2), where $C_{Max}(s)$ is a threshold denoting the maximum cost for the service s. As a consequence, the lower the cost, the lower will be the impact of the feedback gained for those resources. This approach is adopted in order to avoid malicious behavior of peers that try to gain reputation for services having little significance, to be exploited for high value services. The case on which cost is zero (resources are for free) is treated by leaving $\pi_{j,i} = 1$.

Parameter $\beta_{j,i}$ (*Balance*) is computed on the basis of the number of times that a feedback was provided to the same agent, as specified in the left part of Eq. 3.

$$\beta_{j,i}(t) = \begin{cases} 1 & f^S_{j,i} < 0.5 \\ \left(\dfrac{1}{K_{j,i}(t)}\right) & f^S_{j,i} \geq 0.5 \end{cases} \qquad K_{j,i}(t) = \begin{cases} 1 & t = 0 \\ K(t_l) + 1 & (t - t_l) < \Delta t \\ Max\left(1, K(t_l) - \left\lfloor \dfrac{t - t_l}{\Delta t} \right\rfloor\right) & t - t_l \geq \Delta t \end{cases} \qquad (3)$$

The ratio behind computation of parameter β is rather simple. Indeed, it is aimed at limiting activities devoted to obtain a mutual reputation increasing/decreasing. In particular, when $f_{j,i} < 0.5$ parameter β will not affect the final value of f^*. Conversely, when $f_{j,i} \geq 0.5$, i.e. the feedback assumes a "good" value, parameter β will get values from the expression $\left(\dfrac{1}{K_{j,i}(t)}\right)$. The parameter $K_{j,i}$ (defined in the right part of Eq. 3) allows us to limit the impact of potentially collusive users' behavior aiming to increase reputation by providing a number of positive feedbacks with high frequency. In particular, $K_{i,j}(0) = 1$, while, for $t > 0$, $K_{i,j}(t)$ is decreased by 1 each time that for a ΔT and a_j does not give a feedback to a_i. This is achieved by subtracting the factor $\left\lfloor \dfrac{t - t_l}{\Delta t} \right\rfloor$, where t_l is the timestamp of the *last* feedback send by a_j to a_i. It is increased by 1 each time a feedback is released from a_j to a_i with a time elapse smaller than parameter ΔT. We also assume that R is initially set to 0.5, in order to contrast whitewashing strategies [32] without penalizing too new members [24].

Self storage of reputation. We observe that unlike classical reputation systems [10], where reputation is spread by means of a second hand approach by asking to other agents, in FWC peer's reputation is maintained by the Hosted agents running on the users' devices. Indeed, in FWC the traditional approach might fail for the limited connectivity (see Fig. 1) which leads each peer to have a poor and ineffective representation of the reputation into the framework. On the other hand, our approach guarantees the truthfulness of the reputation information because they are updated, signed, stored and spread exclusively by Hosted agents. We remark that these agents are tamper-proof system agents hosted on the peers' devices, and for this reason, they might be considered as local stubs of the Framework agent.

Communication failures or malicious interruption. Suppose that feedbacks cannot be exchanged, due to malicious users' behavior—e.g. avoiding to receive a negative feedback—or communication failure, such that the reputation can not be correctly updated. In the first case, as countermeasure, we expect that the Hosted agent, which monitors the activities of the associate Personal agent, detects a malicious behaviors, it provides to penalize its host by using a system parameter $\eta \in [0, 1[\subset \mathbb{R}$, in order to update reputation as $\left[R_i^{new} = \eta \cdot R_i^{old}\right]$. In the other case, in absence of detected malicious behaviors the Hosted agent can not confirm or exclude the intentionality of the communication failure. A reasonable approach is that the above countermeasure (i.e. penalizing the peers) is applied with a probability which is proportional to the ratio of faults which involved the life of the peers, as recently proposed in [13]. Our approach is pretty simple, as we penalize the peers involved in a communication faults, with probability 0.5. As we discuss in Sect. 4, this is part of the first simulated scenario, denoted in the following as $S1$.

4 Experiments

The FWC reputation model was tested by a set of simulations [14, 18] comprising 10,000 mobile users and 500 residential providers, uniformly distributed into 100 FWC areas. Each experiment consisted of 50 epochs—as we discuss later in this section, this number was enough to show stable and significant trends—and in each epoch 2500 mobile users (i.e. the 25 % of the overall mobile users) interacted with the FWC residential providers. Two different scenarios were simulated, $S1$ and $S2$, as reported in Table 1. Scenario $S1$ includes a number of unreliable peers that release feedbacks closed to 0 to reliable peers, and closed to 1 to unreliable agents. This scenario also includes a probability pf of communication failures. Furthermore, in $S2$ the unreliable peers aim at building positive reputation for services having a low cost. As stated in Sect. 3, positive reputations are exploited to access to significant services, i.e. transactions having high cost. In this respect, the ratio between low to high relevant cost was fixed to 1 : 4. As stated in Sect. 3, a threshold of $\Gamma^* = 0.5$

Table 1 Scenarios simulated and system parameters

Scenario	Untrusted nodes (%)	Unreliable nodes behavior
S1	10	Low performance. Unreliable nodes will cause an interruption of the communication with probability pf
S2	10	Building positive reputations on low cost services for cheating on high cost services. No interruption of communications
$\Delta t = 5,$	$\eta = 0.5,$ $\quad \alpha = 0.5,$ $\quad \Gamma^* = 0.5$ $\quad pf = 0.5,$ $\quad pm = 0.8$	

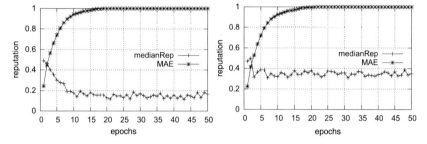

Fig. 2 Median value of reputation (only untrusted agents) and MAE for 50 epochs of simulations. S1 (*left*) and and S2 (*right*)

is used to identify trusted and untrusted counterparts, such that agents assuming a unreliable behavior receive feedbacks less than 0.5 with a certain probability pm (see Table 1). Moreover, at the beginning of each simulation, each agent was set with a reputation of 0.5. Parameters Δt, η, α, Γ^*, pf, pm are reported in Table 1.[1]

As the reputation system is designed to provide basic indexes that reflect the actual trustworthiness of peers, we measured the ratio of untrusted agents which, sooner or later, assume values of reputation reflecting their unreliability. In the following we will refer to the above measure as *MAE* (Malicious Agent Estimation). We also measured the median value of reputation for those agents.

Figure 2 includes two sets of results, which are related to the aforementioned scenarios *S1* and *S2*. Results for scenario *S1*, are shown in the left of Fig. 2. Curve identified by "MAE" shows that, after about 10 epochs, i.e. as the number of interactions among peers becomes relevant, almost the 90 % of the nodes assuming an untrusted behavior show a value <0.5, which is a desirable results. In addition, the median value of reputation (descending curve labeled "medianRep") shows a trend that, starting from initial values of 0.5, gets lower values as simulation epochs increases, which is expected. Right part of Fig. 2 shows results for scenario *S2*. In this case, curve of "MAE" shows results almost identical. It means that, also in presence of dynamic behaviors aimed at gaining reputation for low cost services, the reputation system has a good resilience, showing a stable trend. This is confirmed by the second

---

[1] We planned to present an in depth study related to various combination of these system parameters in a future work.

curve of median value of reputation. Indeed, in spite of values are higher than those of scenario *S1*, reputation of untrusted nodes assume values less than 0.5, which is desirable.

From these preliminary results we can state that the FWC reputation system is effective in quickly identifying all the "malicious" agents (without "false positive") and support the activities of the proposed framework. However, as we discuss in Sect. 6, we planned to perform an in depth study of the reputation model, mainly basing on different combination of parameters shown in the last row of Table 1.

5 Related Work

The relevance of reputation systems is shown by the wide number of researches presented in literature within a wide range of fields [3–5, 16, 17]. However, in this Section we will discuss only those approaches resulting the closest to the FWC reputation model.

Authors of [28] deal with a typical dissemination network of mobile users, equipped with wireless devices, which exchange information when they come into communication range. While users specify what kind of data they need/offer, reputation scores provide information about their trustworthiness. Two reputation schemes are proposed, the first one relies on a centralized trusted party which manages reputation information while the second scheme provides a high user privacy by using a trusted local component, called observer, which exploits cryptographic group signatures to manage/spread reputation information among mobile users.

A lot of related work concerns investigation about reputation on P2P networks [13, 15] by proposing many models, presenting countermeasures for malicious behaviors. For instance, PeerTrust [31] represents an adaptive, reputation-based trust framework. Authors takes account of many different and diffuse types of malicious attacks. Reputation information are locally stored by each peer by Distributed Hash Tables and spread when they interact. However, PeerTrust is designed for usual P2P networks and, therefore, it is not adaptable to our FWC context. In [30] authors try to improve the state of the art of current approaches for mobile P2P scenario, which suffer for the heterogeneous nature of peers, limited-range and intrinsic unreliability of wireless links. They propose an evaluation model, in which a polling protocol and seven metrics provide a real time support to evaluate mobile peers reputation. The heterogeneity of mobile devices and wireless environment implies the necessity, for service providers, to find the best connectivity for mobile users. This issue is addressed by RLoad [2], which is a reputation-based mechanism designed for heterogeneous wireless environments, aimed at selecting suitable networks for mobile users and providing load balancing for network traffic.

For instance, authors of [19] present a reputation model to deal with misbehavior and selfishness in Ad-Hoc network environments. As the latter is difficult to contrast, a mechanism based on reputation to enforce cooperation among nodes is developed by the authors of that work. Each network entity keeps track of other entities' collaboration by calculating their reputation based on various informative sources. Since there is not any advantage to disseminate negative feedbacks, some types of attacks are excluded and not considered at all.

The cited distributed reputation systems for mobile environments deals with some aspects considered in FWC. While some proposal make use of cryptographic techniques to guarantee integrity of sensitive information (reputation and identity), some others include reputation mechanisms on which positive and negative behaviors are rated differently. Finally, in some works, local agents are exploited to manage information about reputation locally, as the already cited [13]. Other considered aspects are reputations of newcomers, as well as service variability and dissemination of reputation scores among the network of peers.

6 Conclusions and Future Work

In this paper we proposed a reputation-based framework, called Federated Wireless Community, capable to support residential users to act as service providers in offering their unused communication and computational resources to mobile peers by means of Wi-Fi hot-spot access points. In order to assist mobile users in consuming residential resources we conceived an agent framework and a distributed reputation model. The reputation model is capable to make ineffective or minimize the impact due to collusive activities and false feedbacks. To verify the performances of the FWC framework we performed a first set of experiments which proved the effectiveness of our proposal in supporting each actor within the scenario we conceived.

As future work we aim to extends experimental results in order to characterize the reputation model basing on various combination of the several parameters involved in the model. We also aim to perform experiments both by means of a dataset collected from real behaviors of users and by comparing FWC with other distributed reputation models.

Acknowledgments This work has been supported by project **PRISMA** PON04a2 A/F funded by the Italian Ministry of University and **NeCS** Laboratory of the Department DICEAM, University Mediterranea of Reggio Calabria.

References

1. Abdul-Rahman, A., Hailes, S.: Supporting trust in virtual communities. In IEEE Computer Society, HICSS, vol 6, Washington (2000)
2. Bi, T., Trestian, R., Muntean, G.: RLoad: Reputation-based load-balancing network selection strategy for heterogeneous wireless environments. In ICNP, pp. 1–3 (2013)
3. Comi, A., Fotia, L., Messina, F., Rosaci, D., Sarnè, G.M.L.: A qos-aware, trust-based aggregation model for grid federations. In On the Move to Meaningful Internet Systems: OTM 2014 Conferences, pp. 277–294. Springer, Berlin (2014)

4. De Meo, P., Messina, F., Rosaci, D., Sarné, G.M.L.: Recommending users in social networks by integrating local and global reputation. In Internet and Distributed Computing Systems, pp. 437–446. Springer International Publishing (2014)
5. De Meo, P., Messina, F., Rosaci, D., Sarné, G.M.L.: Improving grid nodes coalitions by using reputation. In Intelligent Distributed Computing VIII, pp. 137–146. Springer International Publishing (2015)
6. Falcone, R., Castelfranchi, C.: Social Trust: A Cognitive Approach. Kluwer Academic Publishers, Norwell (2001)
7. Gibson, J.: Mobile communications handbook. CRC Press (2012)
8. IxEM (2015). http://www.ixem.polito.it/research/wireless/verrua_savoia_e.htm
9. Josang, A., Gray, E., Kinateder, M.: Simplification and analysis of transitive trust networks. Web Intell. Agent Syst. **4**(2), 139–161 (2006)
10. Jøsang, A., Ismail, R., Boyd, C.: A survey of trust and reputation systems for online service provision. Decis. Support Syst. **43**(2), 618–644 (2007)
11. J. Kaur, S. Saxena, and M. Sayeed. Securing mobile agent's information in ad-hoc network. In 5th International Conference Confluence the Next Generation Information Technology Summit 2014, pp. 442–446. IEEE (2014)
12. Lax, G., Sarné, G.M.L.: Blue: a reputation-based multi-agent system to support C2C in P2P bluetooth networks. In: ICE-B, pp. 97–104 (2006)
13. Lax, G., Sarné, G.M.L.: Celltrust: a reputation model for C2C commerce. Electron. Commer. Res. **8**(4), 193–216 (2008)
14. Messina, F., Pappalardo, G., Santoro, C.: Complexsim: an smp-aware complex network simulation framework. In: Sixth International Conference on Complex, Intelligent and Software Intensive Systems (CISIS), pp. 861–866. IEEE (2012)
15. Marti, S., Garcia-Molina, H.: Taxonomy of trust: categorizing P2P reputation systems. Comput. Netw. **50**(4), 472–484 (2006)
16. Messina, F., Pappalardo, G., Rosaci, D., Santoro, C., Sarné, G.M.L.: A trust-based approach for a competitive cloud/grid computing scenario. Intelligent Distributed Computing VI, pp. 129–138. Springer, Berlin (2013)
17. Messina, F., Pappalardo, G., Rosaci, D., Sarné, G.M.L.: A trust-based, multi-agent architecture supporting inter-cloud vm migration in iaas federations. Internet and Distributed Computing Systems, pp. 74–83. Springer International Publishing, Switzerland (2014)
18. Messina, F., Pappalardo, G., Santoro, C.: Complexsim: a flexible simulation platform for complex systems. Int. J. Simul. Process Model. **8**(4), 202–211 (2013)
19. Michiardi, P., Molva, R.: Core: a collaborative reputation mechanism to enforce node cooperation in mobile ad hoc networks. Advanced Communications and Multimedia Security, pp. 107–121. Springer, Berlin (2002)
20. Misztal, B.: Trust in modern societies. Polity Press, Cambridge (1996)
21. Mui, L., Mohtashemi, M., Halberstadt, A: Notions of reputation in multi-agents systems: a review. In: Proceedings of the 1st International Joint Conference on Autonomous Agents and Multiagent Systems, pp. 280–287. ACM Press, (2002)
22. Perich, F., Undercoffer, J., Kagal, L., Joshi, A., Finin, T., Yesha, Y.: In reputation we believe: query processing in mobile ad-hoc networks. In: Proceedings of the 1st Annual International Conference on Mobile and Ubiquitous Systems: Networking and Services, pp. 326–334 (2004)
23. Rosaci, D., Sarné, G.M.L., Garruzzo, S.: Integrating trust measures in multiagent systems. Int. J. Intell. Syst. **27**(1), 1–15 (2012)
24. Sarvapali, D.R., Dong, H., Nicholas, R.J.: Trust in multi-agent systems. Knowl. Eng. Rev. **19**(1), 1–25 (2004)
25. Shaked, Y., Wool, A.: Cracking the bluetooth pin. In: Proceedings of the 3rd International Conference on Mobile Systems, Applications, and Services, pp. 39–50. ACM Press, New York (2005)
26. Tan, Y., Thoen, W.: An outline of a trust model for electronic commerce. Appl. Artifi. Intell. **14**(8), 849–862 (2000)

27. Vodafone: Vodafone wi-fi community (2015). http://www.vodafone.it/portal/Privati/Vantaggi-Vodafone/Per-i-gia-Clienti/wifi-community
28. Voss, M., Heinemann, A., Muhlhauser, M.: A privacy preserving reputation system for mobile information dissemination networks. In: 1st International Conference on Security and Privacy for Emerging Areas in Communications Networks, pp. 171–181. IEEE (2005)
29. Wang, Y., Vasilakos, A., Jin, Q., Ma., J.: A wi-fi direct based p2p application prototype for mobile social networking in proximity (msnp). In: IEEE 12th International Conference on Dependable, Autonomic and Secure Computing, pp. 283–288. IEEE (2014)
30. Wu, X.: A distributed trust evaluation model for mobile P2P systems. J. Netw. **7**(1), 157–164 (2012)
31. Xiong, L., Liu, L.: PeerTrust: Supporting reputation-based trust for peer-to-peer electronic communities. IEEE Trans. Knowl. Data Eng. **16**(7), 843–857 (2004)
32. Zacharia, G., Maes, P.: Trust management through reputation mechanisms. Appl. Artifi. Intell. **14**(9), 881–907 (2000)

Improving the Weighted Distribution Estimation for AdaBoost Using a Novel Concurrent Approach

Héctor Allende-Cid, Carlos Valle, Claudio Moraga, Héctor Allende
and Rodrigo Salas

Abstract AdaBoost is one of the most known Ensemble approaches used in the Machine Learning literature. Several AdaBoost approaches that use Parallel processing, in order to speed up the computation in Large datasets, have been recently proposed. These approaches try to approximate the classic AdaBoost, thus sacrificing its generalization ability. In this work, we use Concurrent Computing in order to improve the Distribution Weight estimation, hence obtaining improvements in the capacity of generalization. We train in parallel in each round several weak hypotheses, and using a weighted ensemble we update the distribution weights of the following boosting rounds. Our results show that in most cases the performance of AdaBoost is improved and that the algorithm converges rapidly. We validate our proposal with 4 well-known real data sets.

1 Introduction

Ensemble methods have gained considerable attention from the Machine Learning and Soft Computing communities in the last years. There are several practical and theoretical reasons, mainly statistical reasons, why an ensemble may be preferred. A set of learners with similar training performance may have different generalization

H. Allende-Cid (✉)
Escuela de Ingeniería Informática, Pontificia Universidad Católica de Valparaíso,
Valpara íso, Chile
e-mail: hector.allende@ucv.cl

C. Valle · H. Allende
Departamento de Informática, Universidad Técnica Federico Santa María,
Valpara íso, Chile

R. Salas
Escuela de Ingeniería Biomédica, Universidad de Valparaíso, Valpara íso, Chile

C. Moraga
European Centre for Soft Computing, 33600 Asturias, Mieres, Spain

C. Moraga
TU Dortmund University, 44220 Dortmund, Germany

© Springer International Publishing Switzerland 2016
P. Novais et al. (eds.), *Intelligent Distributed Computing IX*,
Studies in Computational Intelligence 616,
DOI 10.1007/978-3-319-25017-5_21

performance, when exposed to sparse data, large volume of data or data fusion. The basic idea of an ensemble learning algorithm is to build a predictive model by combining a set of learning hypotheses instead of laboriously designing the complete map between inputs and responses in a single step. Ensemble based systems have shown to produce favorable results compared to those of single-expert system for a broad range of applications such as financial, medical and social models, network security, web mining or bioinformatics, to name a few.

Boosting algorithms, since the mid nineties, have been a very popular technique for constructing ensembles in the areas of Pattern Recognition and Machine Learning (see [1, 4, 6, 9]). Boosting is a learning algorithm to construct a predictor by combining, what are called, weak hypotheses. The AdaBoost algorithm, introduced by Freund and Schapire [4], builds an ensemble incrementally, placing increasing weights on those examples in the data set, which appear to be "difficult". Ensemble learning is the discipline which studies the use of a committee of models to construct a joint predictor which improves the performance over a single more complex model.

Most of the modern computers nowadays have processors with multiple cores. Adaboost was proposed in a time were the number of cores per machine was more limited. So, it seems natural to use all the resources available to improve the quality of the inference made by this model. We improve the weight estimation phase, which is one of the most important stages in the AdaBoost algorithm. For this we use concurrent computing based in parallel processes. Distributed and Parallel Machine Learning aim mainly to improve the computation times, e.g. when dealing with Large Datasets, while our approach seeks to use the multithread/parallel computation in order to improve the generalization ability of the algorithm. While both use parallel processes, the difference relies on which problem they are trying to solve.

This paper is organized as follows. In Sect. 2, we briefly introduce AdaBoost and Parallel Adaboost approaches. In Sect. 3, we present our proposed model Concurrent AdaBoost. In Sect. 4 we compare the performance of the classic AdaBoost and our proposal. The last section is devoted to concluding remarks and to delineate future work.

2 Adaptive Boosting and Some Parallel Variants

Different ensemble approaches are distinguished by the way in which they manipulate the training data, by the way in which they select the individual hypotheses and by the way in which those are aggregated to the final decision. The AdaBoost Algorithm [4], introduced in 1997 by Freund and Schapire, has a theoretical background based on the "PAC" learning model [14]. The authors of this model were the first to pose the question of whether a weak learning algorithm that is slightly better than random guessing can be "boosted" to a strong learning algorithm. The classic AdaBoost takes as an input a training set $\mathcal{Z} = \{(x_1, y_1) \cdots (x_n, y_n)\}$ where each x_i is a variable that belongs to $\mathcal{X} \subset \mathbb{R}^d$ and each label y_i is in some label set $\mathcal{Y}$. In this particular paper we assume that $\mathcal{Y} = \{-1, 1\}$. AdaBoost calls a weak or base learning algorithm repeatedly in a sequence of stages $t = 1, \ldots, T$. The main idea of AdaBoost is to maintain a sampling distribution D_t over the training set. This sample set is used to train the learner at round t. Let $D_t(i)$ be the sampling weight assigned to the example i on round t. At the beginning of the algorithm the distribution is uniform, that is distribution $D_1(i) = \frac{1}{n}$ for all i. At each round of the algorithm however, the weights of the incorrectly classified examples are increased, so that the following weak learner is forced to focus on the "hard" examples of the training set. The job of each weak or base learner is to find a hypothesis $h_t : \mathcal{X} \rightarrow \{-1, 1\}$ appropriate for the distribution D_t. The goodness of the obtained hypothesis can be quantified as the weighted error:

$$\epsilon_t = Pr_{i \sim D_t}[h_t(x_i) \neq y_i] = \sum_{i:h_t(x_i) \neq y_i} D_t(i). \tag{1}$$

Notice that the error is measured with respect to the distribution D_t on which the weak learner was trained. Once AdaBoost has computed a weak hypothesis h_t, it measures the importance that the algorithm assigns to h_t with the parameter

$$\alpha_t = \frac{1}{2} \ln \left(\frac{1 - \epsilon_t}{\epsilon_t} \right). \tag{2}$$

After choosing α_t the next step is to update the distribution D_t with the following rule,

$$D_{t+1}(i) = \frac{1}{Z_t} D_t(i) e^{-\alpha_t y_i h_t(x_i)},$$

where Z_t is a normalization factor. The effect of this rule is that, when an example is misclassified its sampling weight for the next round is increased, and the opposite occurs when the classification is correct. This updating rule makes the algorithm focus on the "hard" examples, instead in the correctly classified examples. After a sequence of T rounds has been carried out, the final hypothesis H is computed as

$$H_T(x) = sign \left(\sum_{t=1}^{T} \alpha_t h_t(x) \right). \tag{3}$$

AdaBoost loss function is $\ell(y, f(x)) = \exp(-yf(x))$ where y is the target and $f(x)$ is the approximation made by the model. To obtain the exponential loss function, using the weak hypothesis $h_T \in \{-1, 1\}$, one must solve

$$(\alpha_t, h_t) = \underset{\alpha, h}{\operatorname{argmin}} \sum_{i=1}^{n} D_i^{(t)} \exp(-\alpha y_i h(x_i))), \tag{4}$$

for the weak hypothesis h_t and corresponding coefficient α_t to be added at each step and with $D_i^{(t)} = \exp(-y_i H_{t-1}(x_i))$, where H_{t-1} is the strong Hypothesis without the learner h_t.

The solution to Eq. 4 for h_t for any value of $\alpha > 0$ is

$$h_t = \operatorname*{argmin}_h \sum_{i=1}^{n} D_i^{(t)} \operatorname{sign}(y_i h(x_i)), \qquad (5)$$

where I is the indicator function. There are methods that use learners that can learn the weight distribution using them as inputs (reweighting) or methods that use resampling following the weight distribution to select the training set that the learners use in their training process. In [13] the authors state that the latter approach gives better results. Nevertheless it needs to be stated that the weak learner h_t that is selected with this approach is sensitive to the resampling technique.

There have been several approaches to use Parallel Computing together with AdaBoost. In [11, 12] the authors propose two parallel boosting algorithms, ADA-BOOST.PL and LOGITBOOST.PL, which facilitate simultaneous participation of multiple computing nodes to construct a boosted ensemble classifier. The authors claim that these proposed algorithms are competitive to the corresponding serial versions in terms of the generalization performance. They achieve a significant speedup since their approach does not require individual computing nodes to communicate with each other for sharing their data. In [7] a generic, flexible parallel architecture of AdaBoost, which is suitable for all ranges of object detection applications and image sizes is proposed. In [3] a randomised parallel version of Adaboost is proposed. The algorithm uses the fact that the logarithm of the exponential loss is a function with coordinate-wise Lipschitz continuous gradient, in order to define the step lengths. They provide the proof of convergence for this randomised Adaboost algorithm and a theoretical parallelisation speedup factor. The authors in [10] proposed an algorithm, which they call BOOM, for boosting with momentum. Namely, BOOM retains the momentum and convergence properties of the accelerated gradient method while taking into account the curvature of the objective function. They describe a distributed implementation of BOOM which is suitable for massive high dimensional datasets. To the best of our knowledge, all proposed algorithms that use parallel computing, try to improve the computation time, instead of improving the generalization performance. The results obtained by these approaches are at most similar to the ones obtained with classic AdaBoost, because they try to approximate the behaviour of the classic algorithm. In this paper we use concurrent computing, in order to improve the performance of the classic AdaBoost.

3 Concurrent AdaBoost

In this research the main idea is to work with all the processors of the machine in order to use, in other case idle processors, to improve the generalization ability. In most parallel Adaboost approaches, the multiple cores are used to decrease the computation times, by partitioning the dataset in smaller fractions. These approaches try to get an approximation to the model that would have been obtained it had been trained with classic approaches.

In this proposal, instead of using a single weak learner in each Adaboost round, the idea is to use all p the processors available to resample, in a parallel fashion, the original data, and also in parallel train several weak learners. With all the p weak learners we build an ensemble, that is weighted with its training accuracy, and then with the output of the ensemble update the weights of the examples.

In the proposal we choose each h^j according to Eq. 5 where $j = 1, \ldots, p$, and then obtain the ensemble output $E_t(x_i)$ for example x_i in the tth round as

$$E_t(x_i) = \text{sign}\left(y_i \sum_{j=1}^{p} \phi^j h_t^j(x_i) \right), \tag{6}$$

where ϕ^j is the training accuracy of weak hypothesis h_t^j and $\sum_{j=1}^{p} \phi^j h_t^j(x_i)$ is the decision of the p weak learners that were trained in parallel. Then, the weighted error ϵ_t is computed considering the output of the ensemble E_t using $\epsilon_t = Pr_{i \sim D_t}[E_t(x_i) \neq y_i]$. With this, we avoid the necessity of having to select a learner explicitly and use an ensemble of learners instead. Algorithm 1 shows our proposal.

Algorithm 1 Concurrent AdaBoost

 1: Given is the training data set $\mathcal{Z} = \{(x_1, y_1), \ldots, (x_n, y_n)\}$ with n elements, where $x_i \in \mathcal{X}$ and $y_i \in \mathcal{Y} = \{-1, 1\}$.
 2: Initialize the parameters. Pick the parameter p (number of parallel processes) the number of rounds T and $t = 0$.
 3: Initialize the empirical distribution $D_1(i) = \frac{1}{n}$ for each data sample $(x_i, y_i), i = 1..n$.
 4: **repeat**
 5:    Increment t by one.
 6:    Take p bootstrap samples $\mathcal{Z}_t^j$ from $\mathcal{Z}$ with distribution D_t, where j is the number of the parallel process, with $j = 1, \ldots, p$. (In parallel)
 7:    Train p weak learners $h_t^j : \mathcal{X} \rightarrow \{-1, 1\}$ with the bootstraped samples $\mathcal{Z}_t^j$ as the training sets. (In parallel)
 8:    Generate an ensemble E_t with all weak learners h_t^j.
 9:    Compute the weighted error ϵ_t of the ensemble of weak hypothesis E_t as

$$\epsilon_t = Pr_{i \sim D_t}[E_t(x_i) \neq y_i].$$

10:    Compute the empirical importance for the tth ensemble α_t with equation

$$\alpha_t = \frac{1}{2} \ln\left(\frac{1 - \epsilon_t}{\epsilon_t} \right).$$

Algorithm 2 Concurrent AdaBoost (continuation)

11:    Update the empirical distribution as

$$D_{t+1}(i) = \frac{D_t(i)}{Z_t} \times e^{(-\alpha_t \beta(i) y_i E_t(x_i))},$$

where Z_t is the normalization factor.

12:    The strong hypothesis $H_t(x)$ at stage t is given by

$$H_t(x) = sign\left(\sum_{t=1}^{t} \alpha_t E_t(x)\right).$$

13:    Classify the training data set $\mathcal{Z} = \{(x_1, y_1), \ldots, (x_n, y_n)\}$ with the strong hypothesis $H_t(x)$.
14: **until** The stopping criterion is met
15: Output: The strong hypothesis $H_T(x)$

In Fig. 1 we observe the main difference between the Classic Approach and the Concurrent one. In the classic we observe that in each round a single weak learner is trained with a resample of the original data, and the updates of the weights of the distributions are changed using the outputs of the single weak hypothesis. In the case of the concurrent approach, in each boosting round, p resamples obtained in a concurrent fashion are used to train p weak learner to build an ensemble weighted by the training accuracy. This ensemble is then used to update the weights of the distribution.

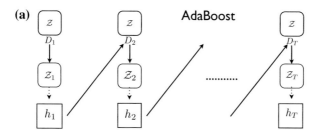

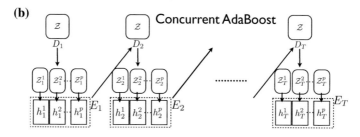

Fig. 1 Classic AdaBoost approach and concurrent AdaBoost. **a** Classic Approach. 1 weak learner h per boosting round. **b** Concurrent Approach. p weak learner h^i per boosting round, where $i = 1, \ldots, p$

4 Experimentation

In this section we validate our proposal with 4 well-known real datasets that can be found in [8]. The datasets are: Liver Disease, Cancer, Ionosphere and Diabetes. The weak learners used were Decision Stumps, and the number of experiments was 20. The data was split in 80 % training and 20 % test. We implemented our proposal in Python Language 2.7.6 and used the latest version of the library for parallel processing, Parallel Python. The experiments were run in an Intel i7 2.6GHz (8 cores) with 16 GbRam running with Ubuntu 14.04 x64. The number of parallel processes in each proposal experiment was 7 (odd number of decisions). The performance measures used to report the results are: Accuracy, Area under the ROC curve, F1 score, Precision and Recall.

In Figs. 2, 3, 4 and 5 we observe the results in the 4 real data sets in terms of test accuracy. We report the results of 20 experimental runs. The best results are observed in the Cancer and Liver Disease data sets. In the Diabetes and Ionosphere data sets, a slight improvement over the Classic approach is observed.

In Table 1 we observe the results of the 4 datasets in terms of Roc-AUC, F1-score, Precision and Recall. We show the results with 5, 10, 15 and 20 boosting rounds. We observe that the proposed approach outperforms the classic one in the majority of the cases. In the cases where our proposal is outperformed the results between both approaches are comparable. The best results are obtained in the Cancer data set, where the Concurrent approach outperforms the classic one in every measure. The worst results are obtained in the Diabetes dataset, nevertheless in some measures the concurrent approach obtains comparable results against the classic one.

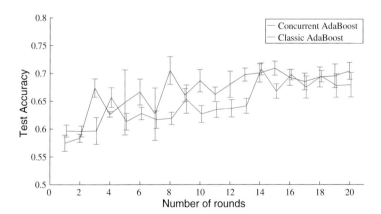

Fig. 2 Liver test accuracy results

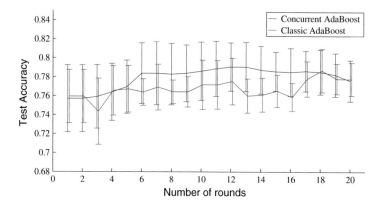

Fig. 3 Diabetes test accuracy results

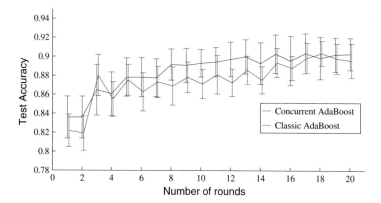

Fig. 4 Ionosphere test accuracy results

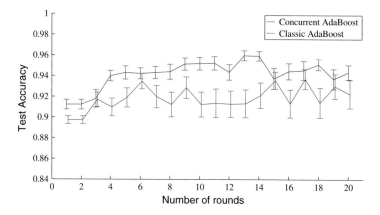

Fig. 5 Cancer test accuracy results

Table 1 Test results of Diabetes, Cancer, Ionosphere, Liver datasets in terms of Roc-auc, F1-score, Precision and Recall performance measures

Rounds	Classic AdaBoost				Concurrent AdaBoost			
	ROC-AUC	F1	Precision	Recall	ROC-AUC	F1	Precision	Recall
Diabetes								
5	0.690	**0.840**	0.788	**0.905**	**0.712**	0.838	**0.804**	0.878
10	0.704	0.841	0.798	**0.893**	**0.739**	**0.847**	**0.823**	0.876
15	0.719	0.833	0.815	0.857	**0.744**	**0.846**	**0.830**	**0.867**
20	**0.737**	**0.840**	**0.826**	**0.860**	0.734	0.839	0.824	0.859
Cancer								
5	0.913	0.889	0.887	0.891	**0.944**	**0.924**	**0.899**	**0.952**
10	0.907	0.881	0.873	0.890	**0.951**	**0.935**	**0.920**	**0.952**
15	0.926	0.910	0.930	0.891	**0.932**	**0.914**	**0.915**	**0.914**
20	0.915	0.894	0.897	0.891	**0.938**	**0.922**	**0.925**	**0.921**
Ionosphere								
5	**0.836**	0.909	**0.867**	0.957	0.834	**0.912**	0.861	**0.973**
10	0.823	0.907	0.856	0.966	**0.853**	**0.922**	**0.876**	**0.976**
15	0.861	0.921	0.887	0.960	**0.872**	**0.929**	**0.894**	**0.967**
20	0.867	0.922	0.891	**0.958**	**0.878**	**0.927**	**0.903**	0.954
Liver								
5	0.597	0.443	0.688	0.337	**0.640**	**0.600**	**0.704**	**0.548**
10	0.617	0.541	0.631	0.476	**0.675**	**0.608**	**0.743**	**0.520**
15	0.659	0.600	0.689	0.536	**0.701**	**0.651**	**0.737**	**0.583**
20	0.673	0.630	0.684	0.584	**0.697**	**0.656**	**0.721**	**0.606**

5 Conclusions

In this work we introduce a concurrent approach to obtain a better estimation of the distribution weights that are updated in each round. We show that using parallel processes and using more than one weak learner per round remarkable results can be obtained. In the classic approach only one hypothesis is used in each boosting round, while neglecting the potential of processors with multiple cores. The parallel approaches in the literature mainly deal with speeding the training process in large data sets, while in this approach we improve the estimations of the distribution weights. In future work we will try this approach in other boosting algorithms and will formalize the approach in terms of convergence.

Acknowledgments This work was supported by the following research grants: Fondecyt 1110854 and DGIP-UTFSM. The work of C. Moraga was partially supported by the Foundation for the Advancement of Soft Computing, Mieres, Spain.

References

1. Bauer, E., Kohavi, R.: An empirical comparison of voting classification algorithms: Bagging, boosting, and variants. Mach. Learn. **36**(1–2), 105–139 (1999)
2. Clemen, R.T.: Combining forecasts: a review and annotated bibliography. Int. J. Forecast. **5**(4), 559–583 (1989)
3. Fercoq, O.: Parallel Coordinate Descent for the Adaboost Problem. In: Proceedings of the 12th International Conference on Machine Learning and Applications (ICMLA), vol. 1, no. 1, pp. 354-358, 4-7 December 2013
4. Freund, Y., Schapire, R.: A decision-theoretic generalization of on-line learning and an application to boosting. J. Comput. Syst. Sci. **55**(1), 119–139 (1997)
5. Freund, Y., Shapire, R.E.: A short introduction to boosting. J. Jpn. Soc. Artif. Int. **14**(5), 771–780 (1999)
6. Kuncheva, L., Whitaker, C.: Using diversity with three variants of boosting: aggressive, conservative and inverse. Lect. Notes Comput. Sci. **2364**(1), 81–90 (2002)
7. Kyrkou, C., Theocharides, T.: A flexible parallel hardware architecture for adaboost-based real-time object detection. IEEE Trans. Very Larg. Scale Integr. (VLSI) Syst. **19**(6), 1034–1047 (2011)
8. Lichman, M.: UCI Machine Learning Repository. University of California, Irvine. School of Information and Computer Science (2013). http://archive.ics.uci.edu/ml
9. Liu, H., Tian, H.Q., Li, Y.F., Zhang, L.: Comparison of four Adaboost algorithm based artificial neural networks in wind speed predictions. Energy Convers. Manage. **92**(1), 67–81 (2015)
10. Mukherjee, I., Canini, K., Frongillo, R., Singer, Y.: Parallel boosting with momentum. Mach. Learn. Knowl. Discov. Databases **8190**(1), 17–32 (2013)
11. Palit, I., Reddy, C.K.: Parallelized Boosting with Map-Reduce. In: IEEE International Conference on Data Mining Workshops (ICDMW), vol. 1, no. 1, pp. 1346–1353, 13 December 2010
12. Palit, I., Reddy, C.K.: Scalable and parallel boosting with mapreduce. IEEE Trans. Knowl. Data Eng. **24**(10), 1904–1916 (2012)
13. Seiffert, C., Khoshgoftaar, T.M., Van Hulse, J., Napolitano, A.: Resampling or reweighting: a comparison of boosting implementations. In Proceedings of the 20th IEEE International Conference on Tools with Artificial Intelligence (ICTAI 08), vol. 1, no. 1, pp. 445–451, November 2008
14. Valiant, L.G.: A theory of the learnable. Commun. ACM **27**(11), 1134–1142 (1984)

Part VI
Intelligent Data Processing

Towards Semi-automated Parallelization of Data Stream Processing

Martin Kruliš, David Bednárek, Zbyněk Falt, Jakub Yaghob and Filip Zavoral

Abstract Current hardware development trends exhibit clear inclination towards parallelism. Multicore CPUs as well as many-core architectures such as GPUs or Xeon Phi devices are widely present in both high-end servers and common desktop PCs. In order to utilize the computational power of these parallel platforms, the applications must be designed in a way that intensively exploits parallel processing. In our work, we propose techniques that simplify the application decomposition process in data streaming systems. The data streaming paradigm may be applied in many data-intensive applications, e.g., database management systems or scientific data processing. In order to employ these techniques, we have developed a data streaming language called Bobolang that simplifies the design of the application. This approach allows the programmer to write strictly serial operators in a traditional language and then interconnect these operators in an execution plan, that presents opportunities for automated parallel processing.

1 Introduction

Streaming systems represent a specific domain of computing environment. These systems operate with data streams, which are basically unidirectional flows of structured tuples. Streams are processed by operators (also denoted functions, kernels, or

M. Kruliš (✉) · D. Bednárek · Z. Falt · J. Yaghob · F. Zavoral
Charles University in Prague, Prague, Czech Republic
e-mail: krulis@ksi.mff.cuni.cz

D. Bednárek
e-mail: bednarek@ksi.mff.cuni.cz

Z. Falt
e-mail: falt@ksi.mff.cuni.cz

J. Yaghob
e-mail: yaghob@ksi.mff.cuni.cz

F. Zavoral
e-mail: zavoral@ksi.mff.cuni.cz

© Springer International Publishing Switzerland 2016
P. Novais et al. (eds.), *Intelligent Distributed Computing IX*,
Studies in Computational Intelligence 616,
DOI 10.1007/978-3-319-25017-5_22

filters) which may have multiple inputs and outputs. These operators transform data from the input streams by performing their built-in functionalities and pass their results into the output streams. The operators are usually implemented in a procedural or object-oriented programming language such as C/C++ and compiled natively.

A streaming application is typically represented as an oriented graph, where the vertices are operators and the edges prescribe the data flow between them. In the remainder of this paper, we will refer to this graph as the *execution plan*. The execution plan is usually described in specialized declarative languages.

There is a large number of existing streaming systems [2–4, 10, 11, 14, 16], while each is designed for a specific purpose or for a different platform. The streaming systems were originally designed for scenarios, where the data naturally occur as a stream (e.g., sensory data) and where continuous processing is required. However, the streaming paradigm can be also used to express parallelism in a more programmer-friendly way. In this context, we recognize two types of parallelism:

- the *inter-operator parallelism* (concurrent processing of multiple operators), and
- the *intra-operator parallelism* (parallelism within one operator).

The *inter-operator parallelism* emerges quite naturally in the streaming systems. It only requires that the operators are truly independent and that there are enough data fragments to keep multiple operators occupied. The *intra-operator parallelism* cannot be achieved automatically by the streaming system task scheduler, since the operators are usually treated as indivisible blocks of code. However, in some cases, we can decompose an operator into multiple sub-operators which perform the same functionality. The decomposed structure leads to a larger execution plan; hence, it presents more opportunities for inter-operator parallelism.

We narrow our focus on shared memory streaming systems which implement the data streams as flows of packets and explicitly expose this implementation to the programmer. Typical representatives of such systems are *The Flow Graph* component from the Intel Threading Building Blocks [13] or the *Bobox* framework [2]. The packet-level processing in shared memory permits certain optimizations, especially when the packets are passed on without modification, or when the data are shallow copied.

In this paper, we address the issues of effective semiautomated parallelization in the streaming systems. These systems separate the design of effective code (reduced to simple serial routines) from the data flow schema where the potential parallelism is expressed along with requirements for implicit synchronization. Furthermore, we introduce techniques how to decompose operators in order to express intra-operator parallelism (within one operator) by the means of inter-operator parallelism (concurrent processing of operators).

The proposed techniques were implemented in Bobolang language [7], which was designed for the specification of execution plans. This language is integrated into Bobox framework [2], which provides a runtime for parallel evaluation of the plans on shared memory systems. This framework is primarily designed for efficient parallel data processing, thus it provides an excellent platform for our experiments.

The paper is organized as follows. Related work is collected in Sect. 2. Section 3 presents our approach towards semi-automated parallelization in streaming systems. The Bobolang language which implements this concept is described in Sects. 4 and 5 concludes the paper.

2 Related Work

Contemporary streaming languages basically differ in their focus designed with a particular intent which significantly influences their syntax and semantics. The languages Brook [3], StreaMIT [14] and StreamC [6] are intended for the development of efficient streaming applications. They introduce a language based on the C/C++ syntax, which allows the programmer to implement the operators and to specify their mutual interconnections. The compiler exploits the streaming nature of the application to perform specific analyses and optimizations designed with a particular emphasis on concurrent execution. The compiler also creates a static mapping of the operators to the execution units such as CPUs, GPUs, or FPGAs.[1]

Another language extension designed for the development of streaming applications is Granular Lucid (GLU) [9]. The operators are implemented in the C language; their structure is described in Lucid. The parallel evaluation of operators is designed in a similar way as in Bobolang.

The X Language [8] is another example of modern streaming language. It is logically similar to GLU, but it uses different syntax (similar to Bobolang) which clearly and explicitly describes the connections between operators. This is especially useful for designing complex algorithms.

The main difference between these languages and our approach is that they lack support for constructions such as multiplication of inputs/outputs (see Sect. 4.1). The absence of this feature requires the use of parallelism inside the procedural code of operators; otherwise, only inter-operator parallelism would be available.

Semiautomatic parallelization is usually studied in a context of particular programming languages such as C (Paralax [15]) or Python (Pydron [12]) where the sequentially programmed source code is transformed into parallelizable pieces of code. FastFlow [1] accelerator supports the easy porting of existing sequential C/C++ applications onto multi-core systems. Code kernels identified by a programmer are offloaded onto a number of additional threads running on the same CPU.

Despite the abundance of parallelism in streaming applications, it is a nontrivial task to split and efficiently map sequential applications to multicore systems. Cordes et al. presents an algorithm [5] which automatically extracts pipeline parallelism from sequential ANSI-C applications. This method employs an integer linear programming (ILP) based approach to automatically control the granularity of the parallelization.

---

[1]Field-programmable gate array.

3 Semi-automated Parallelization

Streaming systems naturally introduce concurrent processing of operators, which is
called inter-operator parallelism. The inter-operator parallelism is usually cooperative (i.e. non-preemptive), associated to sending and receiving data between operators, where an incoming packet triggers the execution of the operator (if it was
suspended after finishing the previous work). To balance the load among available
processors, the operator code is usually allowed to migrate across a pool of worker
threads. In other words, the incoming data packets generate a sequence of tasks which
are assigned to worker threads by the underlying scheduler. This mode of concurrency is often called *pipeline parallelism* and it is sometimes considered a special
case of *task parallelism*.

Although some systems allow parallel code inside individual operators, parallel
programming is difficult for the developer. Instead of implementing intra-operator
parallelism by parallel code inside an operator, we suggest multiplication of the operators in the execution plan and inserting auxiliary operators to dispatch the input
data among the replicas of an operator and to collect the output data. This way, the
available parallelism is explicitly denoted in the execution plan and the individual
copies of the operator may remain sequential.

Of course, the multiplication of an operator must be consistent with its behavior
and the way the data are dispatched. The simplest case, described in Sect. 3.1, is
associated with *stateless* operators which process each packet of data independently.
If the operator depends on its internal state, a copy of the state must be properly
maintained in every replica of the operator. As we will show in Sect. 3.2, there are
situations where the cost of maintaining the replicated state is significantly lower
than the gain of the parallelization.

The cost of dispatching data and maintaining replicated state depends on the cost
of data transfer between the operators. Consequently, our approach aims at systems
which use shared memory for communication between operators and allow cheap
broadcasting and forwarding of packets via sharing memory regions. In addition, we
rely on the ability of the underlying scheduler to balance the load among processors.
Thanks to this ability, the auxiliary operators are not required to balance the load
exactly since minor skewness will be corrected by the scheduler.

In our approach, the designer of the execution plan decides which operators may
be parallelized and marks them as *stateless* or *parallelizable*. In the latter case, the
procedural code of the operator must adhere to the protocol described in Sect. 3.2.
The execution plan will then be automatically transformed by multiplication of selected operators and insertion of auxiliary operators.

Besides the two built-in approaches to parallelization, the plan designer may explicitly invoke the multiplication of operators, using plan annotations described in
Sect. 4.

3.1 Data Parallelism

Data-parallel subproblems would be implemented by a *parallel for* in traditional parallel code. In pipeline systems, such subproblems correspond to stateless operators. Thanks to the absence of internal state, we may split the input stream into several sub-streams and process each sub-stream independently, by identical replicas of the original operator. Two auxiliary operators are required, as shown in Fig. 1: The *dispatch operator* distributes incoming packets to its outputs in round robin manner and the *consolidate operator* interleaves incoming packets. The auxiliary operators have negligible overhead, since they only forward incomming packets.

3.2 Maintaining the Local State

If an operator maintains local state, the parallelization process requires minor modification of the internal function of the operator as well as different data dispatching scheme. The parallelization comes at the cost of performing redundant work. It depends on the nature of the operator whether the cost is acceptable, i.e. lower than the gain by parallelism. The plan designer has to assess the overhead and decide whether an operator should be parallelized this way. Let us consider a typical schema of a general stateful operator with an internal state S:

> $S \leftarrow$ initial state
> **while** not finished **do**
>   *tuples* $\leftarrow$ next part of input (e.g., next packet)
>   process *tuples* whilst using and updating state S
> **end while**

If the processing of the tuples and the update of the state can be effectively separated from each other and the updating of state S takes significantly less time than the processing of tuples, the stateful operator can be effectively parallelized. The concurrency is achieved by replicating the operator while each replica has its own copy of the state. Each replica has a unique index from 0 to $R - 1$ (for R replicas) called *RID* (Replica ID). Each of the replicas then perform the following algorithm:

> $S \leftarrow$ initial state,   *phase* $\leftarrow$ 0
> **while** not finished **do**

Fig. 1 Parallelization of a
stateless operator

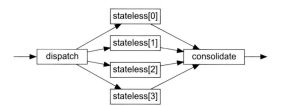

Fig. 2 Parallelization of a
stateful operator (RIDs are in
brackets)

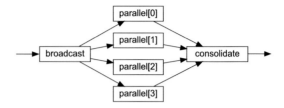

> $tuples \leftarrow$ next part of input (e.g., next packet)
> **if** $phase \bmod N = RID$ **then**
> > process $tuples$ whilst using and updating state S
> **else**
> > update state S using $tuples$
> **end if**
> $phase \leftarrow phase + 1$
> **end while**

We denote operators which are modified in this way as *parallelizable*.

The schema is depicted in Fig. 2. The *broadcast operator* which clones its input for all its outputs is used here instead of the dispatch operator used in the stateless case. The data are efficient shallow copied in the shared memory. All the operator replicas receive identical (shallow) copies of the original stream, but they alternate in the processing of the tuples and in the production of the output stream. The resulting stream is then gathered by the *consolidate* operator as in the case of parallelization of the stateless operators. The body of the operator must be designed by the developer to support this parallelization. On the other hand, many problems allow simple decoupling of the tuple processing and state updates simply by creating conditions in existing code.

4 The Bobolang Language

Bobolang is a declarative language designed for specification of execution plans—together with the procedural code of individual operators, the execution plans forms a parallel application. The application requires a runtime environment, essentially consisting of a dynamic scheduler and streaming support. In our case, the runtime environment is Bobox [2] where the primary procedural language for operators is C++; however, Bobolang itself is independent of both the runtime and the associated procedural language and the same plan may even be used with different implementations of the operators. For type safety, the Bobolang compiler is supplied with a dictionary of *column types* which are provided by the runtime.

The features of Bobolang are shaped by the following objectives:

- Providing a hierarchical decomposition of the execution plan. The operators referenced in the *main* plan may either be *atomic* operators implemented in the associated procedural language or *compound* operators whose interior is defined by a *model* defined in a model library. All the models are again defined in Bobolang, allowing for unlimited (but not recursive) decomposition into a tree-like hierarchy of models with atomic operators at leaves.
- Allowing generic models independent of concrete column types. Similarly to functions in generic programming, generic models infer the number and types of columns at their interfaces from the context of their instantiation.
- Multiplication of inputs, outputs, and operators as a means to provide opportunity for parallelization. The degree of multiplication is either specified directly in the model or set to defaults provided from the outer environment. The defaults are set based on the parallel processing hierarchy of the hardware, considering available threads, cores, CPU sockets, caches, and/or NUMA nodes.
- Automatic multiplication of stateless and parallelizable operators, using insertion of built-in operators *broadcast*, *dispatch*, or *consolidate*.

A definition of a compound operator is illustrated in the following example:

```
operator new_operator(int)->(int,int) {
  split_op(int)->(int),(int) split;
  filter_op(int)->(int,int) filter1, filter2;
  join_op(int,int),(int,int)->(int,int) join;

  input -> split;
  split[0] -> filter1 -> [0]join;
  split[1] -> filter2 -> [1]join;
  join -> output;
}
```

The first line declares a new operator called `new_operator`. The operator has one input (a stream of integers) and one output (a stream of integer pairs). The body of the operator has two parts. The first part contains a list of sub-operators, i.e. instances of nested operators from which the operator is composed of. Each line specifies the operator type together with its input/output data descriptor and declares one or more local identifiers of the nested operator instances.

The second part specifies the connections between operators. Statement `op1 -> op2` defines the connection of `op1` output to `op2` input. The corresponding input and output must have the same data type descriptor. The syntax allows creating chains, so the `op1 -> op2 -> op3` statement is just a shorthand expression for `op1 -> op2` and `op2 -> op3` statements.

In addition to explicitly defined sub-operators, each body implicitly contains two special sub-operators—`input` and `output`. These sub-operators represent the input and the output of the operator `new_operator`.

Operators may have multiple inputs or outputs. The inputs/outputs are indexed by consecutive numbers starting with zero. The index of an output is written in brackets as a suffix of the identifier of the operator, the index of an input is written analogically

Fig. 3 Internal structure of
the `new_operator`

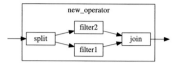

as a prefix. If an operator has only one input/output, the index may be omitted. Note that the `(int)`, `(int)` denotes two streams of integers, whereas `(int, int)` denotes one stream of integer pairs. The resulting internal structure is depicted in Fig. 3.

4.1 Multiplication of Inputs and Outputs

Some operators have the ability to split (or broadcast) their output into N channels while other operators can receive their input from multiple channels. Splitting creates the opportunity for parallelism as the operators between the splitting and merging operators are multiplicated and run in parallel. Bobolang allows the specification of multiplicated outputs and inputs and handles the replication of the operators. On the other hand, the method of splitting (round-robin, hash-based, etc.) as well as merging (ordered, random) is a matter of agreement between the operators involved and must be consistent with the properties of the operators in between.

The main objective of the multiplication mechanism is to allow plan description which does not grow with the degree of multiplication and where the degree of multiplication may be either be specified explicitly or inferred from the environment. The mechanism also allows the propagation of multiplication from the main plan to the lower plans in the model hierarchy, allowing internally parallelized sub-models to communicate via multiplied channels.

In a Bobolang plan, each input and output may be augmented with a *multiplier*, either *explicit* (a number in curly brackets) or *implicit* (an asterisk). In both cases, the multiplier signalizes the ability of the operator to produce or consume multiplicated channels, with the specified or an arbitrary degree. As a result, both channels and operators are multiplicated and the Bobolang compiler tries to find an equilibrium which satisfies the following equation for each edge connecting operator the ith output of the operator op_a to the jth input of the operator op_b:

$$d(op_a) \cdot d(out_{a,i}) = d(in_{b,j}) \cdot d(op_b)$$

Here, $d(op_a)$ and $d(op_b)$ are the degrees of multiplication of the two operators while $d(out_{a,i})$ and $d(in_{b,j})$ are the degrees of output/input multiplication at the edge ends. The products at both sides of the equation correspond to the degree of multiplication of the connecting channel. A part of the output/input degrees may be set by explicit multipliers in the source plan, others are set to 1 where no multiplier is specified. The degrees associated with implicit multipliers (asterisks) as well as the degrees of operators are computed by the Bobolang compiler. Whenever the equations allow a degree of freedom, implicit multipliers are set to default values from the environment. Once all the input/output degrees are fixed, the degrees of operators become uniquely determined by propagation from the main plan borders whose multiplicity degrees are set to 1. Care must be taken when explicit multipliers are used, because the system of equations may become overconstrained and thus no solution may be found.

4.2 Intra-operator Parallelization

We described two types of operators in Sect. 3 and methods of parallelizing them. To make the parallelization process easier, we introduce Bobolang keywords that specify the type of a sub-operator, so the Bobolang interpret can select appropriate parallelization method.

The `stateless` keyword informs the compiler that the sub-operator instance does not have inner state, so it can be always parallelized. The `parallel` keyword denotes operators that are stateful, but have been modified in the way prescribed by our schema and the programmer explicitly requests that they are parallelized.

```
stateless stateless_op()->() op1;
parallel parallelizable_op()->() op2;
```

If we mark the operator as `stateless`, the operator is automatically replaced with the following schema (with respect to the number of inputs/outputs):

```
operator parallelized_stateless (in_type)->(out_type) {
    dispatch(in_type)->(in_type)* disp;
    stateless_op(in_type)->(out_type) op;
    consolidate(out_type)*->(out_type) cons;

    input -> disp -> op -> cons -> output;
}
```

Therefore, when the operator `stateless_op` is used in an execution plan, it is decomposed as shown in Fig. 1. The `parallel` keyword uses very similar schema. The only difference is that it employs a `broadcast` operator instead of `dispatch` operator as depicted in Fig. 2.

5 Conclusion

In this paper, we have presented an elegant concept of semi-automated parallelization designed for data streaming systems. This concept streamlines the implementation of the core functionality of data processing systems whilst providing a safe way how to introduce various forms of parallelism into an application.

Data parallel subproblems with no internal state may be parallelized directly by the means of the replication scheme for stateless operators. The concept of stateless operator is simpler to handle and less error prone, since the programmer designs the internal functionality regardless of the level of parallelism. For more complex cases, we have proposed the concept of parallelizable stateful operator which permits parallelization at the cost of redundant work.

The presented concepts has been implemented in the Bobolang language and successfully applied in the implementation of parallel database engines. The underlying Bobox system is currently being extended to parallel accelerators such as GPUs and Xeon Phi devices. Their unique properties become an impulse for further development of the Bobolang language and the presented parallelization concepts.

Acknowledgments This work was supported by the Czech Science Foundation (GACR) projects P103-14-14292P and P103-13-08195S and by Specific Research SVV-2015-260222.

References

1. Aldinucci, M., Danelutto, M., Kilpatrick, P., Meneghin, M., Torquati, M.: Accelerating code on multi-cores with fastflow. In: Jeannot, E., Namyst, R., Roman, J. (eds.) Euro-Par 2011 Parallel Processing. Lecture Notes in Computer Science, vol. 6853, pp. 170–181. Springer, Berlin (2011)
2. Bednarek, D., Dokulil, J., Yaghob, J., Zavoral, F.: Bobox: parallelization framework for data processing. In: Advances in Information Technology and Applied Computing (2012)
3. Buck, I., Foley, T., Horn, D., Sugerman, J., Fatahalian, K., Houston, M., Hanrahan, P.: Brook for GPUs: stream computing on graphics hardware. ACM Trans. Graph. **23**, 777–786 (2004)
4. Consel, C., Hamdi, H., Réveillère, L., Singaravelu, L., Yu, H., Pu, C.: Spidle: a DSL approach to specifying streaming applications. In: Proceedings of the 2nd International Conference on Generative Programming and Component Engineering, pp. 1–17. Springer, New York, NY, USA (2003)
5. Cordes, D., Heinig, A., Marwedel, P., Mallik, A,.: Automatic extraction of pipeline parallelism for embedded software using linear programming. In: Parallel and Distributed Systems (IC-PADS), 2011 IEEE 17th International Conference on, pp. 699–706 (2011)
6. Das, A., Dally, W.J., Mattson, P., Compiling for stream processing. In: Proceedings of the 15th International Conference on Parallel Architectures and Compilation Techniques, pp. 33–42. ACM, New York, NY, USA (2006)
7. Falt, Z., Bednárek, D., Kruliš, M., Yaghob, J., Zavoral, F.: Bobolang: a language for parallel streaming applications. In: Proceedings of the 23rd International Symposium on High-performance Parallel and Distributed Computing, pp. 311–314. ACM (2014)
8. Franklin, M., Tyson, E., Buckley, J., Crowley, P., Maschmeyer, J.: Auto-pipe and the X language: a pipeline design tool and description language. In: 20th International Parallel and Distributed Processing Symposium. IEEE (2006)

9. Jagannathan, R., Dodd, C., Agi, I.: Glu: a high-level system for granular data-parallel programming. Concurrency—Pract. Expe. **9**(1), 63–83 (1997)
10. Kapasi, U.J., Dally, W.J., Rixner, S., Owens, J.D., Khailany, B.: Programmable stream processors. IEEE Comput. **36**, 282–288 (2003)
11. Mark, W.R., Steven, R., Kurt, G., Mark, A., Kilgard, J.: Cg: a system for programming graphics hardware in a C-like language. ACM Trans. Graph. **22**, 896–907 (2003)
12. Muller, S.C., Alonso, G., Amara, A., Csillaghy, A.: Pydron: semi-automatic parallelization for multi-core and the cloud. In: 11th USENIX Symposium on Operating Systems Design and Implementation (OSDI 14), pp. 645–659. USENIX Association (2014)
13. Reinders, J.: Intel threading building blocks. O'Reilly, Sebastopol (2007)
14. Thies, W., Karczmarek, M., Amarasinghe, S.: StreamIt: a language for streaming applications. In: Compiler Construction, pp 179–196. Springer (2002)
15. Vandierendonck, H., Rul, S., De Bosschere, K.: The paralax infrastructure: automatic parallelization with a helping hand. In: Parallel Architectures and Compilation Techniques, 19th International Conference, Proceedings, pp. 389–400. Association for Computing Machinery (ACM) (2010)
16. Zhang, D., Li, Z.Z., Song, H., Liu, L.: A programming model for an embedded media processing architecture. In: Embedded Computer Systems: Architectures, Modeling, and Simulation, pp. 251–261. Springer (2005)

Optimizing Satisfaction in a Multi-courses Allocation Problem

Ana-Maria Nogareda and David Camacho

Abstract The resource allocation problem is a traditional kind of NP-hard problem. One of its application domains is the allocation of educational resources. In most universities, students select some courses they would like to attend by ranking the proposed courses. However, to ensure the quality of a course, the number of seats is limited, so not all students can enroll in their preferred courses. Therefore, the school administration needs some mechanism to assign the available resources. In this paper, the course allocation problem has been modeled as a Constraint Satisfaction Optimization Problem (CSOP) and two metrics have been defined to quantify the satisfaction of students. The problem is solved with Gecode and a greedy-based algorithm showing how the CSOP approach is able to allocate resources optimizing the students' satisfaction. Another contribution of this work is the allocation of several courses simultaneously, generating feasible solutions in a short time. The allocation procedures are based on preferences for courses defined by students, and on the administration's constraints at Ecole Hôtelière de Lausanne. Ten data sets have been generated using the distribution of students' preferences for courses, and have been used to carry out a complete experimental analysis.

1 Introduction

In many universities and in most business schools, students can choose the courses they would like to attend. Due to quality and sometimes security constraints, the number of seats available in each course is limited, and therefore the demand may be higher than the offer for popular courses, [2, 3]. Different strategies are used by universities to allocate seats to students.

A. Nogareda (✉)
Ecole Hôtelière de Lausanne, Lausanne, Switzerland
e-mail: ana-maria.nogareda@ehl.ch

D. Camacho
Escuela Politécnica Superior, Univ. Autónoma de Madrid, Madrid, Spain
e-mail: david.camacho@uam.es

© Springer International Publishing Switzerland 2016   247
P. Novais et al. (eds.), *Intelligent Distributed Computing IX*,
Studies in Computational Intelligence 616,
DOI 10.1007/978-3-319-25017-5_23

The students' preferences are often submitted using:

- A **Ranking** system, where students rank the courses from their first choice to their last.
- A **Bidding** system that enables students to indicate how much they prefer a course over the others. Each student is credited with a specific amount of points and he bids points on his favourite courses. The number of points given to the different courses depends on the strategy of the student and on his preferences.

With both systems, several allocation methods can be used. Budish and Cantillon compare two mechanisms that use the Ranking system, [1]. The Harvard Business School mechanism consists in allocating one course to each student at each step of the process. For the first step the priority order of students is random and this order is reversed in each subsequent step. The second mechanism is the Random Serial Dictatorship which considers a random order of students and allocates all courses to one student at each step. It is a First-come First-served mechanism where the arrival sequence is defined by a random order.

Sönmez and Ünver analyze mechanisms that use the Bidding system, [8]. The first one is similar to what is used at the University of Michigan Business School, which uses Bidding information to infer students' preferences. Their conclusions show that this method results in an efficiency loss since we may have situations where a student bids more on a popular course and consequently is not assigned to his preferred course.

Diebold et al. compare several matching methods to allocate one bundle of courses to each student using three different metrics, [4]. The *average rank* measures the global welfare of students. A matching will be more *popular* if more students prefer it. The *rank distribution* compares how many students were assigned to their first choice, how many to their second choice, and so on.

In this paper, we compare two methods: the first is based on a greedy approach, the second is based on a Constraint Satisfaction Optimization Problem (CSOP). The generation of our sample data is based on courses offered to students in their final semester at Ecole Hôtelière de Lausanne (EHL); students indicate their preferences with a ranking of all the available courses.

- The **Course Greedy** approach assigns students to courses so that a course is filled with students who ranked it in their first choices.
- The **CSOP** approach assigns students to courses so that the global welfare of students is maximized and the satisfaction of the worst off is optimized.

In addition to the CSOP model, another contribution of this paper is the fact that we do not allocate only one course or one bundle of courses to a student, but several courses.

The rest of the paper is structured as follows. Section 2 describes the course allocation problem and the metrics that will be used to compare different allocations. Section 3 describes the greedy strategy. Section 4 describes the CSOP model. Section 5 presents the results of the two approaches applied to different sets of data. Finally Sect. 6 contains the conclusions.

2 The Course Allocation Problem (CAP)

In most business schools, elective courses with a limited number of seats are allocated to students. Each student has to rank the available courses and will be assigned to courses depending on his preferences. Without loss of generality, we may consider that the number of courses allocated is the same for all students. A formal definition of the problem is presented here. Let $S = \{s_1, \dots, s_n\}$ be the list of students and $G(s)$ the average grade of student s. Let $C = \{c_1, \dots, c_m\}$ be the list of available courses and $K(c)$ the capacity of course c, that means that a maximum of $K(c)$ students can be assigned to course c. Each student must be assigned to γ courses.

Definition 1 The ranking of a student $s \in S$ is a relation $\succeq_s$, such that $\forall c_1, c_2 \in C$:

$c_1 \succ_s c_2 \Leftrightarrow s$ strictly prefers c_1 over c_2
$c_1 =_s c_2 \Leftrightarrow s$ has no preference of c_1 over c_2 or c_2 over c_1

A ranking is said to be total if all courses are ranked by all students.

Definition 2 An allocation μ of courses to students is a function such that

$\forall c \in C : \mu(c) \quad = \{s_1^c, \dots, s_{Card(\mu(c))}^c\}$ is the list of students assigned to c
$\forall s \in S : \mu^{-1}(s) = \{c_1^s, \dots, c_{Card(\mu^{-1}(s))}^s\}$ is the list of courses allocated to s

Definition 3 An allocation μ of courses to students is said to be feasible if

$\forall c \in C \qquad\qquad : Card(\mu(c)) \leq K(c)$
$\forall s \in S \qquad\qquad : Card(\mu^{-1}(s)) = \gamma$
$\forall s_1, s_2 \in \mu(c) \quad : s_1 \neq s_2$
$\forall c_1, c_2 \in \mu^{-1}(s) : c_1 \neq c_2$

In this paper enough seats are available to allocate γ courses to each student, that means

$$\sum_{c \in C}(K(c)) = n \cdot \gamma \qquad (1)$$

The following subsection describes two metrics that will be used to measure the quality of an allocation and to compare two different allocations.

2.1 Comparison of Allocations

The satisfaction of a student s is measured considering his ranking of the courses he is assigned to.

Definition 4 The position $P_s(c)$ of a course $c \in C$ for a student $s \in S$ is defined by

$P_s(c) = Card(c_i \in C : c_i \succ_s c)$
$P_s(c)$ is thus the number of courses that s strictly prefers over c.

Definition 5 A student s has a strict ranking if

$$\forall c_i, c_j \in C \text{ with } i \neq j : c_i^s \prec_s c_j^s \text{ or } c_i^s \succ_s c_j^s \text{ which means } P_s(c_i) \neq P_s(c_j)$$

With a strict ranking, there is no indifference: for any pair of courses, a student always strictly prefers one of them over the other, [6].

Definition 6 For an allocation of courses μ, the satisfaction gap $SatGap_\mu(s)$ of a student $s \in S$ is defined by

$$SatGap_\mu(s) = \sum_{c \in \mu^{-1}(s)} P_s(c) \tag{2}$$

For each course allocated to a student, the satisfaction gap $SatGap_\mu(s)$ counts the number of courses that the student strictly prefers over this course. For example, let's consider a student who is assigned to 2 courses whose positions in his ranking are 5 and 3. That means that there are 5 courses that he prefers over the first course and 3 courses that he prefers over the second course. His satisfaction gap with this allocation is then 8. So the smaller $SatGap_\mu(s)$ is, the happier the student is. If the ranking is strict, then $SatGap_\mu(s)$ has a lower bound: $SatGap_\mu(s) \geq \frac{\gamma \cdot (\gamma - 1)}{2}$.

The quality of an allocation can be measured with different parameters. The metrics used in this paper to analyze the quality of an allocation μ are:

- **Total Satisfaction Gap (TSG)**. Sum of the satisfaction gap of all students.

$$(TSG) = \sum_{s \in S} SatGap_\mu(s) \tag{3}$$

If the ranking is strict, then (TSG) has a lower bound: $(TSG) \geq n \cdot \frac{\gamma \cdot (\gamma - 1)}{2}$.
- **Worst Satisfaction Gap (WSG)**. Worst satisfaction among students. The students whose satisfaction gap is equal to (WSG) are the worst off.

$$(WSG) = \max_{s \in S} SatGap_\mu(s) \tag{4}$$

As we want to have students as satisfied as possible, the objective is to minimize the satisfaction gap, and therefore to minimize both metrics, (TSG) and (WSG).

3 Greedy Strategy: Course Greedy Algorithm (CGA)

To apply this algorithm, students must be ordered using a specific criteria, in the following sections, students are ordered by their grades, ties are randomly broken.

In the first step of the Course Greedy Algorithm, the available seats of each course are allocated to the students who ranked the course first. In step k, only the courses that still have available seats are considered and those seats are allocated to the students who ranked the course in position k. If a student is already assigned to γ courses, he won't be considered in the following steps. This algorithm is sketched in Algorithm 1. To avoid infeasibility, a course becomes mandatory if the number of remaining seats is equal to the number of students who have not been assigned to enough courses.

```
input  : List of students S = [s₁,...,sₙ] ordered by their grades
            ∀i,j ≤ n : i > j ⇒ G(sᵢ) ≤ G(sⱼ)
output: Feasible allocation μ

∀c ∈ C· μ(c) ← ∅
/*Loop per position                                                         */
for k ← 0 to m − 1 do
    /*Loop per course                                                       */
    for j ← 1 to m do
        if Card({s ∈ S : μ⁻¹(s) < γ}) = K(cⱼ) − Card(μ(cⱼ)) then
            /*The course becomes mandatory to avoid infeasibility           */
            μ(cⱼ) ← μ(cⱼ) ∪ {s ∈ S : μ⁻¹(s) < γ}
        end
        if Card(μ(cⱼ)) < K(cⱼ) then
            /*Loop per student                                              */
            for i ← 1 to n do
                if P_{sᵢ}(cⱼ) = k and Card(μ⁻¹(sᵢ)) < γ then
                    μ(cⱼ) ← μ(cⱼ) ∪ {sᵢ}
                end
            end
        end
    end
end
```

Algorithm 1: Course Greedy Algorithm

Example Consider a CAP with $m = 4, \gamma = 2$ and four students $S = \{s_1, \ldots, s_4\}$ with $G(s_1) > G(s_2) > G(s_3) > G(s_4)$. Their rankings are given in Table 1, for example $P_1(c_1) = 0$ means that this course is the favourite course of s_1 together with c_4 since $P_1(c_4) = 0$. As $P_1(c_2) = P_1(c_3) = 2$, s_1 has no preference among those courses, but he prefers c_1 and c_4 over c_2 or c_3. This table contains also the solution obtained when applying the CGA. $\mu(c_i)$ is the list of students assigned to course c_i, for example $\mu(c_1) = \{s_1, s_2\}$ means that s_1 and s_2 are assigned to course c_1. The results are $(TSG) = 6$ and $(WSG) = 2$.

4 The CSOP Model

As defined in [9], a Constraint Satisfaction Optimization Problem (CSOP) is a problem that can be described with three sets and one function (X, D, C, f). $X =$

Table 1 Example

	s	$P_s(c_1)$	$P_s(c_2)$	$P_s(c_3)$	$P_s(c_4)$
Ranking	1	0	2	2	0
	2	0	3	1	2
	3	1	2	0	3
	4	0	2	0	3
Results		$\mu(c_1)$	$\mu(c_2)$	$\mu(c_3)$	$\mu(c_4)$
		$\{s_1, s_2\}$	$\{s_3, s_4\}$	$\{s_3, s_4\}$	$\{s_1, s_2\}$

Ranking: Position of each course for all students. For each student $s \in \{1, 2, 3, 4\}$, $P_s(c_i)$ is the number of courses that s prefers over c_i. Results: Allocation of courses to students and metrics value for CGA: $\mu(c_i)$ is the list of students assigned to course c_i

$\{x_1, x_2, \ldots, x_n\}$ is the set of variables. For each variable x_i, the set of possible values that can be assigned to x_i is defined as its domain D_i. A finite set of constraints $C = \{c_1, c_2, \ldots, c_m\}$ limits the possible combinations of values that can be assigned to the variables. A solution to a CSOP is defined as an assignment of one value to each variable so that the value of one variable belongs to its domain and all the constraints of C are satisfied. The function f maps a numerical value to each possible solution. The objective of the problem is to optimize the value of f.

For the CAP, the variables $X = \{x_j^i : i = 1, \ldots, n, j = 1, \ldots, \gamma\}$ are the courses allocated to each student, e.g. $X_1 = \{x_1^1, \ldots, x_\gamma^1\}$ are the courses allocated to student 1. The domain D_i is the same for all the variables, it is the list of the available courses $\forall i = 1, \ldots, n : D_i = \{c_1, \ldots, c_m\}$. There are two types of constraints: **Unicity**, a course cannot be allocated more than once to a student, and **Capacity**, the number of students assigned to a course cannot exceed its capacity.

$$\forall s \in S : alldifferent(X_s)$$
$$\forall c \in C : Card(x_i^j \in X : x_i^j = c) \leq K(c)$$

The objective function has to take into account the metrics defined in Sect. 2.1 with the following priorities:

Priority 1 Minimize (TSG). The total satisfaction gap must be as small as possible, so that the global satisfaction of students is maximized.

Priority 2 Minimize (WSG). The less satisfied student must be as satisfied as possible.

5 Experimentation

Ten sets of data have been generated for the ranking of 50 students over 10 courses. This ranking is based on the distribution of the preferences of the students over the

Table 2 Experimentation

Course	1	2	3	4	5	6	7	8	9	10
Probability	5	5	6	7	10	10	12	12	15	18

Distribution of preferences (%) of the students over courses

Table 3 50 students, 10 data sets—Comparison CGA-CSOP, Mean ± Standard Deviation of the two metrics (TSG) and (WSG)

	CGA	CSOP—10 s	CSOP—1 min	CSOP—Best known	Best
(TSG)	226 ± 19	217 ± 21	215 ± 21	211 ± 20	CSOP
(WSG)	12 ± 3	8 ± 1	8 ± 1	7 + 1	CSOP

courses given in Table 2: the courses with a higher probability will have a higher probability to be among the first choices of students. For example, course 4 has a probability of 7 % to be the first choice of a student.

All courses have a capacity of fifteen seats, $\forall c \in C : K(c) = 15$. Three courses are allocated to each student: $\gamma = 3$. The generated ranking is a strict ranking, which means that even if a student s is assigned to his first three choices c_1, c_2 and c_3, the position of this courses for s will be $P_s(c_1) = 0$, $P_s(c_2) = 1$, and $P_s(c_3) = 2$. Therefore for any allocation μ and any student $s \in S$ we have:

$$SatGap_\mu(s) \geq 3 \quad (TSG) \geq 150 \quad \text{and} \quad 24 \geq (WSG) \geq 3.$$

Table 3 contains the results. For CGA, the solution is found in less than 1 second. To solve the CSOP, we have used Gecode, [5, 7]. The computation was done with one single processor on an Intel® Core™ i5-3320M CPU 2.60 GHz. Table 3 contains the results after 10 seconds of computation, after 1 minute and the best known solution.

Regarding (TSG), CSOP is on average 4 % better than CGA after 10 s and 6.6 % better for the best known solution. The ten charts in Fig. 1 correspond to the ten data sets analyzed and contain the value of (TSG) for CGA, as this algorithm finds only one solution, the value of (TSG) doesn't change. The charts also contain the results for the CSOP method where we can see for each solution found, the value of (TSG) and the computation time needed to find it.

Regarding (WSG), the first solution obtained with CSOP is always better than with CGA. After 10 seconds, CSOP is 33 % better and even 42 % better for the best known solution.

Time-efficiency. Figure 2 compares the value of (TSG) with CGA and CSOP for the first solution, after 10, 30 s, 1 and 15 min, and includes also the best known result for the CSOP. The first solution obtained with CSOP is not always better than the CGA solution, but after 10 seconds of computation, CSOP outperforms CGA in all data sets.

Scalability. Additional simulations were done with ten data sets of 150 students in order to analyze the possible loss of effectiveness for bigger data sets. Figure 3 compares the results obtained for (TSG) with CGA and the CSOP method for the

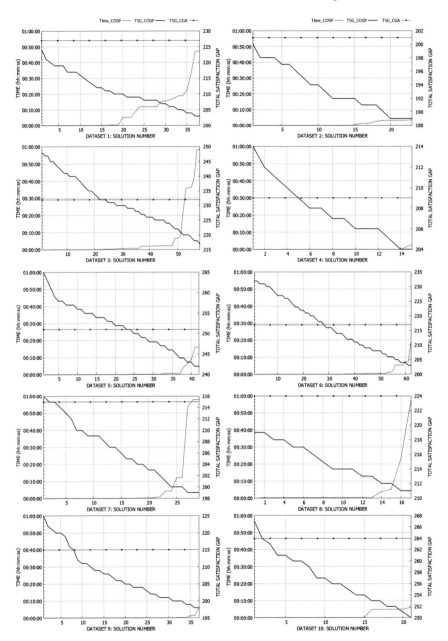

Fig. 1 50 students—Results: Value of the metric (TSG) in the different solutions found by CSOP and time when these solutions are found

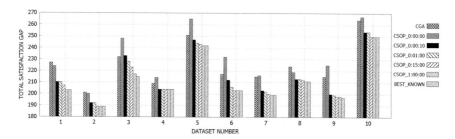

Fig. 2 50 students—(TSG) value for CGA and CSOP for each of the ten data sets in the first solution and after 10 s, 1, 15 min and 1 h of computation

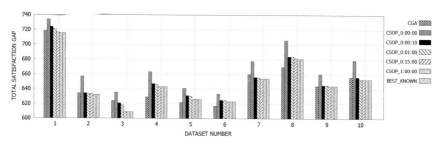

Fig. 3 150 students—(TSG) value for CGA and CSOP for each of the ten data sets in the first solution and after 10 s, 1, 15 min and 1 h of computation

Table 4 150 students, 10 data sets—Comparison CGA-CSOP, Mean ± Standard Deviation of the two metrics (TSG) and (WSG)

	CGA	CSOP—10 s	CSOP—15 min	CSOP—Best known	Best
(TSG)	647 ± 31	652 ± 31	648 ± 31	648 ± 31	CGA
(WSG)	18 ± 5	10 ± 1	10 ± 1	10 ± 1	CSOP

first solution, after 10 seconds, 1, 15 min and 1 h, and also includes the best known result with the CSOP. Regarding (TSG), neither algorithm outperforms the other: CSOP is better in five sets and CGA is better in the other five sets. Regarding (WSG), CSOP always outperforms CGA and on average the results are 44 % better (Table 4).

6 Conclusions and Future Work

In this paper, we have compared two different approaches to solve the problem of allocating several courses to students, where each student gives his preferences with a strict ranking of the available courses. Two metrics have been used to quantify the quality of a solution: the Total Satisfaction Gap (TSG) analyzes the average level of satisfaction of the students and the Worst Satisfaction Gap (WSG) corresponds to the level of satisfaction of the worst off.

The first approach, the Course Greedy Algorithm (CGA), allocates a course to the students who ranked this course first, and if seats are still available, to those who ranked it second and so on. This mechanism is fairer than a simple First-come First-served mechanism.

The second approach is based on a Constraint Satisfaction Optimization Problem (CSOP) and uses the solver Gecode to optimize the value of both metrics. For small instances, the results for both metrics are very quickly much better than with CGA. When the number of students increases, the benefit regarding (WSG) is even more important than for smaller instances. However, the computation time needed to obtain good results for (TSG) increases, and the benefit of this approach is less important than for smaller instances since (TSG) with CSOP is similar to (TSG) with CGA with 150 students. Nevertheless the allocations obtained with CSOP are much fairer for students than the allocation obtained with CGA.

We are currently implementing a third approach using an Ant Colony Optimization (ACO) algorithm in order to compare the results obtained with the three methods with real preferences of students over courses that are offered at Ecole Hôtelière de Lausanne.

Acknowledgments This work has been supported by CIBERDINE Project (S2013/ICE-3095) and by Savier Project (Airbus Defence &Space, FUAM-076915). The authors would also like to thank Vincent Maronnier for his contribution.

References

1. Budish, E.B., Cantillon, E.: (2010). The multi-unit assignment problem: theory and evidence from course allocation at Harvard
2. Cano, J.I., Sánchez, L., Camacho, D., Pulido, E., Anguiano, E.: Allocation of educational resources through happiness maximization. In: Proceedings of the 4th International Conference on Software and Data Technologies (2009)
3. Cano, J.I., Sánchez, L., Camacho, D., Pulido, E., Anguiano, E.: Using Preferences to solve student-class allocation problem. In: Intelligent Data Engineering and Automated Learning-IDEAL 2009, pp. 626–632. Springer, Berlin (2009)
4. Diebold, F., Aziz, H., Bichler, M., Matthes, F., Schneider, A.: Course allocation via stable matching. Bus. Inf. Syst. Eng. **6**(2), 97–110 (2014)
5. Nogareda, A.M., Camacho, D.: Integration of Ant Colony Optimization algorithms with Gecode. In: 20th International Conference on Principles and Practice of Constraint Programming (CP 2014): Doctoral Program Proceedings, 59–64 September 2014
6. Francesca, R., Venable, K.B., Walsh, T.: A Short introduction to preferences: between artificial intelligence and social choice. In: Synthesis Lectures on Artificial Intelligence and Machine Learning 5.4, pp. 1–102 (2011)
7. Schulte, C., Lagerkvist, M., Tack, G.: Gecode. Software download and online material at the website: http://www.gecode.org. Downloaded in May 2014
8. Sönmez, T., Ünver, M.U.: Course bidding at business schools. Int. Econ. Rev. **51**(1), 99–123 (2010)
9. Tsang, E.: Foundations of constraint satisfaction, vol. 289. Academic press, London (1993)

Improving the Categorization of Web Sites by Analysis of Html-Tags Statistics to Block Inappropriate Content

Dmitry Novozhilov, Igor Kotenko and Andrey Chechulin

Abstract The paper considers the problem of improving the quality of web sites categorization using data mining methods. This goal is important for automated systems of parental control. The purpose of such systems is protection from unwanted or inappropriate information. The novelty of the proposed approach is in usage of HTML tags statistics of web pages to improve the categorization of sites that are similar in terms of textual content, but differing in their structural features. The paper describes the architecture of the categorization system, the algorithm of its work, the results of experiments, and assessment of classification quality.

1 Introduction

Data mining methods are known to be used for detecting hidden knowledge in the data: unknown, non-trivial and practically useful. During data analysis it is often necessary to classify the studied object to one of predefined classes; this is the classification problem. Its correct solution is very important and leads to significant progress in many areas. The distinctive features of our time are continuous development and ubiquity of the Internet. In such conditions the importance of automatic classification systems, that distribute web pages by category and block those that are undesirable or offensive, increases. This is extremely important, for example, to protect children from sites containing inappropriate content or to counteract the spread of malware and pirate content.

D. Novozhilov · I. Kotenko (✉) · A. Chechulin
Laboratory of Computer Security Problems, St. Petersburg Institute for Informatics and Automation (SPIIRAS), 39 14 Linija, St. Petersburg, Russia
e-mail: ivkote@comsec.spb.ru

D. Novozhilov
e-mail: novozhilov@comsec.spb.ru

A. Chechulin
e-mail: chechulin@comsec.spb.ru

P. Novais et al. (eds.), *Intelligent Distributed Computing IX*,
Studies in Computational Intelligence 616,
DOI 10.1007/978-3-319-25017-5_24

257

There are many different approaches to the web sites classification. The most efficient and widely used is the analysis of the text content of web pages. However, there are some categories of sites (forum, blog, news) which are almost the same in text content, whereas their structural features are different. In such situations researchers use other classification methods. For example, you can parse URL addresses of the pages or HTML tags of their markup. One of the options in the last approach is to check the presence/absence of specified tags. In the paper we propose an original approach, which, unlike the existing ones, is based on the analysis of statistics of HTML tags, which is the ratio of all occurrences of the tag to the total number of tags on the site. The main objective of the paper is to improve the quality of classification for web sites, which classification by text is difficult, by analysis of web sites based on information about their HTML tags. All stages of this research are reflected in this paper, which has the following structure: at first the review and analysis of related work is conducted, then the proposed solution, the experiments and their results are described. The last section is devoted to conclusions and future plans.

2   Related Works

The most widely applied method is the classification according to the text [1–3]. Another alternative is to move from a consideration of the documents as sets of words to the analysis of their meanings, which are taken from the lexical database [4]. Disadvantage of text classification is that it does not take into account web pages particularities: HTML document is linked by references to other documents; it contains images and other non-text elements. Difficulties are also caused by categories with similar text content, but differing in their structure (for example, blogs, forums, chats). Thus, the method based on the URL analysis was developed. Using it we assume that the web page is rarely visited unless it generates interest among potential readers. That means that the address of the site should somehow reflect its theme [5]. One of the methods here is to split the URL into its component parts, and to analyze the parts obtained. This approach is implemented in [6] for URL analysis to protect from phishing sites. The position of a particular portion in the site address is also important. Another way is to use the length of the host name and the entire URL, count the number of different symbols and analyze URL fragments between these characters. Besides that there are also signs on the basis of information about the host (geographical features, date of registration, the value of TTL, etc.). All these attributes are fed to the input of any classifier [7]. In [8] and [9] there are references to the method associated with the sequence analysis of n-grams, for which the frequency of appearance is calculated. To identify categories, based on structural characteristics, it is necessary to look for other methods, one of which is the usage of HTML tags. The information contained in tags like <title> or <meta> together with the text content may serve as an important source for analysis [10–13]. Our paper discusses a new approach that is based on the analysis not of the content or the number of HTML tags on the page, but their statistics, which is defined as the ratio of all occurrences of the tag to the total number of tags on the site.

3 Models and Implementation

Web pages are known to be different from regular documents primarily by the fact that they are semi-structured using HTML markup tags, interconnected with links, contain fragments of code that is executable on the server side and on the client. Therefore, it is to use methods, which take into account the particularity of the analyzed data. One of possible solutions may be the approaches based on structural features of web pages, i.e. with HTML tags.

The proposed method is also not based on the stored source code of the web pages for later analysis, but works with the statistics of HTML tags instead. We understand the statistics S of tags as the totality of their occurrence frequencies f_i, which is defined as the ratio of the number of instances of this tag n to the total number N of tags on the page, expressed as a percentage. The result is rounded to be more informative: $S = \cup f_i; f_i = \frac{n_i}{N} * 100\%$. We should note that such solution was not found in the beginning of our research, and at first we analyzed simply the number of tags of each type on the page. However, this approach is not quite correct, because, for example, it is incorrect to compare 100 tags <div> on the pages, consisting of 250 and 1000 tags, and they point to a completely different result. The final classifier is based on the Naïve Bayes and Decision Tree algorithms, which basic predictions are combined on the upper level using stacking. As the base models, stacking uses various classification algorithms trained on the same data. Then the meta-classifier is trained on the input data, supplemented by the results of the forecast of the base algorithms. The idea of stacking is that the meta-algorithm learns to distinguish which of the base algorithms it should "trust" on which areas of input data.

To evaluate the quality of classification we use such metrics as precision, recall, accuracy, and the F-measure which combines information on the precision and recall. It should be noted that for the class of systems that perform parental control accuracy metric is of particular importance, since a large number of false positives may cause refusal from application of such systems.

The task of finding the frequency of tags can be solved in several different ways. Let us select two of them: (1) search the tags in the entire HTML document and count the number of occurrences of each of them; (2) use representation of the HTML document as a tree that significantly simplifies solution of the problem, providing a variety of navigation functions and access to its elements. One of the arguments to choose the second approach was that such a tree data structure already exists—it was built for the needs of document parsing and storing its text content to a file without HTML tags.

The key is the tag name, and values are all its representatives, the number of which is counted. If we apply the function to the root of the tree, we shall obtain problem solution. For further analysis, all tags are used which frequency exceeds 2 %. It is empirically set value that allows to exclude from consideration tags that are common to all pages, such as <HTML> , <title> , <head> , <body> etc.

4 Experiments

The experiments were conducted on two data sets ("set1" and "set2"). "Set1" was created basing on the original data from URLBlacklist.com [14], which included the categories: "Books, Hunting, News, Dating, Guns". For each category 1 thousand sites were selected. "Set2" includes content of URLBlacklist.com [14], combined with part of the categories taken from the list of "Shalla Secure Services KG" [15]. Several different catalogs are used to find common features of the sites, as well as by the lack of source data for certain categories within one source and sufficient number of them in another one. Finally there were selected 13 categories: "Chat, Drugs, Forum, Guns, Hunting, Jobsearch, Magazines, Medical, Movies, Music, Press". In each of them there were about 1.5 thousands of sites.

In the course of source data preparation the attention was paid to the boundaries of the categories. Such heterogeneous categories as "radio-tv" or "audio-video" were excluded from consideration, because in fact each of them was divided into two others. "Drugs" and "medical", "guns" and "hunting", on the contrary, were taken specifically to see the work in cases where some specific words and combinations can be common for them. The experimental results for "set1" are presented in Fig. 1. Classification results for tagged and textual features for the first set are shown in Fig. 2. The results of the experiments for the second set are presented in Fig. 3. Classification results for tagged and textual signs for "set2" are shown in Fig. 4. Analysis of the experimental results allows to make conclusion about generally low quality, which does not allow to use this method as the primary tool. The "set 1" gives higher value of accuracy equal to 35.43 %, because it contains less "controversial categories" which may intersect (only "guns" and "hunting"). For the "set 2" the accuracy decreases to 15.08 %, while it is clearly visible that categories "press" and "hunting" became "absorbing".

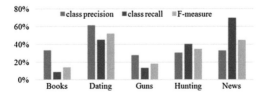

Fig. 1 The values of basic metrics for the set "Set 1"

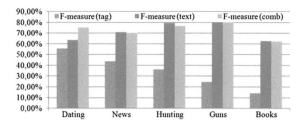

Fig. 2 Categories for set "Set 1", ordered by descending F-measure for classification by tags

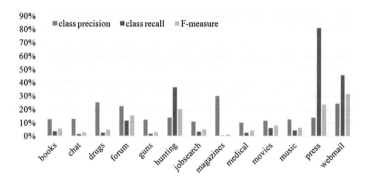

Fig. 3 The values of basic metrics for the set "Set 2"

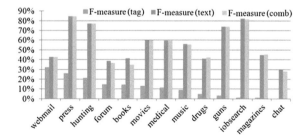

Fig. 4 Categories of set "Set 2", ordered by descending F-measure for classification by tags

Figure 5 shows the values of accuracy for approaches based on the analysis of tags, text and their combinations. The results reflect the improvement of the quality of classification by combining these approaches for 5 and 13 categories. Thus, the studies show that the proposed approach based on the statistics of the HTML tags does not solve the problem of rating by itself, but it can be a good addition in the outlining of categories that differ in their structural features.

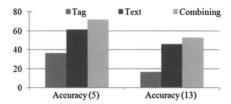

Fig. 5 The accuracy for approaches based on different features and their combination

5 Conclusion

This paper discussed approaches to categorizing web pages that do not have significant differences in text classification, but having different structure. The essence of the proposed method is the use of HTML tags statistics, which is fed to the classifiers. The obtained results show that the quality of the tags based classification is not sufficient to apply this method as a standalone one. But it can be used as a useful complement to existing systems with textual classification. The principles investigated can be applied to improve the quality of systems for protection from information, such as the parental control systems. Further research directions are connected with search for other classifiers and their combinations that will allow to combine textual and statistics tags data analyses, to get rid of "absorbing categories" that are characteristic to decision trees.

Acknowledgment This research is being supported by The Ministry of Education and Science of The Russian Federation (contract # 14.604.21.0147, unique contract identifier RFMEFI60414X0147).

References

1. Joachims, T.: Text categorization with support vector machines: learning with many relevant features. In: ECML-98, LNCS, vol. 1398, pp. 137–142. Springer (1998)
2. Ko, Y., Seo, J.: Automatic text categorization by unsupervised learning. In: Coling'00, pp. 453–459. Morgan Kaufmann (2000)
3. Ntoulas, A., Najork, M., Manasse, M., Fetterly, D.: Detecting spam web pages through content analysis. In: ACM, pp. 83–92 (2006)
4. Kehagias, A., Petridis, V., Kaburlasos, V.G., Fragkou, P.: A comparison of word-and sense-based text categorization using several classification algorithms. J. Intell. Inf. Syst. **21** (3), 227–247 (2000)
5. Attardi, G., Gulli, A., Sebastiani, F.: Automatic web page categorization by link and context analysis. In: THAI'99, pp. 105–119 (1999)
6. Khonji, M., Iraqi, Y., Jones, A.: Enhancing phishing E-Mail classifiers: a lexical URL analysis approach. Int. J. Inf. Secur. Res. **6**, 236–245 (2012)
7. Ma, J., Saul, L.K., Savage, S., Voelker, G.M.: Beyond blacklists: learning to detect malicious web sites from suspicious URLs. In: KDD'09, pp. 1245–1254. ACM (2009)

8. Kan, M.-Y., Thi, H.O.N.: Fast webpage classification using URL features. In: ICIKM 2005, ACM (2005)
9. Geide, M.: N-gram Character Sequence Analysis of Benign vs. Malicious Domains/URLs. Available at http://analysis-manifold.com/ Accessed 24 March 2015
10. Meshkizadeh, S., Masoud-Rahmani, A.: Webpage classification based on compound of using html features and url features and features of sibling pages. Int. J. Adv. Comput. Technol. **2**(4), 36–46 (2010)
11. Patil, A.S., Pawar, B.V.: Automated classification of web sites using naive bayesian algorithm. In: IMECS2012, vol. 1, p. 466 (2012)
12. Riboni, D. Feature selection for web page classification. In: EURASIA-ICT-2002 (2002)
13. Kotenko, I., Chechulin, A., Shorov, A., Komashinsky, D.: Analysis and evaluation of web pages classification techniques for inappropriate content blocking. LNAI **8557**, 39–54 (2014)
14. URLBlacklist.com.: http://urlblacklist.com/ Accessed 24 March 2015
15. Shalla Secure Services KG.: http://www.shallalist.de/ Accessed 24 March 2015

Distributed Architecture of Data Analysis System Based on Formal Concept Analysis Approach

A.A. Neznanov and A.A. Parinov

Abstract This paper describes distributed architecture and data workflow of the analysis system called FCART. Comparing with the similar systems FCART is capable of dealing with various data sources, data preprocessing and interactive analysis, extending functionality by integrating independent web-services and developing plugins. Example of gathering and analyzing data of social networking service is considered.

Keywords Software architecture · Data analysis · Data mining · Graph mining · Formal concept analysis · Knowledge extraction

1 Introduction

Formal Concept Analysis (FCA) [1] has many applications in different fields [2, 3]. In 2012 scientific group headed by S.O. Kuznetsov finished the first stage of crafting software tools for CORDIET project [4]. After that, the group has settled the goal to create an extensible integrated analytic environment for knowledge and data engineers with a set of research tools based on FCA.

This new software system is called "Formal Concept Analysis Research Toolbox" (FCART) [5]. The methodology of using this software is based on modern methods and algorithms of data analysis, technologies for manipulating data collections (including natural language texts), data visualization, reporting and interactive processing techniques. It allows analyst to obtain new knowledge from data with full process tracking and reproducibility of experiments.

A.A. Neznanov · A.A. Parinov (✉)
National Research University Higher School of Economics,
20 Myasnitskaya Ulitsa, Moscow 101000, Russia
e-mail: AParinov@hse.ru

A.A. Neznanov
e-mail: ANeznanov@hse.ru

© Springer International Publishing Switzerland 2016
P. Novais et al. (eds.), *Intelligent Distributed Computing IX*,
Studies in Computational Intelligence 616,
DOI 10.1007/978-3-319-25017-5_25

265

Certainly, there are some other particular software projects like Cubist [6] or Coron [7] or small utilities like Concept Explorer [8]. By now, there are no approaches or integrated program platforms that allow an analyst to use FCA methods and other tools together for solving real life problems. In this paper, we describe new web-based distributed architecture of FCART.

2 Methodology

FCART has been following a methodology of [4] formalizing iterative ontology-driven data analysis process and implementing several basic principles:

1. Iterative process of data analysis using ontology-driven queries and interactive artifacts (such as concept lattice, clusters, etc.).
2. Separation of processes of data querying (from various data sources), data preprocessing (of locally saved immutable snapshots), data analysis (in interactive visualizers of immutable analytic artifacts), and results presentation (in report editor).
3. Extendibility on three levels: customizing settings of data access components, query builders, solvers and visualizers; writing scripts (macros); developing components (add-ins).
4. Explicit definition of analytic artifacts and their types. It allows one to check the integrity of session data and provides links between artifacts for end-user.
5. Realization of integrated performance estimation tools.
6. Integrated documentation of software tools and methods of data analysis.

3 Distributed FCART Architecture Additional Principals

Early version of FCART was constructed as a local native Windows application. Now we suggest a new architecture to advance technological basis to web-based distributed application, to use modern programming techniques, to have a possibility of integrating independent software as components of FCART, and process Big Data. In the foundation of the architectural style there are following additional principals.

- The components of the system are language and framework independent.
- The components communicate with each other using RESTful Web-API [9, 10].
- The components use open standards for two- and three-legged authentication and authorization [11].
- The system use intermediate data storage supporting iterating and paging of big datasets.

First attempt to implement some of these principals had taken place in 2014 [12], and a year after we have checked all project solutions.

4 Structure and Workflows of FCART

In current distributed version, FCART consists of following four parts:

1. FCART AuthServer for authentication and authorization, as well as integration of algorithmic and storage resources.
2. FCART Intermediate Data Storage (IDS) for storage and preprocessing (initial converting, indexing of text fields, etc.) of big datasets.
3. FCART Thick Client (Client) for interactive data processing and visualization in integrated graphical multi-document user interface.
4. FCART Web-based solvers (Web-Solvers) for implementing independent resource-intensive computations.

4.1 Main FCA Workflow

Main data workflow described in details in our previous articles [12]. From the analyst point of view, basic FCA workflow in FCART has four stages (see Fig. 1).

1. Filling Intermediate Data Storage of FCART from various external SQL, XML or JSON-like data sources (querying external source described by External Data Query Description (EDQD).

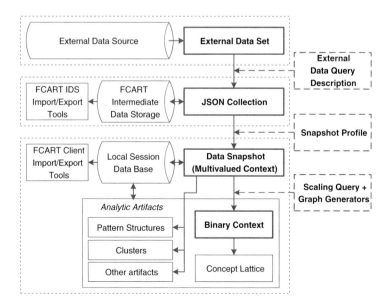

Fig. 1 Main FCA workflow of FCART

2. Loading a data snapshot from local storage into current analytic session (snapshot described by Snapshot Profile). Data snapshot is a data table with annotated structured and text attributes, loaded in the system by accessing external data sources.
3. Transforming the snapshot to a binary context (transformation described by Scaling Query) or graph (by Graph Generator).
4. Building and visualizing concept lattices and other artifacts based on the binary context in a scope of analytic session.

Thick Client. From the end user point of view, Client is responsible for rich end-user interaction. Client has special instruments to visualize and navigate analytic artefacts. For example, a user can work with big formal concept lattices: scale lattice element sizes, use parents-children navigator, filter-ideal selection, separation of focused concepts from other concepts. Thick Client can be extended by plugins and macros. For example, FCART has several drawing techniques for visualizing fragments of big lattices: for rendering focused concept neighborhood and for accenting (by layout and highlighting) selected concepts, "important" concepts, and filter-ideal with different sorting algorithms for lattice levels [12]. All interaction between user and external data storages (except local files) goes through the IDS.

AuthServer and Intermediate Data Storage. IDS is responsible for storing and preprocessing chunks of data from various data sources in common format (JSON collections). In the current release IDS provides a Web-API and uses AuthServer for OAuth 2.0 authorization. There are many data storages, which contain petabytes of collected data. For analyzing such Big Data analyst cannot use any of the tools mentioned above. FCART provides different scenarios for working with local and external data storages. In the previous release of FCART analyst can work with quite small external datasets because all data from external storages convert into JSON files and saves into IDS. In the current release, we have improved strategies of working with external data. Now analyst can choose between loading all data from external data storage to the IDS and accessing external data by chunks using paging mechanism. FCART analyst should specify the way of accessing external data using External Data Query Description (EDQD) [12].

Web Solvers and Web-API implementation. All interaction between Client, Web Solvers and IDS goes through the Web-API. Client construct http-request to the web-service. The http-request to web-service constructed from two parts: prefix part and command part. Prefix part contains domain name and local path (e.g. http://zeus2.hse.ru/lds/). The command part describes operations LDS has to do and represents some function of web-service API. Using described commands FCART client can query data from external data storage to the IDS.

5 Extracting Data from Social Networking Services

In the current release IDS supports scenario to parallel bulk download data from Social networking services (SNS). The bulk download is useful for getting relatively big amounts of data, which takes days and weeks for collecting. For reducing collecting time, we using a parallel architecture of program system, which consists of two types of components: Manager and Agent. In general, social data download process has the following steps (Fig. 2):

1. Manager gets unique identifiers of nodes and saves nodes identifiers to Task Queue.
2. At the same time, Agents listen to the Task Queue. When an identifier appears in the Task Queue, an Agent gets identifier.
3. The Agent checks if there is no such identifier in the Completed Task List and start downloading information about a node to a file on the local computer.
4. The Agent appends the file to the IDS collection.
5. The Agent has downloaded the node information from SNS. It inserts the identifier to the Completed Task List and listens to the Task Queue for the next identifier.

As an experiment of capabilities FCART, we have gathered more than 180,000 person's profiles from social blogging system LiveJournal [13].In LiveJournal,

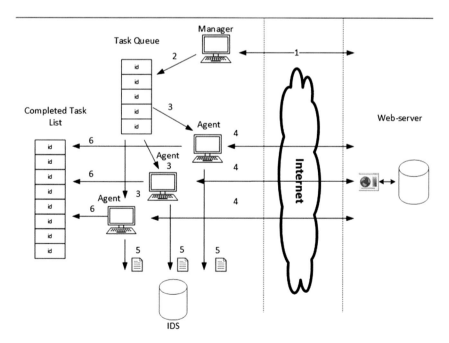

Fig. 2 Bulk downloading of SNS data

friendship relation is not mutual by default. Mutual friendship is a situation when b ∈ a.friends and a ∈ b.friends. For each profile, LiveJournal FCART extracts its friend list and add new nicks into downloading queue. Time of extracting 15 thousands profiles is about 19 h.

The result of "node-to-node" graph query is a digraph, which is saved as an artifact in FCART and can be transformed into binary context. FCART supports interactive browsing of concept lattices. Only ten random person's profiles from LiveJournal dataset (and all their friends) have brought about 5357 vertices in resulting graph and objects and attributes in corresponding context. Time of building this graph is only 0,3 min (on Intel i7 4600 2 GHz).

6 Conclusion

In this paper, we have discussed the distributed architecture of FCART. We believe that a new architecture will improve subsequent development of the system, especially collaborative construction of new solvers and visualizers.

Acknowledgments This work was carried out by the authors within the project "Data mining based on applied ontologies and lattices of closed descriptions" supported by the Basic Research Program of the National Research University Higher School of Economics.

References

1. Ganter, B., Wille R.: Formal Concept Analysis: Mathematical Foundations. Springer, Berlin (1999)
2. Kuznetsov, S.O., Poelmans, J.: Knowledge representation and processing with formal concept analysis. Wiley Interdisc. Rev. Data Min. Knowl. Discov. **3**(3), 200–215 (2013)
3. Poelmans, J., Ignatov, D.I., Kuznetsov, S.O., Dedene, G.: Formal concept analysis in knowledge processing: a survey on applications. Expert Syst. Appl. **40**, 6538–6560 (2013)
4. Poelmans, J., Elzinga, P., Neznanov, A., Viaene. S., Kuznetsov, S.O., Ignatov D., Dedene G.: Concept relation discovery and innovation enabling technology (CORDIET). In: CEUR Workshop proceedings, Concept Discovery in Unstructured Data, vol. 757 (2011)
5. Neznanov A.A., Ilvovsky D.A., Kuznetsov S.O.: FCART: A new FCA-based system for data analysis and knowledge discovery In: Contributions to the 11th International Conference on Formal Concept Analysis, pp. 31–44 (2013)
6. Cubist Project: (http://www.cubist-project.eu)
7. Kaytoue, M., Marcuola, F., Napoli, A., Szathmary, L., Villerd, J.: The coron system. In: Proceeding of the 8th International Conference on Formal Concept Analysis (ICFCA 2010), pp. 55–58 (2010)
8. Yevtushenko, S.A.: System of data analysis "Concept Explorer". In: Proceedings of the 7th national conference on Artificial Intelligence KII-2000, p. 127–134, Russia, (2000) (In Russian).
9. Berners-Lee, T., Fielding, R., Masinter, L.: RFC-3986: Uniform Resource Identifier (URI): Generic Syntax. (http://www.ietf.org/rfc/rfc3986.txt)

10. Hunter II, T.: Consumer-Centric API Design. (https://thomashunter.name/consumer-centric-api-design)
11. The OAuth 2.0 Authorization Framework (RFC-6749): (https://tools.ietf.org/html/rfc6749)
12. Neznanov, A., Ilvovsky, D., Parinov, A.: Advancing FCA workflow in FCART system for knowledge discovery in quantitative data. In: 2nd International Conference on Information Technology and Quantitative Management (ITQM-2014), Procedia Computer Science, vol. 31, pp. 201–210 (2014)
13. LiveJournal: (http://livejournal.com)

Part VII
Machine Learning

Self-Optimizing A Multi-Agent Scheduling System: A Racing Based Approach

Ivo Pereira and Ana Madureira

Abstract Current technological and market challenges increase the need for development of intelligent systems to support decision making, allowing managers to concentrate on high-level tasks while improving decision response and effectiveness. A Racing based learning module is proposed to increase the effectiveness and efficiency of a Multi-Agent System used to model the decision-making process on scheduling problems. A computational study is put forward showing that the proposed Racing learning module is an important enhancement to the developed Multi-Agent Scheduling System since it can provide more effective and efficient recommendations in most cases.

1 Introduction

In the scope of manufacturing systems, scheduling problems are generally complex, large scale, constrained, and multi-objective in nature. Thus, classical operational research techniques are often inadequate to effectively solve them [1]. As such, in this work, different techniques are combined and integrated in order to solve real-world scheduling problems [11].

Due to the complexity and size of this problem, a Multi-Agent System (MAS) is used to model the scheduling environment, so that each job or resource is represented by a single entity (agent). Job and resource agents communicate between them. In this context, MAS emphasize the common behaviors of agents, with some degree of autonomy, and the complexity arising from their interactions. MAS are

I. Pereira (✉) · A. Madureira
School of Engineering from Polytechnic of Porto (ISEP/IPP),
GECAD - Knowledge Engineering and Decision Support Research Center,
Porto, Portugal
e-mail: iaspe@isep.ipp.pt

A. Madureira
e-mail: amd@isep.ipp.pt

© Springer International Publishing Switzerland 2016   275
P. Novais et al. (eds.), *Intelligent Distributed Computing IX*,
Studies in Computational Intelligence 616,
DOI 10.1007/978-3-319-25017-5_26

concerned with the coordination of behaviors of a certain community of agents, in order to share knowledge, capacities, and objectives to solve complex problems [10].

Since scheduling is a NP-complete problem, Metaheuristics (MH) are used to obtain near-optimal solutions in an efficient way, obtaining the optimal solution is not practical, particularly for real world problems. Furthermore, MH have different categorical and/or numeric parameters which makes the tuning procedure a very time-consuming and tedious task. To aid and release the user from this task, a Racing based learning module is proposed.

A brute-force approach is clearly not the best solution to the problem of tuning parameters [3]. A more refined and efficient way can be obtained by streamlining the assessment of candidates and by the release of those who appear to be less promising during the evaluation process. This statement summarizes the Racing approach applied to the problem of tuning parameters on optimization techniques.

The remaining sections of the paper are organized as follows: Sect. 2 presents the manufacturing scheduling problem. In Sect. 3, the Racing architecture is presented, with details about the Multi-Agent Scheduling System and Racing module. The obtained computational results presented in Sect. 4. In Sect. 5 some conclusions are reached.

2 Manufacturing Scheduling Problem

Scheduling problems arise in a diverse set of domains, ranging from manufacturing to transports, hospitals settings, computer, space environments, amongst others. Most of these problems are characterized by a great amount of uncertainty that leads to significant dynamism in the system. Such dynamic scheduling is getting increased attention between researchers and practitioners [1, 11].

The scheduling problem could be stated as: given a set of jobs (composed by a set of operations), a set of resources, and an optimization criteria for performance measure, what is the best way to allocate the resources to the operations considering that precedence relations and the resource availabilities are respected and performance is optimized (maximize or minimized)?

The scheduling problem treated in this work was discussed by Madureira [7], referred as Extended Job-Shop Scheduling Problem (EJSSP), and has some major extensions and differences compared to the classic Job-Shop Scheduling Problem (JSSP), in order to better represent real world problems. JSSP has a set of tasks processing in a set of machines. Each task has an ordered list of operations, each one characterized by the respective processing time and machine where is processed.

EJSSP problems consist in JSSP problems with additional constraints, incorporating scheduling real-world complexity, namely: different release and due dates for each job; different priorities for each job; the possibility that not every machines are used for all jobs; a job can have more than one operation being processed in the same machine; two or more operations of the same job can be processed simultaneously; the possibility of existence of alternative machines, identical or not.

Scheduling problem is included in the NP-complete combinatorial optimization problems class [1]. Their solving methods can be categorized in exact and approximation algorithms [6, 11]. In the former case, an exhaustive solutions space search is made and it is ensured an optimal solution. However these methods are very time consuming. On the other hand, approximation algorithms, including heuristics and MH, do not guarantee the optimal solution but intends to find a satisfactory solution in an acceptable period of time. For this reason, they are used in this work integrated with MAS.

3 F-Race Module Based Architecture

This paper was motivated from many years of research in Scheduling, MH and MAS. A Multi-Agent Scheduling System was proposed in [8, 9] where different Resource agents apply MH to solve the scheduling problem. However, in order to achieve good results, it is necessary to define the parameters for those techniques, which is a very time-consuming task. An idea of using learning techniques came through our mind in order to automate the parameters' tuning and then a Racing module is proposed in this work.

3.1 Multi-Agent Scheduling System

Multi-Agent Systems and MH have been recently used to address and solve different scheduling problems, e.g. [6, 14, 15]. For this reason, in this work, MAS are defined to model a real-time scheduling system, with the objective to use MH for solving the scheduling problem subject to perturbations. The Multi-agent Scheduling System is composed by three kinds of agents (Fig. 1).

Briefly, there are agents representing jobs (or tasks) and agents representing resources (or machines). The system is able to: find optimal or near optimal solutions through MH; deal with dynamism (arrivals of new jobs, cancelled jobs, changing jobs attributes, etc.); switch from one technique to another; change/adapt the parameters of the algorithm according to the current situation [7].

To perform a consistent communication with the user, a User Interface Agent (UI Agent) was implemented. This agent is responsible for the definition of the graphical interface and also for the dynamic generation of the necessary Job and Resource agents, according to the number of jobs and machines (resources) comprising the scheduling problem. This agent is also responsible to assign each task to the respective Job Agent. In the end of the process, it receives the solutions generated by the different Resource agents and applies a repair mechanism in order to check and ensure the feasibility of the final scheduling plan. Job agents process the necessary information about the respective job. They are responsible for the generation of the earliest and latest processing times on the respective job and

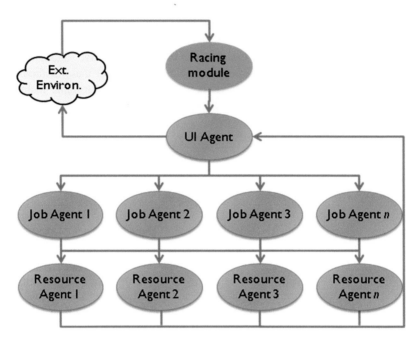

Fig. 1 Multi-agent scheduling system architecture [12]

automatically separate each job's operation for the respective resource. Finally, Resource agents are responsible for scheduling the operations that require processing in the machine supervised by the agent. These agents apply MH in order to find the best possible single-machine schedules/plans of operations and communicate those solutions to the UI Agent.

In order to perform MH self-parameterization used by Resource agents, a Racing module is proposed and described in the next section. The proposed module is embedded on the Multi-Agent Scheduling System architecture.

3.2 Racing Module Proposal

The Racing module described in this section comes from an adaptation of the F-Race method [4]. The impossibility of making a direct implementation of the algorithm for elimination of candidates, since a sufficient number of scheduling instances does not exist, has led to the need of adapting and implementing a solution where the overall operation was similar to the Racing generic algorithm.

The aim of the Racing module undergoes a study of different combinations of MH parameters in optimizing different objective functions (Table 1). Thus, the input parameters refer to the list of objective functions to optimize and the list of MH to validate.

Table 1 Racing algorithm

Step	Description
Step 1	Get problem instances and candidate parameters
Step 2	Create race
Step 4	For each instance For each candidate Create and execute run Remove worst candidates
Step 5	Get best candidate

Traversing the objective functions, the first step involves obtaining a list of candidate parameters of each MH to make a race between them. These races are held in a number of instances. For each instance, each combination of parameters is tested in the system and the data are stored in the database. At each instance, the candidates who obtained the worst results are eliminated. In the end, the best candidate is the one who was able to survive through the process.

The most important of this whole process is the remove candidates' algorithm, described in Table 2, since it is this that decides which candidates will be eliminated from the race. This algorithm takes as input parameters the current race and the list of candidates' parameters, returning the list of survivors' parameters. The first test to perform is check if the list of candidates has more than one combination; otherwise we are looking at the best case. Then we need to check which statistical test to be used. If the candidate list has more than two elements, apply the Friedman test [5]. Otherwise, it uses the Wilcoxon test [5].

Friedman's test requires that there are at least two blocks of results, i.e. it is necessary that all candidates have been carried out in at least two instances. This implies that at the end of the first instance all candidates survive this method. If it is possible to apply the Friedman's test, we should first obtain the ordered rankings of the candidate parameters, and then calculate the sum of each ranking for each candidate. The number of survivors will be the best n (Eq. 1).

$$sobrev = round\left(\frac{1}{\log_2(inst + 1)} \times cand\right) \qquad (1)$$

where $inst$ is the number of instances and $cand$ is the number of candidates

The number of survivors at each step is dependent on the number of instances already executed and the number of candidates. Moreover, it is calculated according to the inverse base-2 logarithm, which gives a similar behavior to the F-race.

Table 2 Remove candidates algorithm

Step	Description
Step 1	Get race instances
Step 2	If more than two candidates, applies Friedman test If only two candidates, applies Wilcoxon test

With this function, it is possible for even a small number of instances, to obtain a quick convergence to the best candidate, unlike what happens with the original F-Race [12].

4   Computational Study

In this section, the set of computational experiments is presented and described, in order to validate the proposed learning mechanism. It is presented the computational results obtained by the framework of this work. Problem instances of academic scheduling taken from the OR-Library [2] have been used.

All results correspond to minimizing makespan, also known as completion time (Cmax). Each instance was run 5 times and the average of the obtained values was calculated. In order to normalize the values, the ratio between the optimal value and the average value of Cmax (Eq. 2) was used to estimate the deviation of the value obtained from the value of the optimal solution.

$$q = 1 - \frac{OptC_{max}}{C_{max}} \tag{2}$$

The objective of the racing based approach is to obtain a suitable parameter for MH from an initial set of parameter combinations. The startup Racing study is to create sets of parameter combinations for each MH, in order to evolve over the training instances and thus obtain the most appropriate parameterization for the technique in question.

Following this methodology, we identify 25 instances for the study, five each dimension is the number of tasks performed. As this work is considered 6 MH, order went through 30 pairs set {MH; dimension}. Since each MH is associated with a different number of parameters (resulting in different numbers of combinations of parameters in each race). It is considered a total of 375 runs per dimension number of tasks which, by multiplying 5 dimensions (10, 15, 20, 30, and 50 tasks), giving a total of 1875 runs at the end of the study. It was possible to identify different parameters from those that gave rise to the previous results.

After obtaining the best combination of parameters for each pair {MH; number of tasks}, could be drawn from the results for test instances, in order to analyze the performance-based approach and Racing conclude as to whether or not a significant advantage compared to the results previously obtained. We intend to compare the results obtained by the Racing module with the previous results in order to obtain information about the advantage of using the learning module based on Racing.

The chart in Fig. 2 allows us to analyze the median and dispersion of data summarized by MH with and without Racing. Table 3 presents the very minimum and maximum values, the mean and standard deviation of such data. From the statistical analysis of the results obtained, it is possible to verify advantage of Racing in use, mainly in Genetic Algorithms and Simulated Annealing, to analyze mean values.

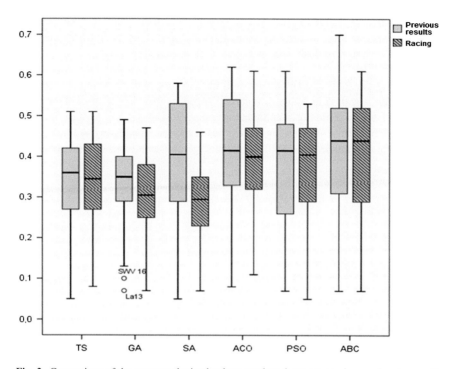

Fig. 2 Comparison of the average obtained values quotient, between previous and racing results

Table 3 Descriptive analysis of average obtained values quotient, between previous and Racing approaches

MH	Results	Min	Max	Avg	Std dev
TS	Previous	0.05	0.51	0.3365	0.12498
	Racing	0.08	0.51	0.3334	0.12226
GA	Previous	0.07	0.49	0.3327	0.10754
	Racing	0.07	0.47	0.3025	0.10654
SA	Previous	0.05	0.58	0.3885	0.15371
	Racing	0.07	0.46	0.2903	0.10556
ACO	Previous	0.08	0.62	0.4238	0.13956
	Racing	0.11	0.61	0.3892	0.11597
PSO	Previous	0.07	0.61	0.3759	0.15305
	Racing	0.05	0.53	0.3669	0.13401
ABC	Previous	0.07	0.70	0.4176	0.16329
	Racing	0.07	0.61	0.4019	0.15480

Analyzing the dispersion of results, it is possible to notice an improvement in the standard deviation in every technique. It is thus possible to state that the parameters resulting from the study of the racing improved results compared with the previous results related to AutoDynAgents system without learning [12, 13].

In order to be possible to analyze the results by Student's t test, the samples are normalized to be able to directly compare the approaches in a comprehensive manner. This normalization was performed by calculating the ratio of average values (Fig. 3).

Analyzing the statistical significance of these results, and observing the values t $(29) = 4.740$, $p < 0.05$ in Table 4, it can be stated, with a degree of confidence of 95 %, there are statistically significant differences between the results obtained initially and the results obtained by Racing based approach, allowing us to conclude the existence of statistically significant advantage of Racing based approach.

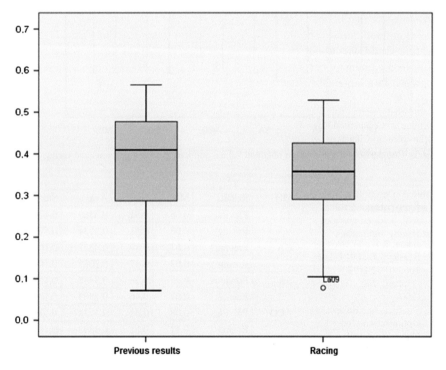

Fig. 3 Comparison of the average obtained values quotient, between previous and Racing results

Table 4 Student's t test results for previous versus Racing approaches		Avg	Std dev	t	DoF	p
	Previous versus Racing	0.03182	0.03677	4.740	29	0.00

5 Conclusions and Future Work

A Multi-Agent Scheduling System was presented in order to assist the user in the decision making process, by suggesting a scheduling plan close to the optimum plan.

A Racing based module was proposed in order to increase the effectiveness and efficiency of the decision making, by providing the scheduling system with the capability of MH parameter tuning. This is essentially the most valuable contribution of this work.

The results showed that the proposed Racing module was a great improvement to the system performance since it can now provide recommendations in a more efficient and effective way. The average deviations were improved in both best and medium values, and the computational time was significantly reduced since the system became more intelligent when choosing a MH (in an effective and efficient way).

Future work includes more research in learning approaches (e.g. Reinforcement Learning and Case-based Reasoning) to improve the performance, as well as a more extensive computational study, to see the influence of a Racing or a hybrid approach in a larger number of optimization problems.

Acknowledgments This work is supported by FEDER Funds through the "Programa Operacional Factores de Competitividade—COMPETE" program and by National Funds through FCT "Fundação para a Ciência e a Tecnologia" under the projects: PEst-OE/EEI/UI0760/2014 and PTDC/EME-GIN/109956/2009.

References

1. Baker, K.R., Trietsch, D.: Principles of Sequencing and Scheduling. Wiley, New York (2009)
2. Beasley J.E.: OR-Library: distributing test problems by electronic mail. J. Oper. Res. Soc. **41**, 1069–1072 (1990)
3. Birattari, M.: Tuning Metaheuristics: A Machine Learning Perspective. Springer, Berlin (2009)
4. Birattari, M., Balaprakash, P., Dorigo, M.: The ACO/F-RACE Algorithm for Combinatorial Optimization Under Uncertainty Metaheuristics, pp. 189–203, Springer, Berlin (2007)
5. Conover, W.J.: Practical Nonparametric Statistics: Wiley, New York (1999)
6. Lau, H.C., Zhao, Z.J., Ge, S.S., Lee, T.-H.: Allocating resources in multiagent flowshops with adaptive auctions. Autom. Sci. Eng. **8**(4) (2011)
7. Madureira, A.: Aplicação de Meta-Heurísticas ao Problema de Escalonamento em Ambiente Dinâmico de Produção Discreta. Ph.D. thesis, Universidade do Minho (2003)
8. Madureira, A., Santos, J., Pereira, I.: MASDSheGATS—scheduling system for dynamic manufacturing environments. In: Ahmed S., Karsiti, M.N. (eds.) Multi Agent Systems.Chapter 17. In-Tech, pp. 333–342. Vienna, Austria. ISBN: 978-3-902613-51-6, (2009a). doi:10.5772/6609
9. Madureira, A., Santos, J., Pereira, I.: A hybrid intelligent system for distributed dynamic scheduling. In: Chiong, R., Dhakal, S. (eds.) Natural Intelligence for Scheduling, Planning and Packing Problems, vol. 250 of Studies in Computational Intelligence, pp. 295–324.

Springer-Verlag, Berlin, Heidelberg. ISBN: 978-3-642-04038-2, (2009b). doi:10.1007/978-3-642-04039-9_12

10. Pan, Q., Wang, L., Mao, K., Zhao, J.-H., Zhang, M.: An effective artificial bee colony algorithm for a real-world hybrid flowshop problem in steelmaking process. Autom. Sci. Eng. **10**(2), 307–322 (2013)

11. Pinedo, M.: Scheduling: Theory, Algorithms, and Systems. Springer, New York (2012)

12. Pereira, I.: Sistema Inteligente para Escalonamento Assistido por Aprendizagem. Ph.D. thesis in Electrical and Computer Engineering. UTAD (2014) (in portuguese)

13. Pereira, I., Madureira, A.: Self-optimization module for scheduling using case-based reasoning. Appl. Soft Comput. **13**(3), 1419–1432 (2013)

14. Vincent, L., Ponnambalam, S.G.: A differential evolution-based algorithm to schedule flexible assembly lines. Autom. Sci. Eng. **10**(4) (2013)

15. Yang, S., Zhang, M.T., Yi, J., Zhang, L., Zheng, L.: Bottleneck station scheduling in semiconductor assembly and test manufacturing using ant colony optimization. Autom. Sci. Eng. **4**(4), 569–578 (2007)

Collaborative Filtering with Hybrid Clustering Integrated Method to Address New-Item Cold-Start Problem

Ferdaous Hdioud, Bouchra Frikh, Asmaa Benghabrit
and Brahim Ouhbi

Abstract Recommender Systems (RSs) are a valuable and practical tool to cope with information overload, as they help users to find interesting products in a large space of possible options. The Collaborative Filtering (CF) approach is probably the most used technique in RSs field due to several advantages as the ease of implementation, accuracy and diversity of recommendations. Despite being much favored over Content-Based (CB) techniques, it suffers from a major problem related to the lack of sufficient data for new-item cold-start problem, which affects recommendations' quality. This paper is focused on resolving issues related to item-side in order to produce effective recommendations. To overcome the above problem, we use a powerful content clustering based on Hybrid Features Selection Method (HFSM), to get the maximum profit from the content. Then, it will be combined side by side to CF under a hybrid RS to improve its performance and handle new-item issue. We evaluate the proposed algorithm experimentally either in no cold-start situation or in a simulation of a new-item cold-start scenario. The conducted experiments show the ability of our hybrid recommender to deliver more accurate predictions for any item and its outperformance on the classical CF approach, which doesn't work as usual especially in cold-start situations.

F. Hdioud (✉)
LTTI Lab, Ecole Supérieure de Technologie, B.P. 2427 Route D'imouzer, Fès, Morocco
e-mail: ferdaous.hdioud@usmba.ac.ma

B. Frikh · A. Benghabrit · B. Ouhbi
LM2I Lab, ENSAM, Moulay Ismaïl University, Marjane II, B.P. 4024 Meknès, Morocco
e-mail: bfrikh@yahoo.com

A. Benghabrit
e-mail: a.benghabrit@gmail.com

B. Ouhbi
e-mail: ouhbib@yahoo.co.uk

1 Introduction

In the last years, the decision-making became more and more difficult with the overwhelming amount of online-information. Therefore, tools are needed to assist users to find what they want quickly and easily. The appearance of Recommender Systems (RSs) tackled this problem, as they facilitate prioritizing information. Thanks to them, the searching time is reduced and the needed or favored information is reached as soon as possible. According to the technique used, RSs can be broadly categorized into four main types [1]: Content-Based (CB) filtering, Collaborative Filtering (CF), Hybrid recommender systems and knowledge-based. The Content-Based RSs are the first conceived, their underlying assumption is using items' *content* and user's profile which contains his past preferences to suggest him new items that are similar to the ones he liked before. Later, these approaches were succeeded by the emergence of CF ones, which are content-agnostic, i.e. they don't exploit any type of item's content in recommendation process. In order to identify the similar users and produce high-quality recommendations, a large amount of opinions (ratings) is required, which presents a big shortcoming. Despite the success of CF approach, it suffers from many serious problems related to data sparsity that affect recommendations' accuracy. The first one is the long tail problem, which concerns items with only few ratings [2]. The second is the cold-start problem [3, 4] which has two variants: (1) new-user when an unknown user joins the system looking forward recommendations [4, 5] and (2) *new-item* (called also *first rater* problem) in which the RS is unable to deliver *new items* without any rating [6]. The new-item problem is a special case of the long tail problem; this paper is focused on resolving issues related to item-side in order to produce effective recommendations. In the CF techniques, the prediction computation is based primarily on the concept of similarity between either users or items; also the quality of recommendation is mainly based on the user ratings, instead of the *content* information, which is completely denied. A *new item,* which is not yet rated, presents a big deal, computing its prediction seems impossible. So, it cannot be recommended to anyone thereafter. However, the highlight of CB systems is that *new items* can be immediately recommended once their attributes are available, then the aforementioned problem is not placed. On the other hand, they achieve a low accuracy compared to CF and rarely used in practical cases [7]. One common thread in RSs research is the need to combine several recommendation techniques to achieve a high-level performance, such systems are called hybrid RSs [8–12]. In this paper we aim to benefit from the *content* to bridge the gap between the *new items* and those already existing by building relationships among them. By doing so, we would be able to overcome the new-item problem. Since, the lack of sufficient ratings will be offset by computing similarities between items based on their whole raw content instead of their ratings.

The rest of this paper is organized as follows: The second section presents the related works while the third section introduces a powerful items' clustering tool based on a Hybrid Features Selection Method (HFSM) and proposes a linear hybrid RS rested on the proposed content clustering algorithm and classical CF approach. The fourth section is devoted to the experimental evaluation. Finally, we offer concluding remarks and summarize the future directions of our research work.

2 Related Works

Due to the limitations of both CF and CB recommender systems, there is few research papers opted to combine many techniques in different ways within a single RS to enhance recommendation quality and tackle some issues. In this section, we describe some papers that use techniques combining content and collaborative approaches to handle cold-start scenario. To combine especially CB filtering and CF to tackle the new item problem, different ways have been proposed and developed. Here we focus only on those using clustering methods. In [3], a predictive feature-based regression model was proposed to tackle cold-start problem; the model exploits all available information of users and items (as user demographic information and item content features). The work conducted in [13], uses clustering method to generate content similarities between items and combined with CF to alleviate the cold-start problem. Later, [14] proposed an extensive hybrid RS using content clustering for both of items and users for the same aim. An Expectation Maximization (EM) clustering method was used in [2] for all items in order to solve the long tail problem. The paper [15] focused on improving the accuracy of RS and minimizing resource consumption by ensuring a dynamic item's clustering. In [16], an enhanced *content* clustering was used to resolve cold-start problem in user-side.

After reviewing these works above, we may notice that they rely only on standard clustering methods such as k-means that is widely used due to its ease and rapidity. These simple clustering techniques are considered as relevant categorization algorithms, when items are characterized by simple enumerated attributes. Therefore it may work successfully in the context of CF (even if items are presented only by their ratings) [16]. Our work surpasses the above approaches since it uses a feature selection method to exploit items' content to detect items correlations accurately. Moreover it introduces semantic as well as statistical aspect feature selection to improve sufficiently similarities' computation. As a consequence, our algorithm operates on a wide content of items using the most relevant content information, this latter will be exploited to detect correlations between items statistically and semantically. By doing so, we generate high quality recommendations and tackle the new item problem.

3   The Proposed System

Our proposed system consists of three modules. The first is content module which ensures the items clustering with Hybrid Features Selection Method (HFSM). The second is the pure item-based CF module depending only on ratings. Both of the two first modules generate an item-to-item matrix similarity. Finally, the third module combines linearly the two generated similarity matrices, which compensate the sparsity of CF similarity matrix by the other based on content clustering. The resulting matrix will be used for the prediction and the recommendation steps. The main operational aspects are depicted in Fig. 1.

3.1   Content-Clustering Module

The number and type of items' attributes need to be correctly described differently from an item to another. Consequently, it is not wise to describe all items by the same small number of attributes with known set of values. To deal with this issue, we use a clustering algorithm that integrated a Hybrid Feature Selection Method (HFSM). Hence, the items are properly represented by using a method that simultaneously selects statistical and semantic relevant features. This makes the similarity between items more accurate, which helps the clustering process better identifying related content items. Once the clustering is performed; similarities between each pair of items are extracted to build the content similarity matrix.

3.1.1   The Hybrid Features Selection Method (HFSM)

The textual description for items is collected from different information sources and stored in large documents. Then to present the documents for the clustering process,

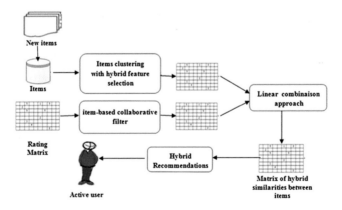

Fig. 1 Overview of our proposed hybrid approach

we use the Vector Space Model. In order to employ it, we first performed a preprocessing step in order to switch into a feature space. Then, the weight of each feature is calculated using the well-known numerical statistic TF-IDF:

$$X_i = \left[tf_1 \log\left(\frac{n}{df_1}\right), tf_2 \log\left(\frac{n}{df_2}\right), \ldots, tf_D \log\left(\frac{n}{df_D}\right) \right]^t \quad (1)$$

where tf_j is the frequency of the term j in the item i, df_j is the number of items that contain the term j, n is the total number of items in the collection, and D is the number of terms.

The Hybrid Feature Selection Method that we proposed in [17] simultaneously selects the most frequent contents by the CHIR statistic [18] and the most pertinent content-based ones by the SIM measure through a weighting model. Hence, the term goodness of a feature w is defined by the following formula:

$$HFSM(w) = \lambda * r\chi^2(w) + (1 - \lambda) * sim(w) \quad (2)$$

where λ is a weighting parameter between 0 and 1. A term w is considered relevant when the value of its $HFSM$ measure is greater than a predefined threshold.

3.1.2 Clustering with Hybrid Features Selection Algorithm (CHFSA)

The CHFSA algorithm [17] alternates between documents clustering and the selection of relevant features that describe the documents. Therefore, the clustering precision is iteratively improved by the selection of relevant features until obtaining a high clustering accuracy at the stability of the process. Detailed algorithm's steps are presented in Fig. 2. To determine the closeness between two items, we used the cosine similarity.

3.2 Collaborative Filtering Module

This module needs to access to the original ratings' matrix and compute similarities between items. As we had already mentioned generating predictions is related heavily on the principle of correlation between either users (in user-based approach) or items (item-based approach). The whole data set must be represented in a suitable format to make the distances' calculation easy later. When it comes to the item-based approach, items must be represented in the users' space, where each item is m-dimensions vector and each dimension is the rating value given by a user to this item. Thereafter, a distance between two items is calculated relatively to ratings given to the both by common users according to a distance's formula. There are several formulas that can be used. We used the Pearson coefficient which is more accurate:

CHFSA Algorithm:

Input: A set S of n items to be clustered
              m: number of distinct terms existing in S
              k: number of clusters
              f: factor in the range of [0,1]
              l: number of selected features
Output: Set of item clusters
              Set of cluster centroids
1: perform the k-means algorithm to get initial clusters and centroids
2: **repeat**
3: **for each** of the m features
4:     calculate the hybrid measure HFSM
5: **end for**
6: rank the terms in a descending order of their criterion function
7: select the top l features from the sorted list
8: **for each** item in S
9:       **for each** feature
10   **if** it is an unselected feature
11:                             reduce its weight by f
12:               **end if**
13:     **end for**
14: **end for**
15: **for each** cluster
16:     recalculate its centroid based on the new weights of the features
17:**end for**
18:**for each** item in the new feature space
19:       **for each** of the k centroids
20:           compute the cosine measure between them
21:       **end for**
22:       assign the item to the cluster that has the closest centroid
23:**end for**
24:**for each** cluster
25:     recalculate its centroid based on the items assigned to it
26: **end for**
27: **until** the centroids no longer move.

Fig. 2 Clustering with hybrid features selection algorithm (CHFSA)

$$PC(i,j) = \frac{\sum_u (r_{ui} - \bar{r}_i)(r_{uj} - \bar{r}_j)}{\sqrt{\sum_u (r_{ui} - \bar{r}_u)^2 \sum_u (r_{uj} - \bar{r}_j)^2}} \qquad (3)$$

– r_{ui}: Is the rating of the item i given by the user u.
– $\bar{r}_i$: Is the mean rating of the item i.

3.3 Hybrid Module

Operations performed by the previous two modules were carried out in parallel. While this module is the core of proposed RS, which is responsible on accessing to the two matrixes generated previously and combine them into a single new complete matrix similarity. Then, predictions of non-voted items are computed. For the hybrid module, two different scenarios are possible. Each one must be treated differently:

- An old item with only few ratings: in this case the number of ratings is insufficient to make accurate recommendations. Then, this module shall compensate this lack through the use of items' content.
- A new item with no ratings: in this case any prediction can't be rested on ratings. The module would be based *only* on *content*.

Therefore, it consists on alternating between content and collaborative information by making certain balance and achieving an acceptable recommendation according to the situation. The total similarity between two items is the combination of the similarities obtained by the two previous modules as follows:

$$Sim_{hybrid} = C.Sim_{content} + (1-C).Sim_{CF} \qquad (4)$$

where $C \in [0, 1]$ is the coefficient of combination, it defines the contribution of each component in the prediction step. We notice that the value of C is variable; it is to be changed according to different situations. If the CF module lacks of sufficient ratings to compute reliable similarities, the content module is valued and C tends to be more important and vise-versa.

Predicting the possible rating's value that a user would give to an item summarizes the prediction's step. The prediction's value of an item i according to a user u is computed as follows:

$$prediction_{ui} = \bar{r}_i + \frac{\sum_{j=1}^{l} (r_{uj} - \bar{r}_j) * sim_{ij}}{\sum_{j=1}^{l} |sim_{ij}|} \qquad (5)$$

- l is the size of the neighborhood of the item i
- sim_{ij} is the hybrid similarity between the item i and his neighbor j as in (4).
- $\bar{r}_i$ and $\bar{r}_j$ are respectively the mean' ratings of the items i and j
- r_{uj} is the rating of the item j given by the user u

When a new item appears in the system, Eq. (5) becomes invalid as the item is unrated by any user (the value $\bar{r}_i$ can't be computed). In this case, the CHFSA is applied solely in order to affect the new item to its closest cluster. Therefore, the prediction is:

$$prediction_{ui} = \frac{\sum_{j \in C} r_{uj} * sim_{ij}}{\sum_{j \in C} |sim_{ij}|} \tag{6}$$

4 Experimental Evaluation

We perform experiments on movie rating data collected from the Movielens[1] web-based RS. The data set contained 100,000 ratings from 943 users of 1,682 movies; each user in the dataset has rated at least 20 items, where rating's value is in 1-to-5 scale. Items are characterized by their id, title and their genre (drama, crime, etc.). An exhaustive textual content of each item is required in order to conduct the CHFSA explained previously. For this reason, we got a detailed synopsis from Wikipedia.[2] For the content module, we picked randomly 20 % of the dataset's items to be considered as new items. The rest of items (the remaining 80 %) will be employed to learn our clustering algorithm. The dataset containing collaborative data (ratings) is divided into a training set and a test set. The 80 % of the data is used as the training set and the rest 20 % is used as the test set.

4.1 Evaluation Metrics

To evaluate the accuracy of a RS, we use: the MAE (Mean Absolute Error):

$$MAE = \frac{1}{|R_{test}|} \sum_{r_{ui} \in R_{test}} |prediction_{ui} - r_{ui}|, \tag{7}$$

and the coverage:

$$Coverage = \frac{n_{pi}}{n_i} \tag{8}$$

where: $prediction_{ui}$: is the prediction and r_{ui}: is the real value's rating, n_i is the number of the items for which we must generate prediction and n_{pi} is the number of items whose prediction was generated by the recommender system.

---

[1]http://grouplens.org/datasets/movielens/.
[2]http://www.wikipedia.org/.

4.2 Experimental Results

Experiment 1: First, we aim to compare the results of the proposed clustering algorithm against baseline item-based CF approach. We evaluate our hybrid approach with borderline weighting cases (C equals respectively to 0 and 1 meaning either the content based on CHFSA or the CF), given different neighborhood sizes. The obtained results are reported in the Fig. 3. We observe that the pure content algorithm outperform the CF and show good results in term of accuracy (MAE don't exceed 0.1), unlike to what is known about classical content approaches that ensure a low accuracy. The high coverage shows the algorithm's ability to compute the prediction for any item regardless of number of ratings, due to the complete matrix similarity generated by the content clustering based on HFSA.

These initial results prove the power of our proposed algorithm to compute accurate similarities between items resting only on a large content which leads to a high quality prediction. Since these similarities are computed once the item's content is present, it allows constructing the fullest similarity matrix as possible, which solves the sparsity problem. It is also proven by a very high coverage. On the other hand, CF doesn't perform well. In most cases it seems unable to calculate prediction, or they are less accurate. This is due to the sparsity of matrix (94 % sparse), as in CF the most we have ratings the most accurate recommendations are generated.

Experiment 2: As the proposed Hybrid RS based on CHFSA is conceived to improve the performance of RS under new-item cold-start situations, we simulate a new-item scenario by keeping only few ratings K per item (K = 2, 5, 10, 30). Then, we conduct several experiments by changing the value of Combination Coefficient C to find out the relation between the number of ratings per item and the contribution of each of the two components (our proposed CB recommendation using CHFSA and the classical item-based CF). The obtained results are presented in Fig. 4; The Fig. 4a shows an extreme new-item cold-start situation, in which the number of ratings K is very small. This proves that CF can't perform better in such cases. On the other side, along with the increasing value of C, the MAE drops until

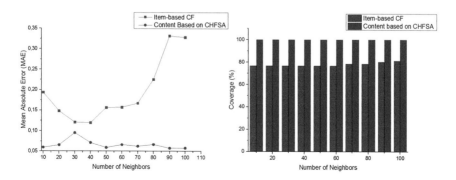

Fig. 3 MAE and Coverage comparison between Item-based CF and content based CHFSA

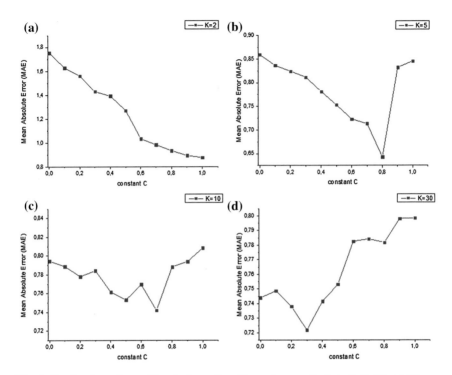

Fig. 4 Simulated new item with only few ratings K equals respectively to 2, 5, 10 and 30

arriving to a lowest value when C = 1 (our content algorithm is dominant). The Fig. 4b shows a different shape than the first one, the lowest value of MAE is reached in C = 0.8. Then, the CF tends to take more importance when the number of ratings K has increased. The Fig. 4c shows a little change compared to Fig. 4b. Even if the number of ratings K has increased, the contribution of content based on CHFSA still emphasized at the expense of CF, the best value of MAE is reached when C = 0.7. The value K = 10 remains negligible compared to the total number of users and items. The last curve in Fig. 4d, emphasizes extensively the importance of CF as the best MAE value is obtained at C = 0.3. Generally, the change in shape over the four curves confirms that: the more items lack ratings, the more our CHFSA Algorithm's contribution is bigger.

5 Conclusion

Incorporating content into CF recommenders, improve significantly the accuracy of the system as it overcomes cold-start situations when we lack of sufficient ratings. The Hybrid Features Selection Algorithm (CHFSA) introduces semantic and statistical dependencies of items to compute similarities between them, which maximizes the profit obtained from their content and upgrade RS's performance. The proposed system overcomes the weaknesses of content and CF approaches, by consolidating one through the other. The conceived RS has for main aim tackling cold-start related to item-side; we conducted many experiments, which simulate new-item situations in which the system worked great and outperformed baseline methods especially in extreme cold-items with no ratings. The proposed solution can be adopted in the context of many recommenders such as: news, web pages, books, articles, and so on, in which items have important content, allowing detecting statistical and semantic connections.

References

1. Adomavicius, G., Tuzhilin, A.: Toward the next generation of recommender systems: a survey of the state-of-the-art and possible extensions. IEEE Trans. Knowl. Data Eng. **17**(6), 734–749 (2005)
2. Park, Y.-J., Tuzhilin, A.: The long tail of recommender systems and how to leverage it. In: Proceedings of the ACM Conference on Recommender Systems, pp. 11–18, (2008)
3. Seung-Taek, P., Wei, C.: Pairwise preference regression for cold-start recommendation. In: Proceedings of the Third ACM conference on Recommender systems, ACM, pp. 21–28 (2009)
4. Hdioud, F., Frikh, B., Ouhbi, B. (2014). Bootstrapping recommender systems based on a multi-criteria decision making approach. In: 2014 Fifth International Conference on Next Generation Networks and Services (NGNS), pp. 209–215. IEEE (2014)
5. Hdioud, F., Frikh, B., Ouhbi, B.: Multi-criteria recommender systems based on multi-attribute decision making. In: Proceedings of International Conference on Information Integration and Web-based Applications and Services, p. 203. ACM (2013)
6. Cremonesi, P., and Turrin, R.: Analysis of cold-start recommendations in IPTV systems. In: Proceedings of the Third ACM Conference on Recommender Systems, pp. 233–236. ACM (2009)
7. Saveski, M., Mantrach, A.: Item cold-start recommendations: learning local collective embeddings. In: Proceedings of the 8th ACM Conference on Recommender Systems. ACM, pp. 89–96 (2014)
8. Jin, R., Luo, S., Chengxiang, Z.: A study of mixture models for collaborative filtering. Inf. Retrieval **9**(3), 357–382 (2006)
9. Prem, M., Raymond, J.M., Ramadass, N.: Content-boosted collaborative filtering for improved recommendations. In: AAAI/IAAI, pp. 187–192 (2002)
10. Basilico, J., Hofmann, T.: A Joint Framework for Collaborative and Content Filtering. In: Proceedings of the 27th Annual International ACM (SIGIR'04), pp. 550–551 (2004)
11. KhanhQuan, T., Ishikawa, F., Honiden, S.: Improving accuracy of recommender system by item clustering. IEICE Trans. Inf. Syst. **90**(9), 1363–1373 (2007)
12. Schein, A.I., Popescul, A.M.: Methods and metrics for cold-start recommendations. In: Proceedings of the 25th Annual International ACM SIGIR Conference (2002)
13. Qing, L.K.: Clustering approach for hybrid recommender system. In: Proceedings of IEEE/WIC International Conference on Web Intelligence WI 2003, IEEE (2003)
14. Sutheera, P., Tsuji, H.: A Multi-Clustering hybrid recommender system. In: CIT 2007, 7th IEEE International Conference on Computer and Information Technology, pp. 223–228 (2007)
15. Wen, J., Wei, Z.: An improved item-based collaborative filtering algorithm based on clustering method. J. Comput. Inf. Syst. **8**(2), 571–578 (2012)

16. Sun, D., Li, C., Luo, Z.: A Content-Enhanced approach for cold-start problem in collaborative filtering. In: Artificial Intelligence, Management Science and Electronic Commerce (AIMSEC). IEEE, pp. 4501–4504 (2011)
17. Benghabrit, A., Ouhbi, B., Frikh, B., Zemmouri, E., Behja, H.: Text Document Clustering with Hybrid Feature Selection. In: Proceedings of International Conference on Information Integration and Web-based Applications and Services, in Vienna, Austria, ACM, pp. 600–605, (2013)
18. Li, Y., Congnan, L., Chung, S.M.: Text clustering with feature selection by using statistical data. IEEE Trans. Knowl. Data Eng. **20**(5), 641–651 (2008)

Applying Reinforcement Learning in Formation Control of Agents

Vali Derhami and Yusef Momeni

Abstract This paper proposes a new Reinforcement Learning (RL) algorithm for formation of agents in regular geometric forms. Due to curse of dimensionality problem, applying RL algorithms in formation problems cannot present suitable performance. Moreover, since the state space in formation problem is large, this leads to long learning time. Here, a multi-agent fuzzy reinforcement learning algorithm is presented that is an extension of fuzzy actor-critic reinforcement learning in a multi-agent environment. The final action for each agent is generated by a zero order T-S fuzzy system. In conventional fuzzy actor-critic RL, there are several candidate actions for consequence of each fuzzy rule and aim of learning is finding the best action among these discrete candidate actions. Here, using the proposed linear interpolation, a continuous action selection for determining the best action for each fuzzy rule is presented. The simulation results show the proposed method can improve the learning speed and action quality.

1 Introduction

Formation control is the cooperating of a group of agents to make a regular shape and maintain the same while moving. Their applications include: search and rescue operation [7], surveying [2] and controlling satellite formation [1]. It should be noted that before moving, agents should first make the formation with definite distances.

In the most previous works, mathematical [10] and traditional control techniques [9] have been applied to control the formation. Formation control by methods based on traditional control or methods combined with fuzzy systems and neural networks have been fully studied [6]. However, there are few papers which have studied this matter through Reinforcement Learning (RL). For example, in [12, 13] RL has been

V. Derhami (✉) · Y. Momeni
School of ECE, Yazd University, Yazd, Iran
e-mail: vderhami@yazd.ac.ir

Y. Momeni
e-mail: yusefmomeni@stu.yazd.ac.ir

© Springer International Publishing Switzerland 2016
P. Novais et al. (eds.), *Intelligent Distributed Computing IX*,
Studies in Computational Intelligence 616,
DOI 10.1007/978-3-319-25017-5_28

297

used to find appropriate behavior, but how the formation is made has not been mentioned. In [13] two different behaviors have been considered for moving, row formation for normal situation and column formation for passing through obstacles. The system finds through Q-learning which of the above-mentioned formations should be chosen in different conditions. In [3], this case has been studied by giving an algorithm and by making use of RL. In [4], RL has been used to create formation in a discrete environment, so that the agents be placed around a table in equal distances. The above methods, due to discretization, have some weaknesses such as low learning rate, curse of dimensionality and low quality performance.

In this paper, using Fuzzy Actor-Critic Reinforcement Learning (FACRL), a new method is proposed to create a formation in a multi-agent environment. Three agents in an environment with a continuous state and action, make an equilateral triangular formation. In classic RL, only cases with finite action space are considered, though in many applications related to the actual world, discretization of action space cannot be so appropriate. Some parts of action space may be more significant than the other parts and in order to obtain a good result, it is necessary that the discretization rate be higher in those parts. Moreover, such important action spaces may not be easy to identify. On the other hand, when the number of discretized actions is increased in an action space, the learning rate is reduced.

To cope with the problems of discrete action selection methods, a solution is given in this paper which searches in the whole action space on a continuous basis to select a proper action for FACRL. One of the major problems in multi-agent RL is curse of dimensionality where, by increasing the number of agents, the memory required for state-action space grows exponentially which decreases the learning rate, considerably. This problem prevents application of RL in actual multi-agent issues, where actions and states are expanded and continuous.

The structure of this paper is as follows: The algorithm of fuzzy actor-critic is introduced in Sect. 2. Then, in Sect. 3 the proposed method for formation is given. In Sect. 4, a method is given for continuous action selection. The result of simulation is in Sect. 5 and finally conclusion is given in Sect. 6.

2 Fuzzy Actor-Critic Reinforcement Learning

Fuzzy actor-critic reinforcement learning is combination of fuzzy system as function approximation and classic actor-critic method. There are two fuzzy systems, one for generation of action (called actor), and other for estimation of value function (called critic). First, describe actor module. Consider a zero-order T-S fuzzy system with n input and one output and R fuzzy rule [5].

$$R_i : \text{if } x_1 \text{ is } L_{i1} \text{ and ... and } x_n \text{ is } L_{in} \text{ then } a_{i1} \text{ with value } w_{i1}$$
$$\text{or} \quad a_{i2} \text{ with value } w_{i2}$$
$$\vdots \qquad \vdots$$
$$\text{or} \quad a_{im} \text{ with value } w_{im} \tag{1}$$

where $s = x_1 \times \cdots \times x_n$ is the vector of n-dimensional input state, $L_i = L_{i1} \times \cdots \times L_{in}$ is the n-dimensional strictly convex and normal fuzzy set of the ith rule with a unique center, m is the number of possible discrete actions for each rule, a_{ij} is the jth candidate action, and weight w_{ij} is the approximated value of the jth action in the ith rule. In every time step for each rule, one of m candidate actions is chosen considering weight of that action. The goal of learning is to update the weights of w_{ij} using reinforcement signal. After that, a greedy policy is implemented. Accordingly, action a_i is a_{ij} with the bigger w_{ij}. In learning stage, any of action selection methods can be applied such as ε-greedy or softmax. The global action is calculated as follows:

$$a(s) = \sum_{i=1}^{R} \mu_i(s) \times a_{ii^+} \tag{2}$$

where $\mu_i(s)$ is the normalized firing strength of the ith rule for state s, and i^+ is the index of the selected action and R is the number of fuzzy rules. For the critic, the same fuzzy inference system as actor is used:

$$\text{if } s \text{ is } S_i \text{ then } v_i \tag{3}$$

this is for ith rule. v_i is the state value for the rule i. Then state value for the state s is obtained:

$$V(s) = \sum_{i=1}^{R} \mu_i(s) \times v_i \tag{4}$$

During learning stage, after action selection of each rule, the global action is calculated due to Eq. 2 and it is applied to the environment, and it goes to state s' and reward r is received. In the new state s', value function $V(s')$ is obtained from Eq. 4, and temporal difference error is obtained from Eq. 5:

$$\bar{\delta} = r + \gamma V(s') - V(s) \tag{5}$$

where $0 < \gamma < 1$ is discount factor. Action weights are updated as:

$$w_{ii^+} = w_{ii^+} + \alpha \times \bar{\delta} \mu_i(s) \tag{6}$$

where α is learning rate and $0 < \alpha < 1$. From Eq. 6 can be seen that the weight of selected action in each rule is updated proportional to its participation in global action. Then state value of each rule v_i, is updated through:

$$v_i = v_i + \beta \times \bar{\delta} \mu_i(s) \tag{7}$$

where β is learning rate and $0 < \beta < 1$.

3 Multi-Agent Fuzzy Reinforcement Learning

In this section, a method is given which is called Multi Agent Fuzzy Reinforcement Learning (MAFRL). One of the important issues in multi-agent reinforcement learning is how to share the reward among agents which is received for their common action. If all agents receive the same reward, this approach is called global reward, but in local reward method every agent receives its reward based solely on its individual behavior [11]. Global reward does not specify which agent has done better or worse behavior. Also, using local reward may cause greedy acts that are not suitable for multi-agent environments which need cooperation of agents. In this paper, actor-critic architecture is applied in condition that the defined reinforcement signal is the same for all the agents and due to the fact that each agent is in a different state, their related critic function will also be different, whereby the agent's policy will be improved. According to Eq. 6 the agent's policy improved with respect to the temporal difference error relating to the state value. This helps that each agent with regard to the rate of its effect, whether positive or negative, on the performance of the whole group, to be able to improve its policy so that there will be no need to be concerned how to share the reward among agents.

4 Linear Interpolation Based Action Selection

In this section, a new method is introduced to select an action in an environment with continuous action space to be used in FACRL. In this method, some candidate actions are considered for each rule, but their values are not fixed and change as will be described, until reaches to its final value. If the interval between the first and the last action is considered permissible for action selection, then for other actions existing in this interval which are not included between candidate actions, some value can be considered with regard to Fig. 1 and a first order linear approximation can be used to calculate its value:

Fig. 1 Continuous action selection for $a_{i,k}$

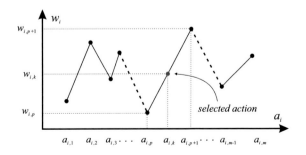

$$w_{i,k} = \left(\frac{w_{i,p-1} - w_{i,p}}{a_{i,p-1} - a_{i,p}} \right) \times (a_{i,k} - a_{i,p}) + w_{i,p} \tag{8}$$

where $w_{i,k}$ weight of the action $a_{i,k}$ and $a_{i,p} \le a_{i,k} \le a_{i,p+1}$ or $a_{i,k} \in [a_{i,p}, a_{i,p+1}]$. In other words k is index of selected action and $p, p+1$ are respectively for before and after selected action. Consequently, $a_{i,k}$ is an action which is located continuously between the two actions $a_{i,p}$ and $a_{i,p+1}$ and its weight is $w_{i,k}$. m is total number of candidate actions in each rule.

The main structure of this action selection method, is based on expansion of soft-max method for continuous environment. When T is high, the possibility for selection of all the points in the given interval is the same and when T decreases, the points with more value, will have more possibility for selection. Possibility of selection $a_{i,j}$ with weight $w_{i,j}$, is obtained by Eq. 9.

$$p(a_i) = \frac{exp(\mu_i w_i(a_i)/T)}{\int_{a_{i,1}}^{a_{i,m}} exp(\mu_i w_i(a_i)/T) da_i} \tag{9}$$

where μ_i is normalized firing strength in ith rule and T is temperature parameter. For action selection, it is necessary to refer to the cumulative distribute function (CDF) relating to Eq. 9. If the structure of the actions and their equivalent weights is considered as in Fig. 1, the CDF can be calculated as follows:

$$CDF = \frac{\int_{a_{i,1}}^{a_{i,p}} exp(\mu_i w_i(a_i)/T) da_i + \int_{a_{i,p}}^{a_{i,k}} exp(\mu_i w_i(a_i)/T) da_i}{\int_{a_{i,1}}^{a_{i,m}} exp(\mu_i w_i(a_i)/T) da_i}$$

$$= \frac{\sum_{j=1}^{p-1} \int_{a_{i,j}}^{a_{i,j+1}} exp(\mu_i w_i(a_i)/T) da_i + \int_{a_{i,p}}^{a_{i,k}} exp(\mu_i w_i(a_i)/T) da_i}{\sum_{j=1}^{m-1} \int_{a_{i,j}}^{a_{i,j+1}} exp(\mu_i w_i(a_i)/T) da_i} \tag{10}$$

Also, with regard to Fig. 1:

$$m_{i,j} = \left(\frac{w_{i,j+1} - w_{i,j}}{a_{i,j+1} - a_{i,j}} \right) \tag{11}$$

$$\delta a_{i,j} = a_{i,j+1} - a_{i,j} \tag{12}$$

As the consequence, Eqs. 13 and 14 are obtained:

$$CDF = \frac{\sum_{j=1}^{p-1} \int_{a_{i,j}}^{a_{i,j+1}} exp\left[\frac{\mu_i}{T}\left(m_{i,j}(a_i - a_{i,j}) + w_{i,j}\right)\right] da_i}{\sum_{j=1}^{n-1} \int_{a_{i,j}}^{a_{i,j+1}} exp\left[\frac{\mu_i}{T}\left(m_{i,j}(a_i - a_{i,j}) + w_{i,j}\right)\right] da_i} +$$

$$\frac{\int_{a_{i,p}}^{a_{i,k}} exp\left[\frac{\mu_i}{T}\left(m_{i,p}(a_i - a_{i,p}) + w_{i,p}\right)\right] da_i}{\sum_{j=1}^{n-1} \int_{a_{i,j}}^{a_{i,j+1}} exp\left[\frac{\mu_i}{T}\left(m_{i,j}(a_i - a_{i,j}) + w_{i,j}\right)\right] da_i} \qquad (13)$$

$$CDF = \frac{\sum_{j=1}^{p-1} \frac{T}{\mu_i m_{i,j}} \left[exp\left(\frac{w_{i,j} + m_{i,j}\delta a_{i,j}}{T}\right) - exp\left(\frac{w_{i,j}}{T}\right)\right]}{\sum_{j=1}^{m-1} \frac{T}{\mu_i m_{i,j}} \left[exp\left(\frac{w_{i,j} + m_{i,j}\delta a_{i,j}}{T}\right) - exp\left(\frac{w_{i,j}}{T}\right)\right]} +$$

$$\frac{\frac{T}{\mu_i m_{i,p}} \left[exp\left(\frac{w_{i,p} + m_{i,p}(a_{i,k} - a_{i,p})}{T}\right) - exp\left(\frac{w_{i,p}}{T}\right)\right]}{\sum_{j=1}^{m-1} \frac{T}{\mu_i m_{i,j}} \left[exp\left(\frac{w_{i,j} + m_{i,j}\delta a_{i,j}}{T}\right) - exp\left(\frac{w_{i,j}}{T}\right)\right]} \qquad (14)$$

To select an action, a normal random number should be chosen between 0 and 1 and with respect to the CDF relating to Fig. 1 which is calculated by Eq. 10, a selected action together with its weight is estimated by Eq. 8. Then the selected action replaces the nearest located action. This has been indicated in Fig. 2.

In order to keep the range of action space unchanged, for the beginning and ending of this range, procedures should be different and these two amounts should remain fixed. If for example, the selected action is closer to $a_{i,1}$ than to $a_{i,2}$, despite its further distance, it replaces $a_{i,2}$, so as to protect changes range of the action space.

Fig. 2 Changes in weights and actions while selecting action $a_{i,k}$

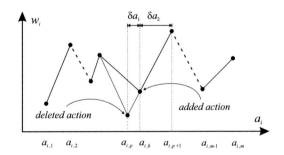

The proposed method, at last makes use of a combination of the above mentioned procedure and the ε-greedy method. That is, with probability $1 - \varepsilon$, the action is selected with maximum value. But with probability ε, Eq. 9 is used to select an action. In order to make use of this algorithm, only the previous action selection methods should be replaced by this method.

5 Simulation

The state space consists of three parameters that are shown in Fig. 3a:

Δm Orthogonal distance of each agent from the bisector of the line sector connecting the other two agents.
Δn Orthogonal distance of each agent from the line passing the other two agents.
Δa The distance between the two other agents.

β is an angle made by a line passing two other agents, and the horizontal line and is used for simplification and symmetrization. Also, the agent is always in the first quadrant of the coordinate plane and in other cases, symmetrization is applied (Fig. 3b).

For the action space, the allowable range is [90, 270] and divided into equal distances and the applied action constitutes the movement in one of these angles. The membership functions of fuzzy system inputs are shown in Fig. 4. The agents will receive a + 10 reward, if they reach the goal and create the formation, otherwise the reward will be −0.01.

$$\text{Reinforcement function} = \begin{cases} +10 & \text{goal state} \\ -0.01 & \text{others} \end{cases} \tag{15}$$

For simulation, agents are randomly placed in an environment of 20×20 dimensions and learning takes place as explained earlier. In this problem $\beta = 0.8$, $\alpha = 0.01$, $\gamma = 0.09$ and for greedy method, ε is equal to 0.2. These cases are applied for both methods of MAFRL and LIBAS. However, in case of using LIBAS method, in addition to the aforementioned cases, T is also needed which is considered equal to 0.1.

Fig. 3 Showing how the state and action of the system is defined

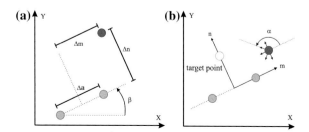

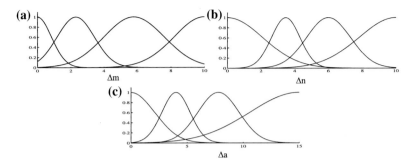

Fig. 4 Membership function of fuzzy system inputs

In testing stage, the agents are randomly placed in 300 different points. After placing the agents in the initial points, an average is taken from total distance which was passed by the three agents, up to the final points. Then, for 60 times more and with the same condition, the system gets learning and again, another average is taken on the whole distance passed by the agents. Finally, the last average is taken once again among the total 60 times and the final result obtained, indicates an average of the total distance passed to reach the goal.

The average distance passed for different number of candidate actions, is given in Table 1 and a comparison has been made for both methods. It should be mentioned that in the table and figures, wherever LIBAS method is mentioned and compared with MAFRL, it means that LIBAS has been applied in MAFRL. Also, in Table 1

Table 1 Comparison between MAFRL and LIBAS methods

Number of candidate action		5	6	9	13	20	30
Compared parameter							
Average of passed distance	MAFRL	24.4	24.9	25.8	27.4	27.5	28.5
	LIBAS	18.7	20.6	18.7	19.1	20.7	21.2
	Improvement ratio	24 %	17 %	27 %	30 %	25 %	26 %
Distance passed while one failed	MAFRL	33.6	33.5	33.8	35.6	35.8	36.4
	LIBAS	22	22.4	22.9	22.1	22.8	24.6
	Improvement ratio	34 %	33 %	32 %	38 %	36 %	32 %
Number of episode to converge	MAFRL	163.1	160.5	164.3	174.3	170.7	176.7
	LIBAS	250.2	220.6	277.5	305	315	351.1
	Improvement ratio	−53 %	−38 %	−69 %	−75 %	−84 %	−98 %

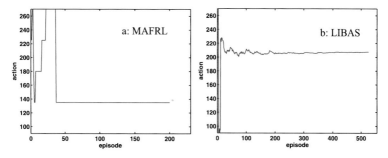

Fig. 5 A sample showing how actions converge

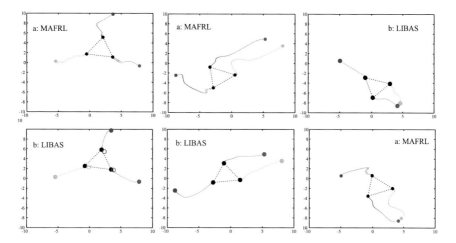

Fig. 6 Behavior of agents to reach the desired formation

improvement percentage means the rate of improvement achieved through application of LIBAS in MAFRL as compared with MAFRL method only. In order to check
the system robustness against failures, it is assumed that one of the agents fails to
move. It is found that the system proves to be robust against failure and the agents
are able to fulfill their duty.

Figure 5 shows convergence for one of the fuzzy rules in one of the agents through
MAFRL and LIBAS methods. It can be seen how LIBAS is able to search the action
space more accurately and find answers not obtainable through discrete action selection method. In Figs. 6 and 7 some pictures are given as examples showing the movement of agents to form triangle formation, whether all three agents are moving or
when one of them fails to move.

Comparing the results, it is clearly understood that, the improvement obtained
through applying continuous action selection method, is greatly considerable.
Although the episode required for convergence has increased, it can be seen that
the average distance passed, has decreased considerably as shown in the table. This

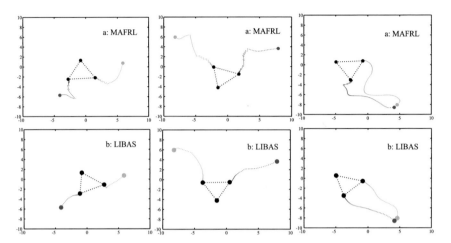

Fig. 7 Behavior of agents to reach the desired formation while one of agents failed to move

shows that usage of LIBAS method has increased reliability of the system. However, decreasing of average distance passed has led to increasing of episodes required for convergence.

6 Conclusion

In this paper, a continuous action selection method developed to be used in fuzzy actor-critic method and was evaluated in creating a formation. The results showed that the continuous action selection method improved the results from the view point of average distance passed by the agents as well as from the view point of system resistance against failure, although it decreased the convergence speed . This speed reduction was due to the fact that the searching space was greater and more time was required for searching. In future works, this method will be preferred to be implemented in single agent environments and also other reinforcement leaning methods.

Acknowledgments This research was financially supported by the Center of Excellence for Robust and Intelligence Systems (CERIS) of Yazd University.

References

1. Bik, J.J.C.M., Visser, P.N.A.M., Jennrich, O.: LISA satellite formation control. Adv. Space Res. **40**, 25–34 (2007)
2. Breivik, M.: Topics in guided motion control of marine vehicles. Ph.D. thesis, Norwegian University of Science and Technology (2010)

3. Chen, G., Cao, W., Chen, X., Wu, M.: Multi-agent q-learning with joint state value approximation. In: Proceedings of the 30th Chinese Control Conference, p. 48784882 (2011)
4. Chen, X., Chen, G., Cao, W., Wu, M.: Cooperative learning with joint state value approximation for multi-agent systems. J. Control Theory Appl. **11**, 149–155 (2013)
5. Derhami, V., Majd, V.J., Ahmadabadi, M.N.: Fuzzy sarsa learning and the proof of existence of Its stationary points. Asian J. Control **10**, 535–549 (2008)
6. Dierks, T., Jagannathan, S.: Neural network output feedback control of robot formations. IEEE Trans. Syst., Man, Cybern. Part B **40**, pp. 383-399 (2010)
7. Feddema, J., Lewis, C., Schoenwald, D.: Decentralized control of cooperative robotic vehicles: theory and application. IEEE Trans. Robot. Autom **18**(5), 852864 (2002)
8. Franco, F.E., Waissman, V.J., Garca, L.J.: Learning the filling policy of a biodegradation process by fuzzy actorcritic learning methodology. Adv. Artif. Intell. **5317**, 243–253 (2008)
9. Izzo, D., Pettazzi, L.: Autonomous and distributed motion planning for satellite swarm. J. Guid. Control Dyn. **30**, 449459 (2005)
10. Macdonald, E.A.: Multi-robot assignment and formation control. M.Sc. thesis, Georgia Institute of Technology (2011)
11. Panait, L., Luke, S.: Cooperative multi-agent learning: the state of the art. Auton. Agents Multi-agent Syst. **11**, 387–434 (2005)
12. Sanz, Y., de Lope, J., Martn, J.A.H.: Applying reinforcement learing to multi-robot team coordination. In: Corchado, E., Abraham, A., Pedrycz, W. (eds.) HAIS 2008. LNCS, vol. 5271, pp. 625632. Springer, Heidelberg (2008)
13. Zuo, G., Han, J., Han, G.: Multi-robot formation control using reinforcement learning, advances in swarm Intelligence. vol. 6145, pp. 667–674. Springer, Berlin (2010)

Context Time-Sequencing for Machine Learning and Sustainability Optimization

Fábio Silva and Cesar Analide

Abstract Computer systems designed to help user in their daily activities are becoming a norm. Specially, with the advent of the Internet of Things (IoT) where every device is interconnected with others through internet based protocols, the amount of data and information available has increased. Tracking devices are targeting more and more activities such as fitness, utilities consumption, movement, environment state, weather. Nowadays, a challenge for researchers is to handle such income of data and transform it into meaningful knowledge that can be used to predict, foresight, adapt and control activities. In order to this, it is necessary to interpret contextual information and produce services to anticipate these conditions. This project aim to provide a system for the creation of information and data structures to generate user models based on activity and sensor based contextual-information from IoT devices and apply machine learning operations to anticipate future states.

1 Introduction

Contextual lifelong information is becoming a reality with the increasing number of devices that allow constant environment and user sensorization. From fitness trackers to household consumption utility monitors, sources of data are diverse and allow different contextual analysis. With concepts and processes from data and information fusion t is possible to not only improve measurement but also improve the quality of information using different heterogeneous sources. Internet of Things (IoT) provides the infrastructure network needed to share data across devices, while growing

F. Silva (✉) · C. Analide
University of Minho, Braga, Portugal
e-mail: fabiosilva@di.uminho.pt

C. Analide
e-mail: analide@di.uminho.pt

© Springer International Publishing Switzerland 2016
P. Novais et al. (eds.), *Intelligent Distributed Computing IX*,
Studies in Computational Intelligence 616,
DOI 10.1007/978-3-319-25017-5_29

309

research areas like smart cities and smart things aim to democratise the use of such infrastructures to, among other things improve living conditions and access to services that may help users in their daily lives.

One application for these concepts can be found in the creation of on-line sustainability reports and the implementation of sustainability indicators on communities of users. In these analysis, sustainable indicators are generally relative to the sustainability dimensions: ecological, social and financial, though, indicators that span across dimensions are also possible. Furthering this idea, the data sources and methods of measuring and assessing sustainability continue non-standardized and dependent on the subject, objective of analysis and area of study. As a result, it seems that no broad framework for dealing with such problem has become relevant, but rather, small implementations specialized implementations dependent on the area of study.

Taking lessons from ambient intelligence, ubiquitous computing and the current availability of data sources as a consequence of the development the interconnection of services and devices through internet protocols, it is possible to think about these problems as computational problems. It means that, computational resources to sense monitor and act upon the environment have become available. This led to the creation of the field of computational sustainability [3]. This field aim to use computational resources to monitor, plan and optimize attributes related to sustainable problems which have application beyond the traditional computational system.

Sustainable problems such as energetic sustainability and energy efficiency are directly affected by human behaviour and social aspects such as comfort. While efficiency is focused on optimization, sustainability is mostly concerned on restrictions put in place to ensure that the devised solution does not impair the future. For instance, a deliberative system that pre-heats a room in the afternoon taking advantage of solar radiation and maintains it warm so it can be occupied in the evening, might conduct an energy efficiency procedure, in order to spend less energy to make that room comfortable later in the day. Considering the cost of heating the same room later in the evening from a lower temperature to a comfortable temperature the energy might be best spent with the previous strategy, but it will not be clear to all users because such action are dependent on context and some user might only see an empty room spending energy on a heating process.

This projects aim to deliver contextual time-sequence actions based on user contextual information. These will use different devices and services connected to data networks to provide organized data input for machine learning models and deliberative tasks. Data operations are based on the streams of time-sequence data which are also defined to according to the learning tasks. The end objective is to provide an automatic work flow that recognises user habits and contexts to control and adapt contextual information to user needs and sustainable objectives.

2 Related Work

2.1 Human Tracking

Different methodologies and procedures exist to keep track of human activities and to make predictions based on previous and current information gathered from these environments. Some of the most common approaches with machine learning techniques involve the use of neural networks, classification techniques, fuzzy logic, sequence discovery, instance based learning and reinforced learning.

Sequence discovery approach is at the heart of learning algorithm in [1], which demonstrates a system that can learn user behavioural patterns and take proactive measures accordingly. The theory developed consists in the discovery of pattern of user behaviour. Such system is composed by three modules, the representation of patterns, the learning patterns and the interacting system. These three modules are predicted to be capable of interacting with the user in natural language, learning patterns of user's behaviour and representing the contents of each discovered pattern in a simple if-then-when rule. Other approach is found in [9] where events are grouped into activities under an ambient intelligence scenario, with sensors spread across a domestic environment. Not only does it target activities of daily living, it also targets the cluster of activities and relationships between them based on mathematical formulations over sets of activities.

Human activity tracking algorithms demonstrate potential for intelligent environments in order to adapt to specific user preferences or environment specific objectives taking in consideration the habits of its users.

2.2 Environment Tracking

Smart cities is a research area under development that aims to provide a technological infrastructure to measure and monitor attributes relevant to city management. Furthermore, it is being used to develop services based on the platforms such as internet of things (IoT), smart grids and public services [6, 8]. These developments enable the constant monitoring of city attributes such as weather, air, living conditions and traffic for example. These are often aggregated in units of time ranging from every few seconds to days, months and years. An historic record is, among other things, used to provide context or even factual data for problems such as global warming, flood control, public health and energy savings.

Every city citizen is subjected to the conditions being monitored and, as such, these services become an interesting context data to analyse behaviours and specially behaviour dependent on context. On a sustainability analysis, this means assessing behaviour impact on sustainability and sustainability indicators.

Our take in environment tracking makes use of dedicated hardware and web-services to gather contextual data as activities are produced. The objective is to create time-sequence strips that can be evaluated on a time basis and used as historic data to perform prediction using models with machine learning technology.

2.3 Machine Learning and Profiling

Models and machine learning often require past evidence to compute, predict and provide insights. The objective is to rely on proof from past events so future occurrences can be estimated ahead of time. This, of course, translates to valuable knowledge, specially when certain outcomes need to be avoided such as uncomfortable environments or unsustainable behaviours. It is in growing demand the need of forecast to prevent and assure compliance.

Projects such as The People Help Energy Savings and Sustainability (PHESS) [10], HVAC control systems based on people estimation [5] and adaptative agents [7], look for means to predict behavioural action of its users based on environment configurations and their current states. To this end, the use ambient sensorization, data fusion techniques, ubiquitous monitoring and smart actuators make it a technological project with benefits for both compliance of policies and an attempt to solve some sustainability assurance. The use of the expression attempt is due to the nature of the definition of sustainability, which for this work was based on the assumption that past and current actions should not impair the future of the execution of such actions in the future. If there is lack of data or information then the analysis may become incomplete, but even worse is the case for uncontrollable factors that impair sustainability without remediation available.

3 User Time-Sequence Context Fusion

The PHESS project, for which this time-sequence context was developed, is intended to, by helping people, make them intervene in the society, in the context and with the purpose of contributing for the energetic sustain-ability of their world. There is perception and desire, in the society and in people, to discuss this matter. In order to make the data design robust enough for consecutive updates, the PHESS system was design to track contextual factor on base units of interest. Raw data is preserved at central data storing nodes with reliable and fault tolerance mechanisms in place and then pre-processed to make meaningful journals of contextual information. This contextual information acts as a middle data preparation which can suffer operations such as slice to decrease contextual information, aggregation to operate on higher base units, and time-sequence to determine the number of days, epoch and period of interest of the analysis. After these operations, the data is ready to become the input of multiple machine learning and ranking algorithms.

Intentionally, the main objective is the re-organization of contextual information so as to provide tailored suggestions to users on sustainable parameters based on evidence form environment parameters and behaviours. In test scenarios, the consumption of utilities electricity was used to perform sustainable assessments. Better yet, the contextual information from structured services inside the PHESS project allow the inclusion of direct context based on what happened and indirect context based on what could have happened if realistic and feasible modifications were made on the base context. This indirect context is dependent on expert knowledge to devise possible scenarios such as the availability of solar energy sources to decrease the environmental and financial cost of grid powered electricity. Such metadata can be used to fuel the assessment of alternative scenarios, which is useful when considering the optimisation or assurance of parameters.

The focus of this project is, more than developing new procedures or algorithms to solving problems, putting these innovations on the hand of the user, with a clear purpose: that these innovative tools should be guided to assist people.

3.1 Data Fusion

The process of data fusion is handled by local central nodes, where data is submitted to data fusion process according to the number of overlapping and complementing sensors. In this regard, there are a number of strategies that can be followed according to context. The first one mentioned is a weighted average of values for the same type of sensors in the same context to get an overview of an attribute with multiple sensors to reduce measurement errors. The weights are defined manually by the local administrator. More sophisticated fusion is employed with complementary sensors which according to some logic defined into the system measure an attribute by joining efforts such as user presence with both RFID readers and wireless connection of personal devices such as smartphones. In this case, the system knows the user is present whenever one or both of the sensors are triggered. In this last example there is the use of heterogeneous data to create attributes with some level of knowledge expertise. Other examples can be found in is the assessment of thermal comfort using default indicator expressed as mathematical formulae such as the PMV index [2] or the more recent physiological estimated temperature (PET) [4]. Other application is the definition of sustainable indicators according to custom mathematical formulae in the platform that shall process some attributes in the system to make their calculation. The PHESS project, in its contextual inference uses convergent data fusion techniques and heterogeneous data sources as income of data. Context can take a form of averaged sensor readings to complex indicators that are directly or indirectly assessed. Operation over datasets are present to filter, group and randomize samples through slice, aggregate and sample operators. A generic visual representation of the time-sequence to represent the evolution of indicators over time is shown by Fig. 1.

Each time-sequence is based on action detection. The action sensors provide the granularity of the analysis as action recognition can be based on user movement (e.g.

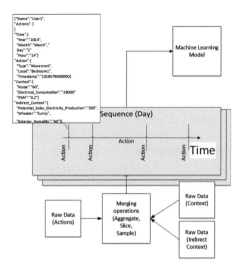

Fig. 1 Daily based time sequence contextual information based on user activity

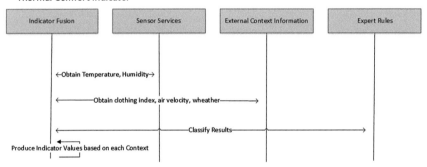

Fig. 2 Indicator Construction in PHESS workflow

from one environment to other), activity detection (e.g. walking, driving, exercising, cooking). Physically, the PHESS platform perform a number of calls to sensor and context services in order to gather the information needed to build the time-sequence. Figure 2 details the list of calls to PHESS internal services to manage and display time-sequences to the user.

The configuration of data fusion steps, the selection of sensors and streams of data is made on the initial step of the system by the local administrator. This flexibility addresses the needs for the system to produce relevant information locally. Also, the historic of time-sequences allows for machine learning tasks and to anticipate user actions based on context or environment response to user behaviour depending on interest.

Table 1 Expert rules
dependent on activity

Attribute	Acceptable range
Artificial light	[50, 2000 lux]
Temperature	[18, 24 °C]
Relative humidity	[20, 60 %]
Ambient noise	[0, 60 db]
Step count	[6000, 10000]
Heart rate	[60, 75 bpm]

3.2 Expert Rules

A predictive system needs to find not only what is expected in the near future but also to guide results according to guidelines extracted from complementary sciences. These are called expert rules due to the fact that they are created from knowledge passed to the system. For instance it is expected that daytime is hooter than nigh time and peak activity occur during daytime while resting period is during night time. As mundane as these rules are, they provide a substantial clues to direct learning efforts, assess sustainable parameters based on living conditions and plan what is ahead.

Other rules such thermal comfort of temperature and humidity or ergonomic lightning conditions provide insight to how much optimization can occur before human comfort and sustainability is affected. A traditional example shall be reduce electrical costs by turning down all electronic equipment. Though, it yields mathematically sound results, the human side is completely forgotten and may enforce hard living condition for today's habits. PHESS platform takes it into consideration as it is belied that no computer centric can endure without taking in consideration usability and comfort.

Table 1 display a simplistic demonstration of expert rules that define acceptable intervals for attributes that can be monitored from environment or users. These showcase what is considered normal ranges in which human behaviour is not negatively effected. If sustainable solution would violate these conditions then it is likely that at least the social dimension of sustainability would be harmed as well. In the system, expert rules act as means to assess indicator results and give them human interpretable meaning.

4 Case Study

The PHESS platform was used to perform experiments validating the theory presented in this research paper. The cases selected were a household environment and a office environment. These environments were set up in the PHESS platform by their physical representation. In each case, fixed and mobile sensors where introduced as in the configuration.

Table 2 User time-sequence results

	Household environment	Office environment	
	User 1	User 1	User 2
Actions detected	53	104	96
Indicators monitored	4	4	4
User notifications	2	4	5

The target was to control, environment comfort, physical condition and compound targets that combine expected contextual conditions with expert rules to design smart notifications to end users. To this end, the creation of time-sequence context was used to gather information about contextual information on each user. Notification were addressed each time.

The fixed sensors used were: an electricity meter to monitor energy consumption, temperature and humidity to monitor environment state. Furthermore, mobile sensors linked to users inside of each space was demonstrated by the use of smartphones, RFID tags and smartwatches data.

The professional environment is composed by a large room inside a office building frequented by a variable number of users. Ergonomic rules are obtained by the required conditions of office work documented by ergonomic studies. On the other hand, the house hold environment is composed of a number of room in which localization determines only weather the user is present in the environment. The sensor reading are also relative to the as generic data about the total building. Expert rules for each case are determined by expert rules as detailed in Sect. 3.2. These rules are transmitted to the PHESS platform which will manage contextual information to extract the maximum benefit for optimization and saving measures while preserving sustainable parameters.

These experiments occurred over in distinct time-frames but they had the duration of a full week of analysis. In Table 2, preliminary result, indicate the number of actions detected and indicators and notifications towards the end user. Indicator design was generated in the PHESS platform and included attributes to monitor sustainable indicators according to the dimensions of sustainability. The indicators are related to the thermal comfort according to the PMV indicator, noise levels, electrical lightning consumptions and number of active steps executed by a user. Notifications are generated according to the compliance of ergonomic rules during the day taking in consideration the last two hours and a proportionate expectation for each indicator. For instance in the case of thermal comfort it is not expected to by violated during the day, but the number of steps is divided by the number of active hours during the day. Electrical optimization is made detecting the need for artificial lightning with weather and solar exposition inputs and their relation to electrical consumption needs.

The context in strips of time-sequence actions, reveals behaviour according to actions and provides historic information to machine learning models. In this case

Table 3 Indicator predicament based on user context

Input	Output	Accuracy (%)
Time & User & Environment	Thermal comfort	83
Time & Action User & Environment	Thermal comfort	84
Time User & Environment	Electrical optimisation	72
Time & Action User & Environment	Electrical optimisation	68
Time User & Environment	Noise comfort	75
Time & Action User & Environment	Noise comfort	86
Time User & Environment	Steps	71
Time & Action User & Environment	Steps	88

predictive action are produced to derive Indicator values according to time or time and action as input. It is possible to see in this example that some routines have more impact on indicator values than other. For instance time and action provides a contextual information that improves the predicament of noise comfort and steps activity during the day. These indicator are clearly more dependent on the activities performed during the day while the other indicator seem to have similar results when only time is considered as input.

This assessment shall be used in the future to suggest activities that improve have impact on the monitored indicators. The strategy employed here might yield good results and also identify which activities have the most success to improve such indicator values according to each user habits (Table 3).

5 Conclusion

Context based on action time-sequences provides context in which actions are made. It is possible to observe which actions are based on context or based on user routines independent of context. The novelty proposed by this work is the middleware and data processing for the creation of sustainable machine learning models based on user activity detection. Furthermore, these models shall support the development of sustainable applications that take advantage of such models and information.

Time-sequence creation from basic aggregate, slice and sample provides a simple work-flow which is powerful enough to deliver meaningful results which can be used to deduce conclusions about future behaviours.After modelling each time strip, con-textual records are ready to be used by machine learning models. Results show that the composed values of indicator can be expressed either by time-sequences, but for some indicators, better results are yield when action is present. In fact, the improvement of the prediction of activities seems to indicate that some activities have more impact on indicator values and thus more prone to be used in suggestion systems.

As future work, there is the need to exploit other models for user context modelling, as well as, the design of better activity detection sensors, based on improved data fusion techniques and heterogeneous devices. The objective is to exploit more sources of data to interpret activity engagement by users. Assessment of time-sequence strips and categorization of sustainable indicators values can be used to fuel both environment actuators and leave notification design system as a fall-back when no actuator can mitigate the impact on indicators.

Acknowledgements This work is part-funded by ERDF - European Regional Devel-opment Fund through the COMPETE Programme (operational programme for com-petitiveness) and by National Funds through the FCT (Portuguese Foundation for Science and Technology) within project FCOMP-01-0124-FEDER-028980 (PTDC/EEI-SII/1386/2012) and project Scope UID/CEC/ 00319/2013. Additionally, it is supported by a doctoral grant SFRH/BD/78713/2011, issued by FCT.

References

1. Aztiria, A., Augusto, J.C., Basagoiti, R., Izaguirre, A., Cook, D.J.: Discovering frequent user-environment interactions in intelligent environments. Pers. Ubiquit. Comput. **16**(1), 91–103 (2012)
2. Fanger, P.O.: Thermal comfort: analysis and applications in environmental engineering. Danish Technical Press, New York (1970)
3. Computational sustainability. IDA. Springer, Berlin (2011)
4. Höppe, P.: The physiological equivalent temperature - a universal index for the biometeorological assessment of the thermal environment. Int. J. Biometeorol. **43**(2), 5–71 (1999)
5. Klein, L., Kavulya, G., Jazizadeh, F., Kwak, J.y.: Towards optimization of building energy and occupant comfort using multi-agent simulation. In: Proceedings of the 28th ISARC, pp. 251–256 (2011)
6. Krčo, S., Fernandes, J., Sanchez, L.: SmartSantandera smart city experimental platform. Electrotech. Rev. 3–6 (2013)
7. Mamidi, S., Chang, Y.H., Maheswaran, R.: Improving Building Energy Efficiency with a Network of Sensing, Learning and Prediction Agents. In: Proceedings of the 11th International Conference on Autonomous Agents and Multiagent Systems, pp. 45–52. International Foundation for Autonomous Agents and Multiagent Systems, Richland, SC (2012)
8. Piro, G., Cianci, I., Grieco, L., Boggia, G., Camarda, P.: Information centric services in smart cities. J. Syst. Softw. **88**, 169–188 (2014). doi:10.1016/j.jss.2013.10.029
9. Rashidi, P., Cook, D., Holder, L., Schmitter-Edgecombe, M.: Discovering activities to recognize and track in a smart environment. IEEE Trans. Knowl. Data Eng. **23**(4), 527–539 (2011). doi:10.1109/TKDE.2010.148
10. Silva, F., Analide, C., Rosa, L., Felgueiras, G., Pimenta, C.: Ambient sensorization for the furtherance of sustainability. In: Ambient Intelligence-Software and Applications, pp. 179–186. Springer, Berlin (2013)

Part VIII
Special Session on Energetic Sustainable Ambient Intelligence (ESAmI 2015)

Intelligent Distributed Systems
for Rural Areas

Luís Frazão, Silvana Meire and António Pereira

Abstract Communication technologies have evolved and grown exponentially in recent years. Unfortunately, this growth is not the case in rural areas, where the small population density does not justify the investment made by Internet providers. The creation and development of projects in rural wireless networks are often an effective alternative to traditional Internet providers. However, these projects tend to disappear soon after their implementation due to the cost of managing and solving problems. This paper presents an intelligent distributed system that can help the network administrators in maintenance tasks by applying automatic intelligent actions and decisions, ease the installation of new network devices for end users, automatically update existing devices, and manage network traffic by applying algorithms that will decide based on intelligence. This is achieved by using an intelligent distributed system executed on every device of the network and is capable of making decisions and actions that will help administrators with problem-solving, automatic configuration of devices, proactive bandwidth and network traffic management, and reduce costs in maintaining the network.

L. Frazão (✉) · S. Meire
Departamento de Informática, Escuela Superior de Ingeniería Informática,
Universidade de Vigo, Ourense 32004, Spain
e-mail: luisfrazao@gmail.com

A. Pereira
INOV INESC Innovation, Institute of New Technologies of Leiria,
2411-901 Leiria, Portugal

L. Frazão · A. Pereira
Computer Science and Communications Research Centre, School of Technology and
Management, Polytechnic Institute of Leiria, 2411-901 Leiria, Portugal

© Springer International Publishing Switzerland 2016
P. Novais et al. (eds.), *Intelligent Distributed Computing IX*,
Studies in Computational Intelligence 616,
DOI 10.1007/978-3-319-25017-5_30

321

1 Introduction

The growth of the number of people with Internet access is amazing. This fast paced evolution has made communication technologies better, faster and cheaper all-around. However, this has not been true for remote areas where the population density is very low and unpopular for the economic interests of the Internet providers. Although the internet backbone bandwidth available for users is increasing, this situation seems to be only happening in urban areas. In rural areas the access to the Internet is most of the cases slow and of poor quality. The actual Internet offer is not equal for everyone due to the non-profitability of implementing Internet services. Wireless network deployments in rural areas are being used as the best solution to provide Internet access [1–4]. Wireless technologies, such as IEEE 802.11, are low cost (unlicensed spectrum), easy, fast to deploy and are able to cover long distances. This has been the most common used technology in wireless rural area networks implemented all over the world.

Prior works [2, 5–12] have identified major problems in the sustainability of rural areas networks. The problem associated with a rural localization is its expensive maintenance due to the distance of the specialized teams. In order to solve these problems and to guarantee the longevity and sustainability of wireless networks in rural areas, we proposed an intelligent system that can automatically act on the network based on previous learned experience, user inputs and monitoring data of the network. Also, it can automatically deploy new configuration to newly installed devices on the network, as well as updating all the devices without the need of local human interaction. Last, the system can collect information of network traffic parameters and decide intelligently on specific problems of the network by applying algorithms created to achieve maximum fairness and quality of experience for the users.

This paper identifies factors that could be used to enhance a network to be autonomous and self-sustainable, and therefore minimizing its cost and maximizing it longevity and sustainability. While the previous researches mentioned are dedicated to specific implementation or specific technology, we aim to achieve a high layer solution that could be used on any wireless network.

The remainder of this paper is structured as follows. In Sect. 2 we briefly describe prior work in rural area wireless networks and the challenges of similar solutions. In Sect. 3 we present an overview of our architecture solution and our conclusions in Sect. 4.

2 Related Work

In the last decade, wireless IEEE 802.11 deployments have been made all around the world in rural areas and developing countries. Some of these deployments focus on providing services for the local community such as health services or education services. However, network management isn't always planned before the deployment of the network. Raman and Chebrolu [5] describe the experiences in using wireless technologies of the Digital Gangetic Plains (DGP) in India. Although they have identified most of the challenges and problems when deploying a rural wireless network, 5 years after the start of the project, Raman et al. says that they "are in the process of putting together a network management tool".

The research and evolution in self-configuring networks are focus on artificial intelligence, creating systems based on multi-agent architectures that allow the main system to be autonomous [13]. Other dynamic configuration systems have been used on wireless local area network (WLAN) like the Wi-Fi Protected Setup (WPS). This method provides a configuration for a small office or home usage that allows a user with no knowledge of network to configure his computer easily. Our goal is to propose an autonomous system that can automatic configure all the devices based on a central unit, contributing to keeping the wireless network scalable and cheap to maintain.

Surana et al. [7] describe their experience in deployments of rural wireless networks and focus on sustainability in projects used by thousands of users. They have deployed rural wireless networks to offer health services in Aravind project [6], and voice communication and Internet in Airjaldi project, but didn't mention solutions for traffic management. This is a main concern in our work. Surana et al. [7] also identified the lack of management in rural area by saying that "remote management solutions for wireless networks that are located in remote rural regions have not received a lot of attention". This is our exactly motivation in creating our work for projects to come.

3 Architecture of the Wireless Rural Area Network

A common rural wireless network infrastructure is similar to the one presented in Fig. 1.

The users of a typical rural wireless network are connected to a wireless router which is connected to other wireless routers. The gateway is often connected to WAN (Wide Area Network)—LAN (Local Area Network) Border Servers, such as web proxy servers or firewalls, which are responsible for securing the local network and providing services for a better Internet use. Other central servers are often used to deploy essential services, such as user authentication or network monitoring. As usual, in rural wireless networks, the administrators are often in a remote location, and have a vested interest in remotely managing the entire network and its services.

In the following Sect. 3.1 we present the monitoring system proposed to manage the network successfully, including intelligent decisions. In the subsequent Sect. 3.2, we present the algorithm proposed to create an automatic configuration of devices and in the final Sect. 3.3, we present a proactive bandwidth management that will focus on applying algorithms to ensure a fair and quality network experience for all the users.

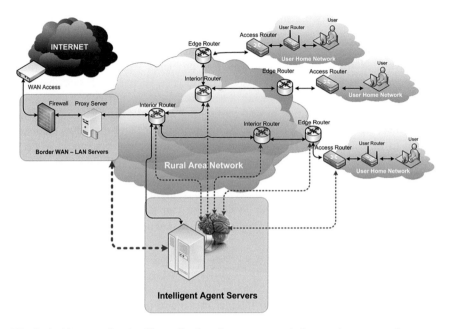

Fig. 1 Architecture of an intelligent distributed system on a wireless rural area network

3.1 Monitoring the Network and Applying Intelligence

A proper network management requires constant monitoring of all active devices and their respective services [5, 8]. A good monitoring task does not only look for yes or no answers. For instance, a router could respond to a ping request, but its CPU may be critical and delaying packets. Therefore, it is very important when monitoring a network device to include decisions based on threshold values of a specific monitored service. This type of monitoring will help the administrators to diagnose and solve network problems. For an excellent diagnose, the administrators must include the use of historical values [9]. When compared to other devices, the administrators could reach a faster conclusion, or recognize a pattern of a known issue. These patterns may help to detect declines or tendencies that could also lead to identify problems before they exist (preventive) [7]. However, the next logical step is to automatically detect these issues and solve them without the need of an administrator [1, 14]. With a richer acquired data, the decision could be better, improving the chances of a successful recovery. Related studies [14] suggest that network management is moving towards the implementation of self-managing network functions with aim of eliminating or drastically reducing human intervention in some complex tasks of network management. One way to achieve this is to create an intelligent system that not only alerts the administrator of possible problems or trends, but also creates actions to solve them. To better clarify our approach, Fig. 2 illustrates the organization and components of the monitoring system.

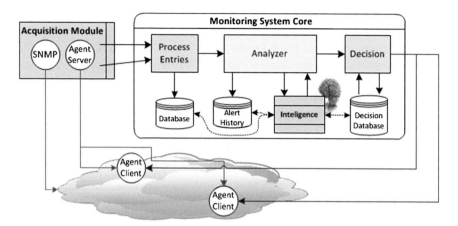

Fig. 2 Monitoring system core

The system for monitoring the network is based on SNMP (Simple Network Management Protocol) queries. When SNMP is available on the device the monitoring system queries the most important variables that could affect the network performance. When SNMP isn't available, an intelligent agent (agent client) is installed in each device that runs local commands in order to get the variables needed by the monitoring system. This module is presented in the figure as the acquisition module and its objective is to receive information every x minutes from the agents on the clients (agent client).

The "Process Entries" module is responsible for gathering all the information received by the acquisition module and insert it in a database. This information is inserted in a raw state, meaning that there is no additional information whether this service is working normally or not.

The Analyzer is the main module of the system. Using the same data inserted on the database, it is responsible for analyzing the received data. This data is analyzed for confirmation of a good or critical condition. If a value is out of its threshold limits, the system will send out alerts. Most of the values defined for the thresholds are inserted by the administrators. For instance, link wireless signals strength for most wireless routers cannot be less than −80 dBm, meaning that, anything below this value should send out an alert.

Other values could be automatic learned by the intelligent system. Using the same example, a wireless link could be working at an average value of −65 dBm in the last 6 months, and all of a sudden, it dropped out to an average value of −78 dBm. Although this value isn't below the threshold for active a warning message, by looking at historical values and its trend, it means that a problem may be on the rise. The intelligent system may decide to send out an alert and take action.

The decision module is responsible for the decision of the actions to be taken, whether an alert to be sent by email, or an action to be sent to the agents in the network devices. Based on the input received by the analyzer, this module decides what actions to take, as configured by the administrators.

The intelligence module it's the brain of the system. It could store information manually configured by the administrator, but also could perform problem prediction and prevention based on previous problems. This can be achieved by comparing the values acquired at a certain time with previously acquired and defined patterns, like the example given in the previous paragraph. This is decision based on history data. If a certain pattern is repeated, actions could be taken in order to avoid the same result as before. Predictions as these can be based on probabilistic or machine learning algorithms for classifications such as Bayesian Network or Naïve Bayes.

3.2   Device Automatic Configuration

Rural area networks are often geographically dispersed over a large area, where the main nodes are interconnected via wireless links to the local user devices. These nodes are often in remote locations of difficult access. Scheduling local visits to these nodes is expensive and hard to accomplish.

Our solution of configuring and updating the devices automatically plans to minimize the cost of maintenance by specialized teams and solve possible problems on non-configured devices.

This architecture uses agents on each device to allow a complete independent solution. An agent is deployed on every device. The agent working on the device it is responsible to keep a good configuration on its device or to deploy a new configuration. The automatic configuration algorithm is given in Fig. 3.

When the devices boots up, it can start in one of two modes: Configuration Mode or Update Mode.

The configuration mode is applied when using the device for the first time. When the device boots up it looks up for a configuration wireless network that it's configured in all devices. This wireless network is often configured as a virtual network on the wireless router, with a hidden service set identifier (SSID) and protected with Wi-Fi Protected Access II (WPA2) encryption using 128 bits password. The device will then choose the best signal of the configuration network it receives, and marks it as its parent device. After establishing a connection to its parent, the agent in the device is able to communicate to the server and receive its personalized configuration file.

The update mode is used every time a device reboots. Before the device initiates its normal operation, the agent will look for new configuration parameters or updates on the server, and apply them immediately.

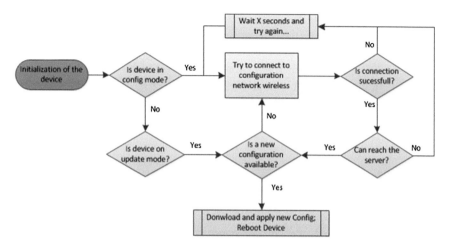

Fig. 3 Algorithm 1: automatic device configuration

3.3 *Proactive Bandwidth Management on Network Traffic*

The automatic system for management bandwidth needs input data to allow it to make decisions and launch alerts with details of the actual state of the network. The network state may be defined as one the following: high congested network, less congested network or no congested network. Each one can be defined as:

> **High Congested**: The network is congested when the packet loss hits the minimum value threshold at a defined timeframe. This timeframe must be enough to the system overcome burst of major traffic rates. A small defined time period may detect many false alarms of congested network.
> **Less Congested**: When there's no network packet lost, the network bandwidth is between 25 and 75 % of the available Internet Service Provider (ISP) bandwidth, and the actual bandwidth it's similar to the average bandwidth registered on historical records with the same timeframes.
> **Not Congested**: The network in not congested when the bandwidth is low (<25 % of the ISP available bandwidth) for a long period of time.

With these 3 defined network states, the system must gather data through time that will enable it to know what the state of the network was at a certain timeframe. For example, considering a timeframe of 24 h, it will be possible to identify on which periods of time a network was highly congested, less congested or not congested. The same will be possible to achieve regarding week periods, month periods, annual periods, and also special periods like holidays and vacation seasons.

The automatic system gathers a variety of information available on the network packets that flow through the network, as well as all the information available on the network devices. After a good collection of data gathering, the system must decide through algorithms which are the acceptable threshold levels for information. This means that the more data the system has, the better the thresholds levels are calculated or adjusted through the comparison of previous similar periods and network states (history data). With these adjustable thresholds levels the system makes decisions about the network actual state and launch alerts and events to improve the network performance. When these fixes are executed, based on the thresholds level defined, the system will then gather the results of its solution to validate its success on the network. This evaluation of the solution provides a valuable feedback and also enables the system to look back to previously actions and learn from them, creating a history of actions and results. Based on this historical data, the system may also anticipate problems on some periods of time and act on them early (proactive).

Not all types of traffic may have the same importance when the delivery priority it's assigned. All traffic must be divided in classes of traffic that represent its importance on the network. This classification will decide the quality, and delay of each type of network flow [15].

Quality of Service (QoS) classification must be defined according to the priority of each type of traffic. The priorities classes will range from "mission critical" (high priority class) to "best effort" (low priority class). Each class of traffic will have in its queues, network packets that have the same attributes and priorities defined by the network administrators. This means that different type of active flows from different application may share the same priority in one class of traffic.

Each class of traffic will be assigned with a reserved bandwidth and also a maximum bandwidth limit. The reserved bandwidth is a guarantee that a minimal bandwidth is being assured. Applications that require no packet loss, low delay and low jitter, such as live video or voice transmission, requires a good amount of reserved bandwidth. Some classes of traffic may have a reserved bandwidth of zero, meaning they operate in any minimum speed and so, no reserved bandwidth is required. This maximum bandwidth enables the class to reach that maximum value of bandwidth available, if it exists, and if is not needed by higher priority classes. The reserved bandwidth for all classes of traffic must be less than the available bandwidth at the ISP. Each class may operate at its maximum reserved bandwidth value at any time. If two or more classes are "fighting" for more bandwidth than their reserved bandwidth, the class with more priority is the first to use it.

With defined classes, the system is able to separate each flow to its respective class type and priority. For this classification the system must identify each type of traffic. This information must be gathered by analyzing the network packets header, network packet payload (by deep packet inspection) and also flow information and flow behavior detection [15]. After this analysis, all the flows must be classified by priority and be assigned to one of the available defined traffic classes on the network. This collection of variables is acquired from packets, flows, QoS classes, network stats, and user activities. All these variables are recorded historically. With a constant monitoring and classification of network traffic the system is able to identify and predict behaviors of the network traffic. With all this information available and stored, the automatic system must learn from it and adjust the QoS configuration. Different machine learning algorithms (Bayesian networks, Decision Trees) could be used for traffic classification. The system is always learning and adjusting dynamic parameters. After a good collection of data, the system must adjust automatically the network QoS [15], by executing some of the following algorithms in Fig. 4.

We have further algorithms that will be presented in a near future extended version of this paper, as well as the intelligence modules tests and results.

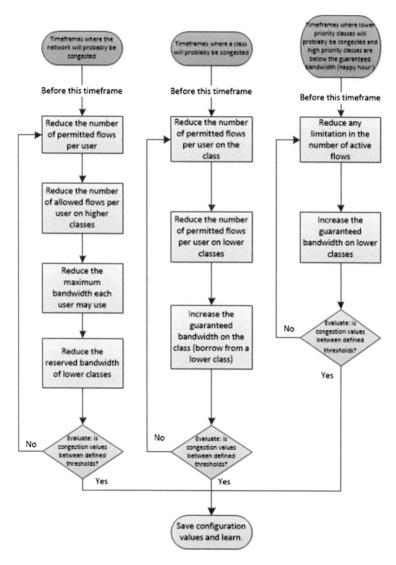

Fig. 4 Algorithm 2—congestion in specific timeframes

4 Conclusions

Rural areas with low population density can take advantage of wireless networks. Unfortunately many projects have failed due to the lack of planning for maintenance and sustainability over the years. By looking at some existing problems in these types of networks we were able to focus our effort on creating alternatives to those problems.

The solution presented permits to prevent the usual problems associated with rural areas wireless network by applying intelligence while executing monitoring functions that will ensure the administrators a safe and stress free maintenance of the network. Also, the solution permits that a new device in the network may be installed without the need of the presence of an administrator, therefore reducing costs and maximizing the network sustainability. This solution also allows the devices to be constantly updated. Lastly, the solution permits to enhance the network performance by applying intelligent decisions based on the actual state of the network variables that will create a fair usage and improve the quality of experience for all the users.

The advantages that these intelligent systems can bring to rural areas wireless networks are focus on reducing the costs in installation, maintenance and improving the network performance. This allows the network to grow sustainable and creating conditions for the users to keep using the network at their maximum performance.

For future work, we plan to implement the proposed solution in a real rural wireless network environment, test all the services and functions described and adjust it accordingly.

References

1. Subramanian, L., Surana, S., Patra, R., Nedevschi, S.: Rethinking wireless for the developing world. In: Hotnets-V, Irvine, California (2006)
2. Brewer, E., Demmer, M., Du, B., Ho, M.: The case for technology in developing regions. Computer (Long. Beach. Calif.) 25–38 (2005)
3. Bernardi, G., Buneman, P., Marina, M.K.: Tegola tiered mesh network testbed in rural Scotland. In: Proceedings of the 2008 ACM Workshop on Wireless Networks Systems for Developing Regions, WiNS-DR '08, p. 9 (2008)
4. Thakur, A., Hota, C.: Sustainable wireless internet connectivity for rural areas, In: 2013 International Conference on Advanced Computing and Communication, pp. 1335–1340 (2013)
5. Raman, B., Chebrolu, K.: Experiences in using WiFi for rural internet in India. IEEE Commun. Mag. **45**, 104–110 (2007)
6. Surana, S., Patra, R., Nedevschi, S., Brewer, E.: Deploying a rural wireless telemedicine system: experiences in sustainability. Computer (Long Beach, California). **41**, 48–56 (2008)
7. Surana, S., Patra, R., Nedevschi, S.: Beyond pilots: keeping rural wireless networks alive. In: Proceedings of the 5th USENIX Symposium Networked Systems Design and Implementation, NSDI'08, pp. 119–132. San Francisco, California (2008)

8. Brewer, E., Demmer, M., Ho, M.: The challenges of technology research for developing regions. Pervasive Comput. IEEE **05**, 15–23 (2006)
9. Surana, S., Patra, R., Brewer E.: Simplifying fault diagnosis in locally managed rural WiFi networks. In: Proceedings of the 2007 Workshop on Networked Systems for Developing Regions, NSDR '07, p. 1. ACM Press, New York, USA (2007)
10. Ishmael, J., Bury, S., Pezaros, D., Race, N.: Deploying Rural Community Wireless Mesh Networks. IEEE Internet Comput. **12**, 22–29 (2008)
11. Johnson, D.L., Pejovic, V.: Traffic characterization and internet usage in rural Africa. In: Proceedings of 20th International Conference on Companion World Wide Web (WWW 11), pp. 493–502. ACM, Hyderabad, India (2011)
12. Salvador, N., Filipe, V., Rabadão, C., Pereira, A.: Management model for wireless broadband networks. In: The Third International Conference on Systems and Networks Communication, pp. 38–43. Sliema, Malta (2008)
13. Gavalas, D., Greenwood, D., Ghanbari, M., O'Mahony, M.: Hierarchical network management: a scalable and dynamic mobile agent-based approach. Comput. Netw. **38**, 693–711 (2002)
14. Ranganai C.: UniFAFF: a unified framework for implementing autonomic fault management and failure detection for self-managing networks. Int. J. Netw. Manag. **9**(4), 17–31 (2007)
15. Este, A., Gargiulo, F., Gringoli, F., Salgarelli, L., Sansone, C.: Pattern recognition approaches for classifying ip flows. In: Structural, Syntactic, and Statistical Pattern Recognition. Lecture Notes in Computer Science, vol. 5342, pp. 885–895, (2008)

Intelligible Data Metrics for Ambient Sensorization and Gamification

Artur Quintas, Jorge Martins, Marcos Magalhães, Fábio Silva
and Cesar Analide

Abstract The interaction between and people is being defined by technology. New concepts appearing in our society such as Internet of Things allow common devices to be connected to the internet and sharing data with other devices in the environment. The flow of data and information available today can be so overwhelming that it can lose significance by their complexity if not handled properly. This proposal details the use of environmental and behavioural information to produce intelligible data metrics on driving events that can be aggregated and understandable in meaningful manners. Furthermore, their application for the promotion of better behaviours is exemplified with techniques extracted from gamification.

1 Introduction

Ambient Intelligence (AmI) is a research area that concerns with augmenting the environment in a proactive manner to support people interaction with the physical environment. AmI can act in different ways, for instance, keeping a space comfortable, assisting user in their activities such as driving or motivating users to acquire better habits and behaviours.

A. Quintas (✉) · J. Martins · M. Magalhães · F. Silva · C. Analide
Department of Informatics, University of Minho, Guimaraes, Portugal
e-mail: arturffq@gmail.com

J. Martins
e-mail: jorgemartins_10@hotmail.com

M. Magalhães
e-mail: marcos_leonardo@msn.com

F. Silva
e-mail: fabiosilva@di.uminho.pt

C. Analide
e-mail: analide@di.uminho.pt

© Springer International Publishing Switzerland 2016
P. Novais et al. (eds.), *Intelligent Distributed Computing IX*,
Studies in Computational Intelligence 616,
DOI 10.1007/978-3-319-25017-5_31

333

Ambient sensorization fits in the intelligent systems field and uses sensor technology and communications to obtain data to create decisions. As an example of such projects, where the goal is to contribute for the good of society, we can refer PHESS (People Help Energy Savings and Sustainability) [11, 12], or SmartSantander [9].

The use of sensors and other methods for the acquisition of data and relevant information is common for most projects. Moreover, the need to present such data and information to the user is motivated by studies that state that user awareness influences the way a user uses a system. Therefore, using data obtained from sensors it is possible to gather data and to develop intelligent systems that have impact on user behaviour. One example is the subject of intelligent cities [7], which has been implemented on some cities with purpose of offering services that are aiming to reduce unnecessary expenses and improving management of available resources. For instance, we could present the project executed in the city of Santander [9], Spain. Besides being profitable, it allows people to access the data that are being captured, such as the location of public transports, traffic control, illumination and water control.

In these projects, a person can interact with these systems either by being presented with useful information gathered or challenged based on motivational strategies to engage user into certain behaviours or actions. In the first case, while data received from sensors can be informative to a machine, it is often ill suited to inform the person. It is therefore important to present to the target audience in a meaningful and understandable manner. Intelligible Data Metrics (IDM) aim to extract features from sensor data and present them in an intelligible manner. The metrics representing features should therefore have a justification that is not opaque to the user. The second case uses data collected to motivate users towards beneficial behaviours in a society. Gamification is a common case in these scenarios as through user competition, behaviours and actions that benefit the system are promoted. The objectives, challenges and points are often rewarded by monitoring system, which in the case of ambient sensorization mean sensor networks.

Taking the example of driving inside cities, our developments aim to use an existent platform to demonstrate the use of data fusion techniques to produce intelligible data metrics. After the exploration of sensorization methods, this article details the developed a prototype to launch services related to sustainable driving and user comfort. It leans on the PHESS platform [11, 12] for service composition and to implement user awareness and motivational strategies towards the management of traffic. More specifically, it is intended the development of intelligible data metrics that support the detection of comfort on the driving domain of the PHESS project. This extension also includes the use of intelligible data metrics to support a gamification engine that manages motivation between users and aims to reduce risky behaviours and discomfort.

2 Related Projects

Intelligible Data Metrics' main concern is to bring context into local data, and create metrics that the user can identify easily. In our project, these metrics concern sustainability regarding driving behaviour. In this section, a review of process for data fusion, driving metrics and gamification elements are presented, as they shall be used in the service our team have developed. Sustainability metrics can be branched out into three areas, which while meaningful on their own, intersect with each other in several ways.

2.1 Data Fusion

In terms of Data Fusion the system can be qualified in several manners. One that seems to present an intuitive assessment is Dasarathy's Classification as presented in [3], which is according to [1] one of the most well known means of classifying a data fusion system. This classification has five levels based on combinations of input/output between three levels named as data, features, and decisions. Where a process can output either information of the same level or of the next level from the input. Consequently they can be named as Data In-Data Out (DAI-DAO), Data In-Feature Out (DAI-FEO), Feature In-Feature Out (FEI-FEO), Feature In-Decision Out (FEI-DEO), and Decision In-Decision Out (DEI-DEO).

In relation to data fusion, the paradigm used for the context of this work goes according to the classification system of data fusion introduced by Dasarathy [3]. In terms of application,the process used were DAI-DAO for the association and filtering of the data, followed by the process DAI-FEO which utilizes the brute data of the data sources to extract characteristics that describe one entity on the environment.

The prototype developed, enables reception of all information that is being monitored by environmental and behavioural sensors. The point is to first enrich the data in reference to a space-time location, and then, apply rules, in order to, create a justifiable score on each of the sustainability factors. As such it is important not only to relate all data to a space, but also a relative and absolute time.

2.2 Driving Event Classification

In order to produce information about traffic flow and route safety it is necessary to gather information about relevant driving patterns in city areas. The focus of our analysis was derived from indicators accepted from the literature and implemented on the PHESS Driving system [10]. Among other characteristics, this system allows the recording of driving trips and assess to sensor values such as GPS, accelerometer, light and gyroscope from user trips. There is also, behavioural classification based

Fig. 1 Data fusion model

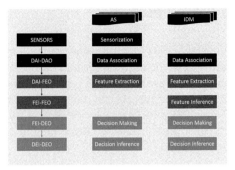

on accelerations and curve driving which are also available. These are based on three level classification from green to red based on the level of risk. Currently, the data used comes directly from sensors, but with the development of new intelligible data metrics, this is information is fine tuned with information fusion from cartographic data and other contexts (Fig. 1).

In projects of a relevant nature to this one authors have used both approaches for differing concerns nevertheless acceleration appears to be a common metric in this type of evaluation of road traffic, [4] uses accelerometer based data to find out whether the road has potholes as well as whether the user is manipulating the device or not, while GPS based motion is used to infer traffic flow. In [8] it is concluded that GPS does provide a reliable source for speed and uses it to address driving behaviour, [10] takes into account the frequency of critical accelerations to determine the aggressiveness and the GPS to determine where turns happen.

2.3 Comfort Assessment and Predicted Mean Value Indicator

Ambient sensorization can be carried out by projects that implement a sensor networks. To this end, projects like the PHESS project enable the creation of rapid ambient intelligence prototypes through its robust sensor network design implemented over multi-agent platforms [11]. With such middleware and access to sensor that monitor environmental and behavioural attributes is possible to determine levels of comfort for a various attributes such as the thermal level or the noise level. Thermal comfort is typically assessed through indicators, in this case there are several indicators in the literature, the current prototype uses the Predicted Mean Value (PMV) index. The PMV index is the average vote estimated from a set of individuals in the environment, and the method which determines this indicator has been developed by Fanger [2]. This choice for the first prototype was based on familiarity with this indicator but other alternative indicators in the literature exist, such as the Physiological Equivalent Temperature (PET) [5]. Fanger, expert in the field of thermal comfort and the perception of environments. With the PMV value, using the scale $[-3, 3]$ it is possible to find the level of thermal comfort using the specified indicator. An increasing negative score mean more cold discomfort while an increasing positive score mean hot discomfort. The goal is to score around the value zero.

2.4 Gamification

User awareness is useful for diagnosing the state and impact of user behaviour, however, to encourage behavioural modification, there are strategic methods that involve communities of users, competition and user rewards within computational systems. One of such methods is gamification, which is seen as a simple use of game mechanics to artificially engage users in activities that otherwise they would not be involved in. The addition of mechanics such as points, rewards and badges are used in activities as trivial for the promotion of desirable behaviours. These mechanical, often take the form of a virtual reward system which may include: points, badges, levels and virtual coins. Nowadays, there are many organizations from schools, software companies, pharmaceutical companies, government organizations, among others, using gamification to train their workers, solve problems and generate new ideas and concepts [6].

Elements of gamification are chosen as strategy to along the use of intelligible data metrics lead people to diagnose and alter their behaviour to become more secure in regard to their driving habits.

3 Service Implementation

The services developed in the context of this work aim to improve data fusion process with the creation of contextual analysis with intelligible data metrics and use gamification elements to make users improve their driving behaviours. The goal of intelligible data metrics to the context of classifying driving behaviours and comfort requires capture of environmental and behavioural attributes which is a demanding task. For this purpose, the implementation and management of such metrics in based on the PHESS project which provides middleware able to cope with diverse data and sources of information. This project provides the necessary infrastructure of data collection through sensor networks and the centralized storage of data. It also facilitates the development of data fusion techniques by providing dedicated web-services and a multi-agent based system to autonomously collect data from different sources. The gamification is built from the access to the PHESS services implemented in the scope of this work.

For ease of understanding the complete implementation of this system is depicted in the following sections: data sources, intelligible data metrics and gamification.

Fig. 2 Prototype with
temperature, humidity and
sound sensor

3.1 Data Sources

To obtain environment attributes it is necessary to use different sensors and a platform of prototyping that allows to collect data to be exported and treated. On Fig. 2
it is shown the Arduino prototype for the detection of the comfort levels inside environments. It is composed by two sensors, one sensor that measures temperature and
humidity, and other that captures sound intensity. Coupled with portable batteries as
power sources and wireless communication it becomes possible to make ubiquitous
sensorization.

The sensor portrayed in this prototype are DHT11 which allows the reading of
temperature and humidity of the environment and a noise sensor. Both are used to
assess thermal comfort the first with the help of the PMV indicator and noise levels
based on personal preferences defined by user input.

In terms of behavioural analysis, for the case study selected, the PHESS project
provides raw information about driving trips and raw access to mobile sensors used
such as GPS and accelerometer. The expectation is that by gathering information
of several trips through the same location the system can learn relevant information about the location itself. And that this data can then be used to produce intelligible metrics concerning sustainability factors of said location. The types of data
present are Time, GPS, Velocity, Acceleration, Weather, Date, and Location. In this
approach, raw driving data is obtained from the PHESS driving project [10].

On the other hand projecting GPS points into the road network using map-
matching while more heavy in terms of resources and only available in mapped
regions, it provides for a more interesting context of the data that takes into account
actual roads and characterizes in a more specific and intelligible manner. It also
allows for measurements against road types and speed limits, which may prove useful. In Sect. 3.2 a detailed explanation on the services to produce this outcome are
presented. Its impact becomes evident when noisy GPS data is translated to street
points. The development of classification procedures using the cleansed data is more
familiar for the user than just GPS points.

Weather comes in the form of information concerning temperature, wind, precip-
itation, visibility, and others. It is generally not as localized as other features and may
depend on availability of data for each space-time location.

3.2 Intelligible Data Metrics

For IDM, most computation in the server application side can be performed in an event driven way. Considering the dependency on web latency as well as computation times that a system of this type is contingent on. An event driven system allows for less waste of computational resources while the server awaits on the various responses from the differing resources both over the web or from memory with slower access times. Therefore, the server consists of a set of modules that connect to APIs from various services reliant and hardware in order to gather different types of data on the PHESS project.

The data fused will take different forms, from GPS coordinates to accelerometer entries, along with data about time, weather, and holidays. Features such as proximity to points of interest and decisions such as user input rules will also be used in order to classify user behaviour, as well as, user scoring.

Data received on the IDM server is not always representative of sensors with the same frequencies scope or range. Therefore a need arises to associate data based on their timestamps. Given the higher frequency of the acceleration data than that of the GPS data, a way has to be found to associate acceleration records to GPS coordinates and/or a location on the road network. This can be accomplished in several ways, the IDM intends to implement and compare association of a set of acceleration entries to each GPS entry based on timestamps. A currently employed way to do this is to divide time between each two GPS points and bin the accelerations according to which time interval they are in, the pseudo-code of this binning strategy is given in (1).

$$
\begin{aligned}
lowerBoundT(0) &= 0; \\
lowerBoundT(i) &= (time(i-1) + time(i))/2; \\
upperBoundT(i) &= if \quad (exists(i+1)) \\
&\qquad then \quad (time(i) + time(i+1))/2; \\
&\qquad else \quad presentTime(); \\
filter(i) &= t >= lowerBoundT(i) \cap t < upperBoundT(i); \\
bin(t) &= yield \quad i \quad in \quad gpsItems \quad where \quad filter(i)
\end{aligned}
\tag{1}
$$

This can allow for spatial information enrichment in cases where GPS points skip road network edges where other attributes may belong. But it has one of the downfalls of the projection based approach, it depends on the map-matching success and the accuracy and completeness of the underlying map. There can be no assumption that every trip submitted will be successfully matched, therefore a mode to fall back on must always be present.

While the fusion process intends to offer the enriched information, some of the fusion does not happen on the Intelligible Data Metrics side, but rather in external services, this includes Map-Matching which in this case is limited to the association of a sequence of GPS points with time stamps to a road network, namely OpenStreetMaps (OSM). Weather and Timezone retrieval are also computed and retrieved from external services. The Intelligible Data Metrics layer is not one that deals with controlling sensors directly, but it does deal with the coordinated use of sensor data.

3.3 Gamification Design

The gamification component works in parallel with the project PHESS Driving and the creation of Intelligible Data Metrics to improve driving behaviours.

A user visit to its profile triggers a query to the servers responsible for the PHESS Driving database and Intelligible Data Metrics construction. The game mechanics include points awarded for each trip will dependent on metric performance and various factors such as time and travel distance, average speed and color of the trip.

The use of these game mechanics increase the competitiveness factor between users of the system, that because users may have access to the performance of other users, motivating and encouraging users to improve bad driving habits in order to aspire to reach the top of rankings. The mechanism is inspired in a gamification implementation over intelligent environment developed for the PHESS project and adapted to the driving scenario.

4 Results

This section will describe results in each component of the system developed. They are integrated as a work flow that spreads results from one component to the next that independently obtain different results and pursue different objectives.

Environmental attributes captured by the arduino sensors are recorded trough PHESS services. The values recorded include a rating of the environment's comfort level based on sound and PMV indicators. During the period of 6 days of testing, uncomfortable environment was detected trough each day to a cumulative 2 hours and 39 min due to either PMV index or noise discomfort. An important aspect to retain is that noise discomfort was only verified by a period of 8 min. Considering those statistics it can be said that the environment is mostly considered a good environment most of the time and that the temperature and humidity were more disrupting to the comfort in this context. These values were recorded inside an environment, due to restrictions on their actual use during driving. Although this fault shall be corrected in a near future, these results demonstrate the possible context information that can be obtained.

The use of Intelligible Data Metrics was based on the localization and map-matching of comfort assessment and driving trips. The point data regarding the GPS, even if not originally connected to the corresponding road on the map, seems to follow the road's shape most of the time, this suggests that [8] was correct in its findings that GPS points may present, on their own, a good way to make an approximation

Fig. 3 Point projection
from complex path shape

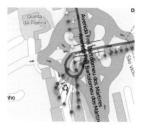

of the vehicle speed. Furthermore one issue became readily apparent, the matching algorithm's way of projecting points into the closest point in the matched path line is too naive to properly grasp some of the more complex path shapes as can be seen in Fig. 3.

The gamification game mechanics are implemented based on the reading of the Intelligible Data Metrics which are used to attribute points for each local with positive Intelligible Data Metrics found in the system. As the metrics are associated to locals, the verified result is users generally avoiding such areas to increase their score when such is possible. As a point of reference, this observation refers to users inside cities which mostly take short trips.

5 Conclusions and Future Work

This project presents a system that operate on different abstraction levels regarding their objectives, but execute compounded tasks to perform the creation of Intelligible Data Metrics and their actual use, in this case, for comfort assessment, driving assessments and gamification to improve user driving behaviours. Initial results provide a proof-of-concept with a working set of services that inform and encourage the user towards sustainable driving. It also provides evidence on the usefulness of the PHESS project to the rapid development of services and data collection and fusion processes.

Concerning future work, will include comfort assessment inside vehicles and the development of a model that allows prediction of comfort levels as opposed to the current classification analysis. In the case of Intelligible Data Metrics future work will encompass further exploration of error and to try to balance context awareness with the least possible error.

Acknowledgements This work is part-funded by ERDF—European Regional Devel-opment Fund through the COMPETE Programme (operational programme for com-petitiveness) and by National Funds through the FCT (Portuguese Foundation for Science and Technology) within project FCOMP-01-0124-FEDER-028980 (PTDC/EEI-SII/1386/2012) and project PEst-OE /EEI/UI0 752/2014. Additionally, it is also supported by a doctoral grant, with the reference SFRH/BD/78713/ 2011, issued by FCT.

References

1. Castanedo, F.: A review of data fusion techniques. Sci. World J. **2013** (2013). doi:10.1155/2013/704504
2. Charles, K.E.: Fanger's Thermal Comfort and Draught Models Fanger 's Thermal Comfort and Draught Models IRC Research Report RR-162, p. 29, October (2003). DOI IRC Research Report RR-162
3. Dasarathy, B.V.: Sensor fusion potential exploitation-innovative architectures and illustrative applications. Proc. IEEE **85**(1), 24–38 (1997). doi:10.1109/5.554206
4. Dolui, K., Mukherjee, S., Datta, S.: Traffic status monitoring using smart devices. In: 2013 International Conference on Intelligent Interactive Systems and Assistive Technologies (IISAT), pp. 8–14 (2013). doi:10.1109/IISAT.2013.6606431
5. Höppe, P.: The physiological equivalent temperature—a universal index for the biometeorological assessment of the thermal environment. Int. J. Biometeorol. **43**(2), 71–75 (1999)
6. Kapp, K.: The Gamification of Learning and Instruction: Game-based Methods and Strategies for Training and Education. Pfeiffer essential resources for training and HR professionals. Wiley (2012)
7. Longo, M., Roscia, M., Lazaroiu, G.C.: Innov. Multi-agent Syst. Appl. Smart City **7**(20), 4296–4302 (2014)
8. Saiprasert, C., Pattara-atikom, W.: Smartphone enabled dangerous driving report system. In: 2013 46th Hawaii International Conference on System Sciences (HICSS), pp. 1231–1237 (2013). doi:10.1109/HICSS.2013.484
9. Sanchez, L., Galache, J., Gutierrez, V., Hernandez, J., Bernat, J., Gluhak, A., Garcia, T.: Smart-Santander: the meeting point between Future Internet research and experimentation and the smart cities, pp. 1–8
10. Silva, F., Analide, C., Novais, P.: Assessing road traffic expression. Int. J. Artif. Intell. Interact. Multimed. **3**(Regular Issue), 20–27 (2014). doi:10.9781/ijimai.2014.313
11. Silva, F., Analide, C., Rosa, L., Felgueiras, G., Pimenta, C.: Ambient sensorization for the furtherance of sustainability. In: Ambient Intelligence-Software and Applications, pp. 179–186. Springer, Berlin (2013)
12. Silva, F., Analide, C., Rosa, L., Felgueiras, G., Pimenta, C.: Social networks gamification for sustainability recommendation systems. In: Distributed Computing and Artificial Intelligence, pp. 307–315. Springer, Berlin (2013)

Hierarchical Architecture for Robust People Detection by Fusion of Infrared and Visible Video

José Carlos Castillo, Juan Serrano-Cuerda, Antonio Fernández-Caballero
and Arturo Martínez-Rodrigo

Abstract Robust people detection systems are nowadays using heterogeneous cameras. This paper proposes an hierarchical architecture which is focused on robustly detecting people by fusion of infrared and visible video. The architecture covers all levels provided by the INT^3-Horus framework, initially designed to perform monitoring and activity interpretation tasks. Indeed, INT^3-Horus is used as the development environment where the approach starts with image segmentation in both infrared and visible spectra. Then, the results are fused to enhance the overall detection performance.

1 Introduction

This paper introduces a hierarchical architecture for robust people detection inspired in the fusion of infrared and visible video. The proposal is focused on human detection starting from image segmentation. The main idea consists on performing human detection in the infrared and visible spectra by two different nodes connected to each

J.C. Castillo (✉)
Department of Systems Engineering and Automatic, University Carlos III of Madrid,
Madrid, Spain
e-mail: jocastil@ing.uc3m.es

J. Serrano-Cuerda
Instituto de Investigación en Informática, Universidad de Castilla-La Mancha,
Albacete, Spain
e-mail: jserranocuerda@gmail.com

A. Fernández-Caballero
Departamento der Sistemas Informática, Universidad de Castilla-La Mancha,
Albacete, Spain
e-mail: Antonio.Fdez@uclm.es

A. Martínez-Rodrigo
Instituto de Tecnologías Audiovisuales, Universidad de Castilla-La Mancha,
Albacete, Spain
e-mail: arturo.martinez@uclm.es

© Springer International Publishing Switzerland 2016
P. Novais et al. (eds.), *Intelligent Distributed Computing IX*,
Studies in Computational Intelligence 616,
DOI 10.1007/978-3-319-25017-5_32

343

camera. Then image fusion is performed in a central node to enhance the detection performance in each spectrum separately. Indeed, in the current context of increased surveillance and security, more sophisticated and robust surveillance systems are needed [20]. The current proposal for a robust human detection system has been developed from the general framework INT^3-Horus [8]. This framework is conceived as a developing environment for every kind of system which performs monitoring and activity interpretation tasks.

Thermal infrared images have a number of unique features in comparison with the visible spectrum images. The objects' intensity is mainly determined by their temperature and radiated heat, and is independent from the current lightning conditions. Given that, a detection system based on this spectrum might be equally applied in day and night conditions. However, human's thermal signature does not always satisfy this condition [12]. Although an efficient segmentation may also be performed with high temperature in the environment using background segmentation techniques, it is hard to carry out a shape-based classification or to distinguish people based on the features of the human body. On second term, most infrared images have lower spatial resolution and sensitivity than visible-light spectrum images, mostly due to the technological limitations of the cameras which acquire them. As a consequence, these images usually have low quality and contrast with the background, as well as a great amount of noise.

The apparent color of an object is mainly influenced by two physical factors such as the spectral energy distribution of the lighting source affecting it and the reflective properties of the object's surface [19]. The representation of a complete human by a single color model is usually too restrictive, even if the model includes several modes. Thus, spatial information has been recently included in approaches such as correlograms, in which a co-occurrence matrix expresses the probability of pixels of two different colors placed to a certain distance between each other [1], texture and color information to generate a high multidimensional space [17] or contours to detect borders in those regions of interest corresponding to human in the scene [16].

Some researchers are performing image fusion by using visible and infrared images together to enhance the performance of people monitoring [13]. The underlying idea is to take advantage of the strengths of both visible and infrared spectra. So, different image fusion techniques are arising, although the fusion of infrared and visible images is not trivial [6]. An example is found in [5] where a background-subtraction technique which fuses contours from infrared and visible imagery for persistent people detection in urban settings is presented.

The rest of the article is organized as follows. Section 2 describes the INT^3-Horus levels selected to implement the new people detection architecture. Some initial results validating the approach are offered in Sect. 3. Finally, some conclusions are provided in Sect. 4.

2 INT³-Horus Levels for Robust People Detection

INT³-Horus is a multi-sensor framework to carry out monitoring and activity interpretation (e.g. [3, 8]). The framework establishes a set of levels with some clearly defined input/output interfaces to provide a hierarchy to the processing. If thinking of several sensors that provide input information at the lowest level (the acquisition level), several modules, each one responsible for the acquisition of a type of sensor, are located. For each level, the framework provides a set of inputs and outputs to be met by the modules. The greatest advantage of the INT³-Horus framework resides in enabling the adaptation of a flexible set of levels to a particular final system [2, 4]. Though, a whole set of levels are proposed to cover every step of a generic multi-sensorial activities interpretation system [7, 14, 15].

The most suitable levels for the needs of the current proposal, robust people detection by infrared and visible spectra video fusion, has been selected. These are: *Acquisition, Segmentation, Fusion, Identification* and *Tracking*. The structure of these levels, as well as their inputs and outputs, are described next in detail.

2.1 Acquisition Level

In the first level, two acquisition nodes are working in parallel, grabbing images from an infrared and visible-light camera, respectively.

Both cameras are placed in parallel and focused to a common point of the same scenario, since our objective is to obtain two similar views of the same scene. Rear and front views of our installation can be observed in Fig. 1a, b.

Infrared test images are acquired from a *FLIR A-320* camera, with a sensitivity from −40° to 70°. The camera grabs frames at a resolution of 320×240 pixels with a frame rate of 5 frames per second. Dynamic temperature range is enabled, that is,

(a)   **(b)**

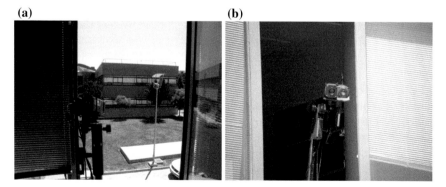

Fig. 1 Installation for simultaneous acquisition in the infrared and visible spectra. **a** Back view. **b** Front view

the brightest pixel always corresponds to the maximum detected temperature in the scene, and the darkest pixel is associated to the minimum temperature registered in the scenario.

Visible image acquisition is realized with a *SONY FCB-EX780bp* camera grabbing at 384 × 288 pixels and a frame rate of 5 frames/second. This frame rate has been forced to synchronize the acquired frames with those acquired from the infrared camera.

2.2 Segmentation Level

In this level, two modules run simultaneously to detect human candidates in both spectra. It is important to point out that the *Segmentation* level includes a set of segmentation algorithms implemented to capitalize the specific features of each spectrum, as well as to cope with different monitored scenarios. Thus, our proposal allows to empirically choose the more suitable segmentation algorithms for a given environment. Notice that in this *Segmentation* level false positives are also discarded. To do so, a set of shape restrictions is imposed (i.e. their shape, location and height/width proportion).

The detected human candidates are later refined in the *Fusion* level. A rule-based system is implemented to obtain a more robust human detection from infrared and color segmentation.

2.2.1 Segmentation in Infrared Spectrum

The main advantage of using the infrared spectrum for human segmentation is that people usually appear with greater intensity in a frame. This property gains importance when the ambient temperature is not too high or when the scenario is poorly illuminated. An important problem in infrared imaging segmentation appears when the ambient temperature is too high, e.g. summer, and humans appear colder than the environment or even at the same temperature. This problem hardens their distinction even by the human eye, as shown in Fig. 2a.

Different approaches have been studied and implemented (e.g. [9, 18]) for infrared segmentation in INT^3-Horus. The first technique only uses the intensity information of the latest acquired frame, taking advantage of the assumption of humans being warmer than other objects in the image. Motion information is later added to this method. At this point, we would like to point out that our main intention is to establish which proposal is more suitable for a particular monitored environment.

(a) (b)

Fig. 2 Image acquired at a high ambience temperature. **a** Infrared spectrum. **b** Visible spectrum

2.2.2 Segmentation in Visible Spectrum

Although many segmentation algorithms can be found for object detection in visible video, only a few of them are specifically focused on human detection. Besides, on many occasions the detected objects are classified as humans on a higher level, *object tracking*. Different approaches are tested on this level too. On this occasion, the motion history of the objects found in the scene has been used as well as the comparison between the current frame and a background or reference image (e.g. [10, 11]). The most suitable approach is chosen based on the results obtained on each scenario.

2.3 Fusion Level

Next the information from the two segmentation nodes arrives to the Fusion node, which is in charge of performing a *Fusion* of those human candidates detected on visible and infrared spectra. This level is one of the most interesting proposals of the current work. Its role is especially important since both spectra have a series of limitations and strengths which will be taken into account to elaborate a strong and reliable fusion algorithm. A *region level fusion* approach has been chosen since it allows using intelligent fusion rules regarding the specific features of each spectrum. Thus, sharpness problems are mitigated in dark and warm environments thanks to the use of complementary information from both cameras combined with intelligent fusion rules.

The *Fusion* algorithm requires images from both cameras in the same coordinates system as a requirement to fuse information from both spectra. Therefore, as the first fusion step, it is necessary to perform an initial calibration to have a common field of view on both images. In our proposal, the visible frame is used as a reference image and a series of geometrical transformations are applied to the infrared image. The result is shown in Fig. 3.

Fig. 3 Result of the
calibration of the infrared
and visible images

The fusion process starts after the calibration has been performed. The main objective is to find the humans in the scene using the regions of interest (ROIs) associated to the human candidates found in the previous *Segmentation* level. First, humans who are found in the common zone to both spectra are unified. The result is a set of ROIs similar to each other and detected on both spectra. As an example, Fig. 4 shows a human detected in both segmentations. It was predictable that these results would not be perfect. For example, while the feet of the human are not detected in the infrared spectrum (as shown in Fig. 4a), part of his head is not found in the visible one, whilst part of the lawn next to the human feet is also included in the region of interest, as appreciated in Fig. 4b. As a result, a new ROI is formed, as seen in Fig. 4c.

Apart from the human candidates detected in both spectra, results only belonging to the infrared or visible one are also considered, depending on some relevant image features. These are the mean illumination and the standard image deviation for the infrared spectrum, while the average intensity is our main cue in the visible spectrum.

Fig. 4 Fusion of regions of
interest in infrared and
visible spectra. **a** Infrared
ROI. **b** Visible ROI. **c** ROI
achieved by the fusion
process

(a)   **(b)**   **(c)**

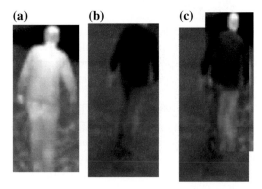

3 Data and Results

The selected test environment is an outdoor scenario. The scenario does not have any predefined access, so that a pedestrian enters into the scene from the lower limits as well as at the left or right sides of the image. A platform constructed of concrete is located in the lower part of the scene. This material quickly absorbs the temperature of the environment. The same property is also present in the building placed in the scene background. The building shows additional problems for infrared human detection. The reason is that the infrared camera automatically performs thermal attenuation, which results in the lack of accuracy in obtaining far objects' temperature. The attenuation causes the thermal readings of pedestrians to be confused with the temperature of the building, this way hardening their isolation from the scene background.

   Figures 5 and 6 show the performance of applying fusion techniques to the cases in which the IR and color spectra do not work properly. This cases cover some of the most challenging cases. For instance, sometimes the infrared spectrum does not only work better than the visible information under low lightning conditions, but also in zones covered by shades (see Fig. 5a). In this case, the infrared spectrum image detects a human candidate (see Fig. 5b), and this detection is used as the final detection results, just as depicted in Fig. 5c. The opposite case is also possible. In Fig. 6b, the human detected in the scene is very difficult to distinguish from the background in the infrared spectrum. Yet, he/she can be easily detected in the visible spectrum, as shown in Fig. 6a. Thus, the detection in the visible spectrum is assigned to the the final result is shown in Fig. 6c.

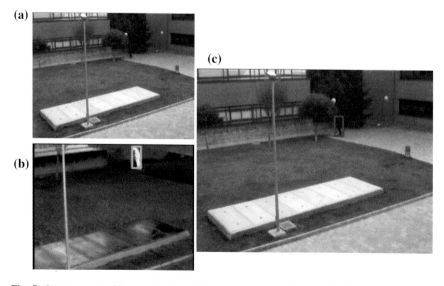

Fig. 5 Improvement of human detection after a poor segmentation result in the visible spectrum. **a** Segmentation in the visible spectrum. **b** Segmentation in the infrared spectrum. **c** Fusion results

Fig. 6 Improvement of human detection after a poor segmentation result in the infrared spectrum. **a** Segmentation in the visible spectrum. **b** Segmentation in the infrared spectrum. **c** Fusion results

4 Conclusions

This paper has introduced a hierarchical architecture for robustly detecting people through infrared and visible video fusion. The architecture is based on the INT³-Horus framework, initially designed to perform monitoring and activity interpretation tasks.

The main stages of the proposed architecture for video fusion have been described. First, the suitability of the approach for robust human detection is presented. Next, the levels chosen from that framework were detailed, explaining in depth the objective of each one of them. After an initial *Acquisition* stage, human candidates from both spectra are detected in the *Segmentation* level. In this level different algorithms were implemented, using visible and infrared information, to choose the most appropriated approaches for the objectives and environment where our system is deployed. Results from the segmentation on both spectra (detected by the chosen algorithms) are unified in a *Fusion* level, obtaining a single list of possible humans.

Acknowledgments This work was partially supported by Spanish Ministerio de Economía y Competitividad/FEDER under TIN2013-47074-C2-1-R grant.

References

1. Capellades, M., Doermann, D., DeMenthon, D., Chellappa, R.: An appearance based approach for human and object tracking. In: Proceedings of IEEE International Conference on Image Processing, vol. 2. IEEE, pp. 85–88 (2003)
2. Carneiro, D., Castillo, J.C., Novais, P., Fernández-Caballero, A., Neves, J.: Multimodal behavioral analysis for non-invasive stress detection. Expert Syst. Appl. **39**(18), 13376–13389 (2012)
3. Castillo, J.C., Carneiro, D., Serrano-Cuerda, J., Novais, P., Fernández-Caballero, A., Neves, J.: A multi-modal approach for activity classification and fall eetection. Int. J. Syst. Sci. **45**(4), 810–824 (2014)
4. Costa, Â., Castillo, J.C., Novais, P., Fernández-Caballero, A., Simoes, R.: Sensor-driven agenda for intelligent home care of the elderly. Expert Syst. Appl. **39**(15), 12192–12204 (2012)
5. Davis, J., Sharma, V.: Background-subtraction using contour-based fusion of thermal and visible imagery. Comput. Vis. Image Underst. **106**(2–3), 162–182 (2007)
6. Elguebaly, T., Bouguila, N.: Finite asymmetric generalized gaussian mixture models learning for infrared object detection. Comput. Vis. Image Underst. **117**(12), 1659–1671 (2013)
7. Fernández-Caballero, A., Castillo, J.C., Rodríguez-Sánchez, J.M.: Human activity monitoring by local and global finite state machines. Expert Syst. Appl. **39**(8), 6982–6993 (2012)
8. Fernández-Caballero, A., Castillo, J.C., López, M.T., Serrano-Cuerda, J., Sokolova, M.V.: Int³-horus framework for multispectrum activity interpretation in intelligent environments. Expert Syst. Appl. **40**(17), 6715–6727 (2013)
9. Fernández-Caballero, A., Castillo, J.C., Serrano-Cuerda, J., Maldonado-Bascón, S.: Real-time human segmentation in infrared videos. Expert Syst. Appl. **38**(3), 2577–2584 (2011)
10. Fernández-Caballero, A., López, M.T., Carmona, E.J., Delgado, A.E.: A historical perspective of algorithmic lateral inhibition and accumulative computation in computer vision. Neurocomputing **74**(8), 1175–1181 (2011)
11. Fernández-Caballero, A., López, M.T., Saiz-Valverse, S.: Dynamic stereoscopic selective visual attention (dssva): Integrating motion and shape with depth in video segmentation. Expert Syst. Appl. **34**(2), 1394–1402 (2008)
12. Goubet, E., Katz, J., Porikli, F.: Pedestrian tracking using thermal infrared imaging. Infrared Technol. Appl. XXXII **6206**, 797–808 (2006)
13. Leykin, A., Hammoud, R.: Pedestrian tracking by fusion of thermal-visible surveillance videos. Mach. Vis. Appl. **21**(4), 587–595 (2008)
14. Pavón, J., Gómez-Sanz, J., Fernández-Caballero, A., Valencia-Jiménez, J.J.: Development of intelligent multisensor surveillance systems with agents. Robot. Auton. Syst. **55**(12), 892–903 (2007)
15. Rivas, A., Martínez-Tomás, R., Fernández-Caballero, A.: Multiagent system for knowledge-based event recognition and composition. Expert Syst. **28**(5), 488–501 (2011)
16. Rodriguez, M.D., Shah, M.: Detecting and segmenting humans in crowded scenes. In: Proceedings of the 15th International Conference on Multimedia. ACM, pp. 353–356 (2007)
17. Schwartz, W.R., Kembhavi, A., Harwood, D., Davis, L.S., Human detection using partial least squares analysis. In: IEEE 12th International Conference on Computer Vision, pp. 24–31 (2009)
18. Sokolova, M.V., Serrano-Cuerda, J., Castillo, J.C., Fernández-Caballero, A.: Fuzzy model for human fall detection in infrared video. J. Intell. Fuzzy Syst. **24**(2), 215–228 (2013)
19. Yilmaz, A., Javed, O., Shah, M.: Object tracking: a survey. ACM Comput. Surv. **38**(4) (2006)
20. Zin, T.T., Takahashi, H., Toriu, T., Hama, H.: Fusion of infrared and visible images for robust person detection. In: Image Fusion. Intech, pp. 239–264 (2011)

Part IX
Special Session on Cognitive Models and Emotions Detection for Ambient Intelligence (COMEDAI 2015)

Recommendations with Personality Traits Extracted from Text Reviews

Antonella Di Rienzo and Asana Neishabouri

Abstract It is well known that human reasoning and decision-making are strongly influenced by psychological aspects. Recent works explore the adoption of personality traits to provide personalized recommendations. In this article, we report experimental results obtained with implicit recognition of Big Five personality traits from users' text reviews. Hence, we present a personality-based recommender system with the analysis of the overall users' satisfaction regarding the list of recommended items, showing promising results.

1 Introduction

Recommender systems have obtained great success as an intelligent information system to help deal with the information overload problem, especially in the field of e-commerce. Previous studies on recommender systems mainly consider user preference information (e.g., user ratings, users' past behavior), item properties (e.g., price), or user demographic information (e.g., gender). For instance, collaborative filtering approaches are methods able to make automatic predictions about the user's interests by collecting his preferences, while content based filtering approaches are based on the description of the item and a profile of the user's preference. Other information (e.g., contexts, tags and social information) are also used in the implementation of recommender system [25]. Recently, some studies have considered the recommendation based on users' psychological characteristics [5, 10].

Personality can be defined as a set of characteristics possessed by a person that uniquely influences his or her cognitions, emotions, motivations, and behaviors in various situations. Personality is also an effective factor for decision making. People with similar personality characteristics are more likely to have similar interests and

A. Rienzo (✉) · A. Neishabouri
Politecnico di Milano, Piazza Leonardo da Vinci 32, Milano, Italy
e-mail: antonella.dirienzo@polimi.it

A. Neishabouri
e-mail: asana.neishabouri@mail.polimi.it

© Springer International Publishing Switzerland 2016
P. Novais et al. (eds.), *Intelligent Distributed Computing IX*,
Studies in Computational Intelligence 616,
DOI 10.1007/978-3-319-25017-5_33

355

preferences [2]. One of the biggest obstacle to the widespread adoption of personality based recommender systems is the difficult task of eliciting personality traits from users by means of standard quizzes [9]. Such quizzes contain around one hundred questions and users are negatively affected by the task of answering to all of them. The risk is for the user to provide random answers, thus leading to wrong personality traits. On the contrary users are willing to write reviews about products or services (e.g. TripAdvisor, Amazon, Yelp). Recently a number of works [1, 11] suggest to extract personality traits from free text.

In this work, we propose to use the automatic extraction of personality trait from text reviews as an input to the recommender system that works in cooperation with a more traditional hybrid collaborative filtering + user based filtering + demographic based recommender system. Thanks to the blending of these technologies it is possible to boost traditional recommender systems with the power of a personality based recommender system without the need of bothering the users with complex quizzes. Our study has been organized into 2 steps: in the first step we have implemented a system to extract personality traits from text reviews and we have validated the quality of the personality traits by comparison with outputs from a standard questionnaire; in the second step we have implemented a cascaded hybrid recommender system (personality based, demographic based and collaborative filtering) and we have validated the quality of the recommendations through users' feedbacks. The structure of the paper is the following: in Sect. 2 we provide the related works; in Sect. 3 we propose our personality elicitation system; in Sect. 4 we proceed with our personality based recommender system; in Sect. 5 we will make a conclusion and discuss possible further works.

2 Related Works

The most commonly used human psychological aspects in recommender systems include personality traits [5, 10], demographic information [19], emotion [7], temperament [13] and lifestyle [12]. Nunes and Hu [17] propose a personality-based recommender system to provide a better personalized environment for the customer. They claim that one interesting outcome of introducing a psychological dimension into the recommender system could be the possibility of products categorization based not only on their attributes (price, physical parameters, etc.), but also on the effect they may have on the consumer. Lin and Mcleod [13] propose a temperament-based filtering model incorporating human factors, especially human temperaments (Keirsey's theory), into the processing of an information recommendation service. Their model categorizes the information space into 32 temperament segments. Combining with the content based filtering technique, their method aims at recommending the information units which best matched both users' temperaments and interests. Even though the system utilizes personalities to model user profiles, they don't really take the psychological relation between human personalities and information items into account. In [20], the authors apply the relation between musical preferences and personality traits found in [22] to recommend music. Recio-Garcia et al. [21] introduce a novel method of generating recommendations to groups based on existing techniques of collaborative filtering and taking into account the group personality composition. They test their method in the movie recommendation domain and experimentally evaluated its behavior under heterogeneous groups according to the group personality composition. In [23], the authors present a prototype of personality based recommender system in which personality traits are extracted from text reviews. Although inspiring, their work is limited for two reasons: the lack of validation of the extracted personality traits and the lack of integration with other sources of information such as user's demographics item attributes and users' ratings.

Researchers have proposed various approaches to deal with the personality elicitation problem. In 2005 a pioneering work by Argamon et al. [1] classified neuroticism and extroversion using linguistic features such as function words, deictics, appraisal expressions and modal verbs. Oberlander and Nowson [18] classified extraversion, stability, agreeableness and conscientiousness of blog authors using n-grams as features and Naive Bayes as learning algorithm. Mairesse et al. [14] ran personality recognition in both conversation (using observer judgements) and text (using self assessments via Big5). Iacobelli et al. [11] used as features word n-grams extracted from a large corpus of blogs, testing different extraction settings, such as the presence/absence of stop words or inverse document frequency. Chen et al. [3] considered the role of personal values in social media texts using the Linguistic Inquiry and Word Count (LIWC) text analysis software.

We propose an alternative way of constructing a personality based recommender system, which considers users' demographic features, restaurants ratings given by users and automatic recognition of users' personality traits from their text-reviews.

3 Personality Elicitation

The Big5 factor model, introduced in psychology by Norman [16], emerged from empirical analysis of rating scales, and has become a standard over the years. The five bipolar personality traits, namely Extraversion, Neuroticism, Agreeableness, Conscientiousness and Openness, have been proposed by Costa and Mac-Crae [15]. We annoted users' personality by means of our personality elicitation system, that makes use of correlations between written text reviews and the Big5 personality traits. Personality Elicitation from Text (PET henceforth) consists in the automatic classification of authors' personality traits from pieces of text they wrote. This task, that is partially connected to authorship attribution, makes usage of techniques from linguistics, psychology, data mining and communication sciences.

3.1 Methodology

Our methodology makes a review sentiment calculation process (positive or negative) based on Hu and Bing Liu opinion word lexicon [8] determining whether the polarity of each opinion is positive or negative and combining this prediction with linguistic cues to find the most probable personality trait, which is one of the big five personality factors extracted from Tripadvisor user's restaurant review-text. This is based on the assumption that one user has one and only one complex personality, and that this personality emerges at various levels from written text reviews.

In this manner, we try to extract linguistic cues, which are useful to determine important information features and consequently the user personality, such as: pronouns, verb tenses, articles and etc. This allows us a possibility to create a precise user profile that better matches with user characteristics. Hence, in this preprocessing phase the unstructured text data is converted into numerically structured data matrix. The text transformation process comprises of: text to lowercase conversion, tokenization, stop words removal, stemming, text tagging and finally the generation of the Feature Vector Matrix (FVM) for representing the text on the basis of its characteristics. In the processing phase the system generates the personality user profile according to the Table 1.

3.2 Validation

We run a temporary online experiment with 27 participants, who had to explicitly mention their demographic data (age, gender) and who were asked to write 15 restaurants reviews in English, rating each restaurant on a 1 to 5 star scale. Since research suggests that the same five-factor structure of personality can be found in multiple countries [24] and our academic environment and campuses are international, we made our study with an heterogeneus set of people. The participants were Iranian, Ukrainian, Chinese, Palestinian, Egyptian and Italian master students from Politecnico di Milano, in the field of Computer, Management or Mechanical Engineering

Table 1 Linguistic cues with respect to personality traits

Personality traits	Linguistic cues
Agreeableness	Positive emotions, first person pronouns, present verbs, word longer than 6 letters
Extrovert	Positive emotions, third person pronouns
Neuroticism	Negative emotions, present verbs, first person pronouns
Openness	Negative emotions, present verbs, future verbs, third person pronouns, second person pronouns, prepositions, articles
Conscientiousness	Positive emotions, present verbs, prepositions

Table 2 Confusion matrix

	27 participants	Text reviews				
		Agreeableness	Extrovert	Neuroticism	Openess	Conscientiousness
Quizzes	Agreeableness	9				
	Extrovert		3			
	Neuroticism	1		2	1	
	Openess	2			3	1
	Conscientiousness				1	4

aged between 24 and 35, 12 males and 15 females. Although they gave their consent to take part to the study, participants were not previously informed about the personality aspect until they faced the Big5 personality test, which appeared after writing the reviews. All of them decided to perform it and this could ensure that users were neutral, not knowing how text reviews were going to be processed and allowing us to achieve more reliable results. To evaluate the effectiveness of our implicit personality recognition methodology, we compared the output implicitly extracted from text reviews, which is one trait of the Big-Five Factor per each person, with the result of the Big-Five Factor Markers extracted from the International Personality Item Pool, developed by Goldberg [6], filled by our participants explicitly. The trait with highest score from questionnaire result has been considered as the most effective representation of the user's personality and has been compared with the result produced by our algorithm in order to understand if the latter was affected by misclassification. The confusion matrix resulting from the experiment is shown in Table 2. The rows correspond to traits extracted explicitly from the psychological test, and the columns correspond to personality isolation results implicitly extracted from text reviews. The numbers on the diagonal are telling us how many users have their trait, extracted explicitly by questionnaire, identical to the personality trait extracted from text reviews. Based on the confusion matrix, we can clearly see the accuracy and conflict of our algorithm: Agreeableness and Extroversion have 100 % accuracy; Openness has 50 % accuracy, 33 % conflicts with Agreeableness and 17 % conflicts with Conscientiousness; Conscientiousness has 80 % accuracy and 20 % conflicts with Openness; Neuroticism has 50 % accuracy and 25 % conflicts with Openness and Agreeableness.

4 Personality-Based Recommender System

We propose a unique cascading hybrid recommendation approach which refines, in a staged process, the recommendation from among the candidate set by combining the users' personality features, the users' demographic information and the users' ratings.

4.1 Application

Our application has been implemented in Matlab and it works as follows.

When TripAdvisor users, who have more than 3 reviews in restaurant domain, enter their username, they can see:

- their demographic features such as age range and gender, which are recorded in the trip advisor database
- and their personality, extracted by our algorithm.

In addition the system provides them with a list of 10 restaurant recommendations, based on the user personality, gender and age range similarities with other reviewers who gave restaurants ratings greater than or equal to 4. If the user has no account in trip advisor or has less than 3 reviews, he could type at least 250 words producing 3 reviews (or copy 250 words of his text reviews) in the comment box, then specifying his age range and gender, he could receive the recommendations.

4.2 Validation

To validate our new hybrid recommender system, we used the perceived Accuracy, Novelty and Satisfaction measures. Standard metrics (e.g., error metrics and accuracy metrics), which are evaluated just by statistical methods developed in the fields of information retrieval and machine learning (e.g., hold-out or k-fold cross-validation), defines objective quality. However, this is not always a suitable indicator of the potential for persuasion of a recommender system.

Our results are based on how the users see a recommender system with the characteristics related to subjective elements. Therefore to get an initial feedback from the users, in this second step we conducted an offline evaluation and we invited 21 participants to give answers to questions about novelty, perceived accuracy and satisfaction of the given recommendations.

1. Novelty measures were collected as follows: for each recommended item we first showed its title (with no other information) and asked the question Have you ever been on this restaurant? The answer (yes/no) was then used to compute the First Order Novelty (FON). A value of 1 was given if the user had never been on the restaurant and 0 otherwise. If the user had never been on the restaurant, we proceeded with an indepth exploration to assess Second Order Novelty (SON). We asked the user if he/she had ever heard about the restaurant, inviting him/her to explore related information (location, picture) to help him/her refresh his/her memory. If a user answered yes (or if FON was 0) the SON value was set to 0, otherwise it was set to 1.
2. Perceived accuracy measures were collected as follows: for each recommendation, if the user had gone to the restaurant, he/she was asked to rate how much

he/she liked or disliked it (on a 1 to 5 scale). If the user had not gone to the restaurant, we invited to see a review of the restaurant and the picture; we asked him to rate the degree of potential interest for the restaurant. For each user, we calculated the average of the values of Novelty, and Perceived accuracy that were assigned to each recommendation.

3. User opinion measures were collected as follows: users were asked to provide a global judgment regarding the list of recommended restaurants in binary format [1-0] (yes/no).

Our results were reasonable and depicted that the restaurant recommender system based on personality had 88 % of novelty, could predict user preferences with 82 % of accuracy and could achieve 79 % user satisfaction.

We also applied classic information retrieval metrics to evaluate our engine, using Precision, Recall, Fallout and F-measure (F1). Ideally, a good algorithm should have high recall and low fallout, it means should recommend interesting items and avoid recommending items of no interest to the user.

$$Precision = \frac{\#Relevant\ Recommended\ Items}{\#Recommended\ Items} \qquad (1)$$

Precision is a measure of the ability of the system to return only relevant results to the user. If the user's rating was greater than 3, we considered the restaurant as a relevant item for recommendation. In Fig. 1 the users' precision measure is shown. The overall precision of our personality based recommender system in average is about 82 %.

$$Recall = \frac{\#Relevant\ Recommended\ Items}{\#Existing\ Relevant\ Items\ on\ Database} \qquad (2)$$

Recall is the proportion of relevant recommendations to the user that appear in top recommendations. In Fig. 2 each user recall measure is shown. In our personality based recommender system the average of the recall is 61 %.

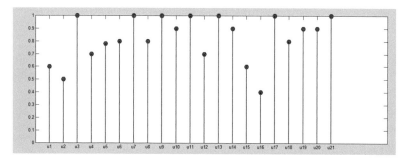

Fig. 1 Precision

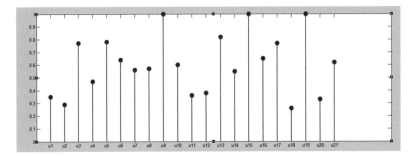

Fig. 2 Recall

$$Fallout = \frac{\#Irrelevant\ Items}{\#Existing\ Irrelevant\ Items\ on\ Database} \qquad (3)$$

The fallout is a measure of the aptitude of the system to suggest a restaurant given it is irrelevant for the user. In Fig. 3 each user fallout measure is shown. In average the overall fallout is 23 %.

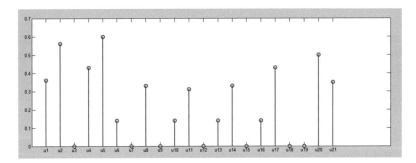

Fig. 3 Fallout

$$F1 = \frac{2 * Precision * Recall}{Precision + Recall} \qquad (4)$$

The F1 score can be interpreted as a weighted average of the precision and recall [4]. It indicates an overall utility of the recommendation list where the best score has its value at 1 and worst score at the value 0.

In Fig. 4 the F-measure is shown. The average F1 measure is about 67 %.

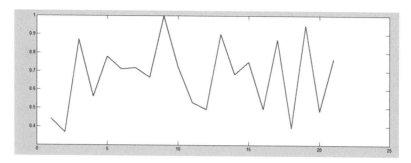

Fig. 4 F-measure

5 Conclusion and Future Work

We have shown that personality traits can be extracted from reviews through linguistic cues. Therefore, this experiment has successfully obtained the type of personality that can be used to suggest restaurants based on personality type similarity. In particular we made two main contributions. First, we conducted our analysis in a realistic scenario and the obtained results were corroborated by users' satisfaction, proving that the hybrid recommender system, in restaurant domain, could benefit from the personality study. Most importantly, we could improve the state of the art, building an algorithm that implicitly infers people personalities from their reviews. In future we expect to increase the number of our participants, taking into account users' disappointment while filling out the time consuming questionnaires that we needed for evaluating both our personality recognition algorithm and our recommender system. To this purpose, we are planning our recommender system to work in an online setting. Finally, we intend to build a user model that considers all big five dimensions of user personality traits simultaneously instead of selecting the highest of them.

Acknowledgments Special thanks to our participants for their cooperation.

References

1. Argamon, S., Dhawle, S., Koppel, M., Pennebaker, J.: Lexical predictors of personality type. In: Proceedings of the Joint Annual Meeting of the Interface and the Classification Society of North America, (2005)
2. Cantador, I., Fernandez-Tobas, I., Bellogn, A., Kosinski, M., Stillwell, D.: Relating personality types with user preferences in multiple entertainment domains. In: UMAP Workshops, Citeseer (2013)
3. Chen, J., Hsieh, G., Mahmud, J.U., Nichols, J.: Understanding individuals' personal values from social media word use. In: Proceedings of the 17th ACM Conference on Computer Supported Cooperative Work and Social Computing, pp. 405–414. ACM (2014)

4. Cremonesi, P., Garzotto, F., Turrin, R.: Investigating the persuasion potential of recommender systems from a quality perspective: an empirical study. ACM Trans. Interact. Intell. Syst. (TiiS) **2**(2), 11 (2012)
5. Dunn, G., Wiersema, J., Ham, J., Aroyo, L.: Evaluating interface variants on personality acquisition for recommender systems. In: Houben, G.J., McCalla, G., Pianesi, F., Zancanaro, M. (eds.) User Modeling, Adaptation, and Personalization. Lecture Notes in Computer Science, vol. 5535, pp. 259–270. Springer, Berlin (2009)
6. Goldberg, L.R.: The development of markers for the big-five factor structure. Psychol. Assess. **4**(1), 26 (1992)
7. Gonzalez, G., de la Rosa, J., Montaner, M., Delfin, S.: Embedding emotional context in recommender systems. In: IEEE 23rd International Conference on Data Engineering Workshop, pp. 845–852 (2007)
8. Hu, M., Liu, B.: Mining and summarizing customer reviews. In: Proceedings of the Tenth ACM SIGKDD International Conference on Knowledge Discovery and Data Mining, pp. 168–177. ACM (2004)
9. Hu, R.: Design and user issues in personality-based recommender systems. In: Proceedings of the Fourth ACM Conference on Recommender Systems, vol. 10, pp. 357–360. ACM, New York, NY, USA (2010)
10. Hu, R., Pu, P.: A study on user perception of personality-based recommender systems. In: De Bra, P., Kobsa, A., Chin, D. (eds.) User Modeling, Adaptation, and Personalization. Lecture Notes in Computer Science, vol. 6075, pp. 291–302. Springer, Berlin (2010)
11. Iacobelli, F., Gill, A.J., Nowson, S., Oberlander, J.: Large scale personality classification of bloggers. In: Affective Computing and Intelligent Interaction, pp. 568–577. Springer (2011)
12. Lekakos, G., Giaglis, G.M.: Improving the prediction accuracy of recommendation algorithms: approaches anchored on human factors. Interact. Comput. **18**(3), 410–431 (2006)
13. Lin, C.H., McLeod, D., et al.: Exploiting and learning human temperaments for customized information recommendation. In: IMSA, pp. 218–223 (2002a)
14. Mairesse, F., Walker, M.A., Mehl, M.R., Moore, R.K.: Using linguistic cues for the automatic recognition of personality in conversation and text. J. Artif. Intell. Res. **30**(1), 457–500 (2007)
15. McCrae, R., Costa, P.: The Neo Personality Inventory Manual. Psychological Assessment Resources, Odessa (1985)
16. Norman, W.T.: Toward an adequate taxonomy of personality attributes: replicated factor structure in peer nomination personality ratings. J. Abnorm. Soc. Psychol. **66**(6), 574 (1963)
17. Nunes, M.A.S., Hu, R.: Personality-based recommender systems: an overview. In: Proceedings of the Sixth ACM Conference on Recommender Systems, p. 56. ACM (2012)
18. Oberlander, J., Nowson, S.: Whose thumb is it anyway?: classifying author personality from weblog text. In: Proceedings of the COLING/ACL on Main Conference Poster Sessions, Association for Computational Linguistics, pp. 627–634 (2006)
19. Pazzani, M.: A framework for collaborative, content-based and demographic filtering. Artif. Intell. Rev. **13**(5–6), 393–408 (1999)
20. Perik, E., De Ruyter, B., Markopoulos, P., Eggen, B.: The sensitivities of user profile information in music recommender systems. Proceedings of Private, Security, Trust, pp. 137–141 (2004)
21. Recio-Garcia, J.A., Jimenez-Diaz, G., Sanchez-Ruiz, A.A., Diaz-Agudo, B.: Personality aware recommendations to groups. In: Proceedings of the Third ACM Conference on Recommender Systems, pp. 325–328 ACM (2009)
22. Rentfrow, P.J., Gosling, S.D.: The do re mis of everyday life: the structure and personality correlates of music preferences. J. Personal. Soc. Psychol. **84**(6), 1236 (2003)
23. Roshchina, A., Cardiff, J., Rosso, P.: A comparative evaluation of personality estimation algorithms for the twin recommender system. In: Proceedings of the 3rd International Workshop on Search and Mining User-Generated Contents, pp. 11–18. ACM (2011)
24. Triandis, H.C., Suh, E.M.: Cultural influences on personality. Annu. Rev. Psychol. **53**(1), 133–160 (2002)
25. Zheng, N., Li, Q.: A recommender system based on tag and time information for social tagging systems. Expert Syst. Appl. **38**(4), 45754587 (2011)

A Framework for the Automation of Multimodalbrain Connectivity Analyses

Paulo Marques, Jose Miguel Soares, Ricardo Magalhaes, Nuno Sousa and Victor Alves

Abstract In neuroscience research, there has been an increasing interest in multimodal analysis, combining the strengths of unimodal analysis while reducing some of its drawbacks. However, this increases complexity in data processing and analysis, requiring a big amount of technical knowledge in image manipulation and a lot of iterative processes requiring user intervention. In this work we present a framework that incorporates some of this technical knowledge and enables the automation of most of the processing in the context of combined resting-state functional Magnetic Resonance Imaging (rs-fMRI) and Diffusion Tensor Imaging (DTI) data processing and analysis. The proposed framework presents an object-oriented architecture and its structure reflects the nature of three levels of data processing (i.e. acquisition level, subject level and study level). This framework opens the door to more intelligent and scalable systems for neuroimaging data processing and analysis that ultimately will lead to the dissemination of such advanced techniques.

P. Marques (✉) · J.M. Soares · R. Magalhaes · N. Sousa
Life and Health Sciences Research Institute (ICVS), School of Health Sciences,
University of Minho, Campus Gualtar, 4710-057 Braga, Portugal
e-mail: paulo.c.g.marques@ecsaude.uminho.pt

J.M. Soares
e-mail: josesoares@ecsaude.uminho.pt

R. Magalhaes
e-mail: ricardomagalhaes@ecsaude.uminho.pt

N. Sousa
e-mail: njcsousa@ecsaude.uminho.pt

P. Marques · J.M. Soares · R. Magalhaes · N. Sousa
ICVS/3Bs - PT Government Associate Laboratory, Braga/guimares, Portugal

P. Marques
Clinical Academic Center Braga, Braga, Portugal

V. Alves
Department of Informatics, University of Minho, Braga, Portugal
e-mail: valves@di.uminho.pt

© Springer International Publishing Switzerland 2016
P. Novais et al. (eds.), *Intelligent Distributed Computing IX*,
Studies in Computational Intelligence 616,
DOI 10.1007/978-3-319-25017-5_34

365

1 Introduction

With the advent of non-invasive neuroimaging techniques, the neuroscience research field has put a lot of effort in studying the human brain in-vivo. Recently, one of the main focus of research has been the study of the brain as a network [1]. Techniques such as functional Magnetic Resonance Imaging (fMRI) enable researchers to map brain activity both at rest and during task execution. In particular, Functional Connectivity (FC—statistical dependencies in brain signals along time) analyses reveal networks of segregated regions, commonly designated as Resting State Networks (RSNs), which display coordinated activity in rest conditions (Fig. 1a) [2]. On the other hand, Diffusion Weighted Imaging (DWI) MRI acquisitions enable the reconstruction of the main white matter axonal tracts responsible for the brains wiring through Diffusion Tensor Imaging (DTI) tractography (Fig. 1b) leading to the concept of Structural Connectivity (SC) [3]. The combination of the above mentioned techniques (Fig. 1c) is of major interest for the scientific and clinic research enabling to understand what structural alterations underlie functional alterations [4].

The power of the combination of the previously mentioned techniques also brings some drawbacks. Both SC and FC analysis have a fair amount of complexity that has been overcome with the development of specially tailored software tools. However, combining both methodologies adds a new layer of complexity to the analysis, involving the manipulation of several software tools from different providers, knowledge regarding image manipulations and time consuming procedures that are not familiar to most of researchers and clinicians interested in such analysis. More-

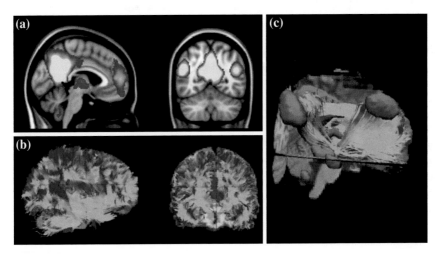

Fig. 1 Resting state fMRI analysis enable the identification of RSNs (**a**), while DTI tractography enable the reconstruction of the main brain tracts (**b**). The combination of both analyses ultimately leads to identification of the tracts underlying the functional patterns of the RSNs (**c**)

over, the number of procedures that need to be repeated and parameters that need to be set increase the chance of error and decrease the reliability of the results [5, 6].

Although Ambient Intelligence (AmI) systems are commonly perceived as intelligent homes or devices that would self-adjust to the owner's needs, interacting with them and providing care and assistance, this kind of systems could also be of major importance in clinical practice and research. Ideally, in the neuroimaging context, AmI systems would adjust the processing procedures to the data available and the user intentions. These systems need to be context aware, personalized and adaptive, all of them characteristics of AmI [7, 8]. To the best of our knowledge, such systems are scarce, if not inexistent. In the present work, we present a simple framework developed with the goal to create a layer of transparency regarding image manipulation procedures, enabling the automation of several processing steps and reducing the need for user intervention in the process.

2 Framework Overview

The proposed framework is intended to facilitate the access of less experienced users to complex analysis such as the ones presented. Without such a framework, users would need to have knowledge regarding image file types, image processing algorithms, different software manipulation and even some programming skills. For instance, without such a framework the user would typically use software such as dcm2nii[1] to convert images from raw DICOM [9] format into NIfTI format [10]. Then he would preprocess the fMRI acquisition with several consecutive steps in order to increase data quality with other software tools (e.g. SPM [11], FSL [12], AFNI [13], BrainVoyager [14]) and repeat the same steps for every subject enrolled in the study. Afterwards, he would gather all the preprocessed images and perform group Independent Component Analysis (ICA) in order to identify RSNs using tools such as MELODIC[2] and GIFT.[3] Next he would need to isolate clusters with yet another software tool with GUI or simply by command line. Regarding the DTI data, preprocessing would need to be performed in a similar manner, repeating the process for every subject and using tools such as FSL, AFNI or TRACULA.[4] Then the user would perform the tractography on the preprocessed DWI data. Available tools for such purpose include, among others, DTIStudio [15], TrackVis [16] and 3DSlicer [17]. Finally, the isolated clusters would need to be combined with the tractography. For this purpose, the user would need to understand the concepts of standard space (where the clusters were formed) and native space (where the tractography is performed) and understand how to move images from one space to another again using different software tools, implementing different algorithms, each one involv-

---

[1] http://mccauslancenter.sc.edu/mricron/dcm2nii.htlm.

[2] http://fsl.fmrib.ox.ac.uk/fsl/fslwiki/.

[3] http://icatb.sourceforge.net/.

[4] https://surfer.nmr.mgh.harvard.edu/fswiki/Tracula.

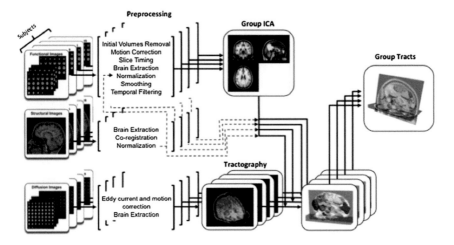

Fig. 2 Illustration of the complexity of the data processing workflow in combined FC and SC analyses

ing tradeoffs. A schematic overview of the complexity of data processing in such analyses is presented in Fig. 2.

With the implementation of the framework here presented, we aim at creating systems that have most of the knowledge necessary to such analyses embedded in its structure, adjusting the necessary procedures to the data and user intentions This represents a layer of abstraction that deals with software compatibility, different file formats, different spaces of image processing, pipelining and data structure. In fact we developed a software tool named BrainCAT [18] that implements the framework here presented. With this tool, users only need to arrange their data in an intuitive manner and the software will only require the user interaction when strictly necessary adjusting the pipeline according to the needs.

3 Implementation Details

In the case of combined FC and SC analysis, three different kinds of MRI acquisitions are typically necessary: a resting state fMRI acquisition in order to estimate the RSNs, a DWI acquisition necessary for the tractography and a structural acquisition that is needed for intermediate steps such as the removal of non-brain tissue from the images and spatial normalization procedures. Since each acquisition requires specific processing steps, independent of the other acquisitions, and have different native image spaces, this is reflected in the Object-Oriented architecture of the framework through the creation of three different classes (Mri, Dti, Mri, respectively), which implements the methods and functions necessary for the handling and processing of the data from each acquisition.

Besides dealing with different kinds of acquisitions, studies combining SC and FC are typically group studies, involving several different subjects, each one undergoing the same acquisition protocol. This was is also reflected in the frameworks architecture trough the creation of a higher-level class called Subject. This class holds three attributes, one for each of the Fmri, Dti and Mri classes thus holding the necessary tools for processing all of the acquisitions of each subject. Additionally, subjects data and methods requiring the combination of more than one acquisition are implemented at this level.

Finally, a class name Study holds information regarding the study details. This class sits at the highest level, whose main attribute is an array of Subject objects. This class is also responsible for the analysis itself, holding the information necessary for the pipelining of the processing steps. All the image-processing procedures are implemented with freely available and previously validated command line tools pro-

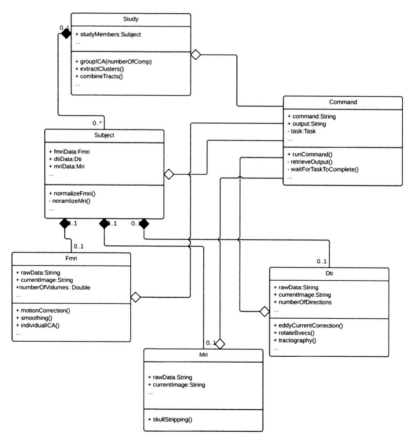

Fig. 3 Unified Modeling Language (UML) representation of the frameworks architecture presenting its main classes

vided with MRICron [19], FSL [12] and TrackVis [16]. An additional class named Command handles commands invocation and output retrieval, thus each command that needs to be run is an actual object. An overview of the Frameworks architecture can be visualized in Fig. 3.

4 Main Classes

As previously mentioned, the Command class was created in order to deal with running and retrieving output from the command line tools that need to be run. One of the main requirements for this class is the need for the commands to be run sequentially since, in the implemented workflow, the output from one step of data processing is the input of the following step of the pipeline. Since most of the programs run in the bash environment, bash environment variables were also added to the runtime environment during the programs execution. Objects of this class are used at all levels of the framework in order to run the command line tools.

The Fmri class implements the necessary methods to handle fMRI data. Besides holding the getters and setters for the input images and images generated throughout the workflow, it holds a state variable that holds the image currently being used. Except when explicitly defined otherwise, steps of the processing pipeline are always applied to this image, thus the input of most steps is also the output of the same steps. Analyses that depend only on the fMRI acquisitions are also implemented in this class. Examples of such procedures include motion correction, smoothing and temporal filtering, as well as single-subject ICA.

Similarly to the Fmri class, the Dti class implements the functions necessary to the manipulations and processing of DWI acquisitions. This includes the required getters and setters, data processing steps and subject specific analyses that only require the DWI acquisition. Similarly, to the Fmri class, a state variable holds the image currently being used in enabling pipelining. Distortions correction and tractography are examples of processing steps which this class implements.

The Mri class implements the methods needed to process the structural acquisitions, which, apart from getters and setters. In the context of the BrainCAT application, only one method for the removal of non-brain structures was implemented. However other methods could be implemented for other needs and so it also holds the last generated image during processing in an appropriate attribute.

As previously described, the Subject class contains three main attributes, each one belonging to one of the previously described classes (i.e. Fmri Dti, Mri), and implements the methods that require the manipulation of images from more than one acquisition. Whenever a processing step is required involving individual acquisitions, the necessary functions are called in the corresponding attribute. Whenever more than one acquisition is required, the most recent images are retrieved from the necessary attributes of the Subject class and the processing is performed within this class. At the end, the state variables of the attributes are updated. This is the case of the spatial normalization of the fMRI data where both fMRI and structural MRI images are used. Additionally, some steps are quite common in image processing and would be repeated if the analysis were to be made separately for each acquisition. In this case, objects of this class check if such procedures were already performed throughout the processing pipeline and, if so, it does repeat the process, thus saving precious computation time. This is also the case of normalization procedures where spatial transformations are only computed once.

The Study class is the class responsible for the implementation of the whole work-flow. The main attribute of this class is an array of objects of the Subject class hold-ing the data from each subject belonging to the study. Almost every neuroimaging workflow incorporates a preprocessing stage where data quality is enhanced through a series of sequential procedures applied separately for each subject. As so, in such procedures, the array of Subject objects is run sequentially for each step. Thus, each step is run independently for each subjects dataset, opening an opportunity for paral-lelization and/or distribution of processing procedures. This class is also the respon-sible for the implementation of group-wise procedures and analysis. For example, in the case of combined FC and SC analyses, Group-ICA, RSNs extraction and mer-gence of each subjects tracts is performed in this level.

5 Discussion

The presented framework was developed in the context of combined FC (i.e. RSNs) and SC (i.e. tractography) analysis and allowed to create a software tool that auto-mates the entire data analysis (BrainCAT). In this particular case the user sets a work-ing directory with the images of several subjects participating in the study, starts the application and it automatically recognizes the subjects whose data is suitable for analysis. The inclusion/exclusion of certain processing steps is left at the users choice. However, some steps (e.g. spatial normalization) are highly dependent on other steps (e.g. removal of non-brain structures). As such, if the user opts for one analysis where spatial normalization is strictly required, the system automatically recognizes it, performs the spatial normalization using the necessary acquisitions and, if the removal of non-brain tissue was not performed, it performs it before apply-ing the normalization.

This kind of framework has, however, the potential to be extended to basically any neuroimaging analysis where different modalities are combined. In fact, the same framework was used in order to combine volumetric analysis derived from the struc-tural MRI acquisitions with DTI tractography. Similarly structural and functional whole brain FC and SC connectivity analyses have also been performed. As long as the necessary functions are implemented at the required levels, virtually any com-bination of analysis is possible. Moreover, most of the software tools used in the neuroimaging field are still serial computing tools. The proposed architecture has the potential to be parallelized and distributed in order to analyze large amounts of data. It should be noted that such studies take tenths or hundreds of hours of process-ing time, obviously depending on the number of subjects and size of the datatsets.

Future developments of the framework could envisage a learning module that would gather usage data and adapt the systems setting to the habits of the user. This system would enable a richer user experience and would facilitate even further the process of setting the analyses.

6 Conclusions

Ideally, in the neuroimaging context, AmI systems would adjust the procedures to the data available and the user intentions. These systems would need to be context aware, personalized and adaptive, all of them characteristics of AMI systems.In this sense, proposed framework enables the creation and expansion of AmI systems in neuroimaging research, leading to the generation of more intelligent, user friendly and capable systems that may have the potential to disseminate complex neuroimaging techniques to researchers and clinicians without strong technical knowledge. All the procedures are applied using freely available software running in bash shell and thus can be implemented in any UNIX system. BrainCAT, the software developed using the presented framework, is freely available.[5] New modules are being added to the software based on the same framework.

Acknowledgments This work has been supported by FCT—Fundao para a Cincia e Tecnologia within the Project Scope UID/CEC/00319/2013. PM was supported by the SWITCHBOX project through the grant SwitchBox-FP7-HEALTH-2010-grant 259772-2 and RM is supported by the Portuguese North Regional Operational Program (ON.2 O Novo Norte) under the National Strategic Reference Framework (QREN), through the European Regional Development Fund (FEDER) by a fellowship from the project FCT-ANR/NEU-OSD/0258/2012 funded by FCT/MEC (www.fct.pt) and by FEDER.

References

1. Gong, G., He, Y., Concha, L., et al.: Mapping anatomical connectivity patterns of human cerebral cortex using in vivo diffusion tensor imaging tractography. Cereb Cortex **19**(3), 524–536 (2009). doi:10.1093/cercor/bhn102
2. Damoiseaux, J.S., Rombouts, S.A., Barkhof, F., et al.: Consistent resting-state networks across healthy subjects. Proc. Natl. Acad. Sci. U.S.A. **103**, 13848–13853 (2006). doi:10.1073/pnas.0601417103
3. Basser, P.J., Pajevic, S., Pierpaoli, C., et al.: In vivo fiber tractography using DT? MRI data. Magn. Reson. Med. **44**(4), 625–632 (2000)
4. van den Heuvel, M.P., Mandl, R., Luigjes, J., et al.: Microstructural organization of the cingulum tract and the level of default mode functional connectivity. J. Neurosci. **28**, 1084410851 (2008). doi:10.1523/JNEUROSCI.2964-08.2008
5. Hasan, K.M., Walimuni, I.S., Abid, H., et al.: A review of diffusion tensor magnetic resonance imaging computational methods and software tools. Comput. Biol. Med. **41**, 10621072 (2011). doi:10.1016/j.compbiomed.2010.10.008

---

[5]http://www.icvs.uminho.pt/research-scientists/neurosciences/resources/braincat.

6. Haller, S., Bartsch, A.J.: Pitfalls in FMRI. Eur. Radiol. **19**, 2689–2706 (2009). doi:10.1007/s00330-009-1456-9
7. Vasilakos, A., Witold, P.: Ambient Intelligence, Wireless Networking, and Ubiquitous Computing. Artech House, Inc (2006)
8. Rech, J., Klaus-Dieter, A.: Artificial intelligence and software engineering: Status and future trends. KI **18**(3), 5–11 (2004)
9. Digital imaging and communications in medicine (DICOM): National Electrical Manufacturers Association (1998)
10. Cox, R.W., Ashburner, J., Breman, H., et al.: A (sort of) new image data format standard: nifti-1. Human Brain Mapp. **25**, 33 (2004)
11. Penny, W.D., Friston, K.J., Ashburner, J.T. et al (2011) Statistical Parametric Mapping: The Analysis of Functional Brain Images: The Analysis of Functional Brain Images. Academic press
12. Smith, S.M., Jenkinson, M., Woolrich, M.W., et al.: Advances in functional and structural MR image analysis and implementation as FSL. NeuroImage **23**, S208–S219 (2004)
13. Cox, R.W.: AFNI: software for analysis and visualization of functional magnetic resonance neuroimages. Comput. Biomed. Res., Int. J. **29**(3), 162–173 (1996)
14. Goebel, R.: Brainvoyager: a program for analyzing and visualizing functional and structural magnetic resonance data sets. Neuroimage **3**(3), S604 (1996)
15. Jiang, H., van Zijl, P.C., Kim, J., et al.: DtiStudio:resource program for diffusion tensor computation and fiber bundle tracking. Comput. Methods Programs Biomed. **81**(2), 106–116 (2006)
16. Wang R, Benner T, Sorensen AG et al (2007) Diffusion toolkit: a software package for diffusion imaging data processing and tractography. Proc. Intl. Soc. Mag. Reson. Med. **15**(3720)
17. Pieper S, Halle M, Kikinis R (2004) 3D Slicer. In: IEEE International Symposium on Biomedical Imaging: Nano to Macro, 2004, pp. 632–635. IEEE
18. Marques, P., Soares, J.M., Alves, V. et al. (2013) BrainCAT-a tool for automated and combined functional magnetic resonance imaging and diffusion tensor imaging brain connectivity analysis. Frontiers Human Neurosci. **7**
19. Rorden, C., Brett, M.: Stereotaxic display of brain lesions. Behav. Neurol. **12**, 191200 (2000)

Emotion Effects on Online Learning

Ana Raquel Faria, Ana Almeida, Constantino Martins and Ramiro Gonçalves

Abstract Learning is understood as an educational activity which aims to help develop the capacities of individuals. These capabilities make individuals able to establish a personal relationship with the environment in which they are inserted. For the learning process to develop the individual has to use its sensory, motor, cognitive, affective and linguistic. Thus, this work aims at studying the effect of emotion in learning systems in online learning environments, analysing the extent to which emotional state can influence the thinking, decision making and learning process.

Keywords Learning styles · Student modeling · Adaptive systems · Affective computing

1 Introduction

The impact of emotions on learning process and especially in online learning has recently grown as shown in literature [1, 2]. The increasing examination of the complex set of parameters related to online learning discloses the significance of the emotional states of learners and especially the relationship between emotions and

A.R. Faria (✉) · A. Almeida · C. Martins
GECAD - Knowledge Engineering and Decision Support Research Center Institute of Engineering, Polytechnic of Porto (ISEP/IPP), Porto, Portugal
e-mail: arf@isepp.ipp.pt

A. Almeida
e-mail: amn@isepp.ipp.pt

C. Martins
e-mail: acm@isepp.ipp.pt

R. Gonçalves
INESC TEC and University of Trás-os-Montes e Alto Douro, Vila Real, Portugal
e-mail: ramiro@utad.pt

© Springer International Publishing Switzerland 2016
P. Novais et al. (eds.), *Intelligent Distributed Computing IX*,
Studies in Computational Intelligence 616,
DOI 10.1007/978-3-319-25017-5_35

affective learning [2]. This causes an emotional gap between student and his online learning environment. To attempt to minimize this problem by proposing the creation of a new framework that can assess the emotional state of the student using affective computing techniques and developing a proper response to induce a positive stimulus in order to facilitate the learning process and improve the student's learning results. This study began with the premise the student's negative emotions have negative effects on his behaviour, attention, motivation and ultimately in the learning process outcomes. On the other hand positive emotions would benefit the learning process. This is corroborated in some previous studies [2, 3]. Another premise was made regarding the influence of the student learning preferences and personality in his learning process. This premise was also established by previous studies [4–6]. For this study of emotions in online learning systems, the system built must be able to capture the student's emotional state. The capture of an emotional state cannot be invasive therefore Affective Computing technologies will be used. Anticipating the creation of a new model could combine one or more techniques of Affective Computing. One of the challenges will be to correctly identify and with a certain degree of reliability the student's emotional states. This process will also include the study of emotional states that can be recognized, its duration in time, what its influence on the teaching and learning process. Another important issue is the recognition of the emotional state that provides a positive emotion for learning and what incentives can be introduced to reverse a negative emotional state.

2 Affective Computing

The study of affect is included in different fields of science such as psychology, cognitive science, neuroscience, engineering, computer science, sociology, philosophy, and medicine. This has contributed to different understanding of basic terms related to affect such as emotions, feelings, moods, attitudes, affective styles, temperament, motivation, attention, reward, and so many others [7]. One of the problems in studying affect is the definition of what emotion is. Moreover, the definition and the terms associated to it. Nearly hundreds of definitions of emotion have been registered since 1981 [8]. To analyse emotion, there are several theories, which attempted to specify the interrelationships of all the components involving an emotion and the causes, the reasons and the function of an emotional response. Although there are several works summarizing these approaches, some of these theories are very controversial among the intellectual community. Nevertheless, they were the starting points for most of the research works done today in affect recognition. There are several types of emotion recognition like: Facial Expression Recognition, Eye tracking, Emotional Speech Recognition, recognizing emotion from brain activity and detection of emotion in text.

A facial expression is the result of the movements or positions of face muscles. Facial expression recognition plays an important role in natural human-machine communication [9]. Most of Facial Expression Recognition (FER) research was influenced by the early theory of emotion in with Ekman's work was based [10]. His work was based on the assumption that the emotions are universal across individuals as well as human ethnics and cultures [11]. Nowadays advances in facial recognition software make the recognition of basic facial expressions like anger, disgust, fear, happiness, sadness surprise, and others possible. A study carried out by Ekman and Friesen [12] developed the Facial Action Coding System (FACS), which is a manual technique for the measurement of facial movement FACS can code practically any anatomically possible facial expression, decomposing it into the specific Action Units (AU) that identifies independent face motion in a temporal order. Manually coding a segment of video is a timely and expressive method performed by highly trained human experts. Studies have been made trying to automate this method, see [13, 14].

Eye tracking is the process of measuring either the point of gaze (what one is looking at) or the motion of an eye in relation to the head position. The analyses of eyes properties can be used to measure the human response to visual, auditory or sensory stimuli [15, 16]. Emotion is measured through the eye tracking hardware and a statistical program which determines the excitement level to a visual image [15]. Parameters analyses include: pupil size, blink properties and gaze. The pupil size is related to an emotional reaction, a study performed by Partala and Surakka [16] indicates there is an alteration in pupil size when a subject faces a positive or negative stimulus opposite when facing a neutral stimulus. Blink has also been associated with emotional responses, for example with defensive reactions, the eye is modulated blink startle [17]. Finally, gaze patterns have been linked to emotional reactions [18].

Speech recognition consists in the ability of a machine or program in identifying words or phrases from the spoken language. The main application of speech recognition resides in assisted technology to help people with disabilities. The vocal aspect of a communication also carries information about speech emotional contents. So, speech can be divided in two parts: an explicit message, consisting of what was said; and an implicit emotional expression, entailing how the message was said. Speech Emotion Recognition (SER) aims to recognize the user emotional state in his speech signal [19]. Most acoustic features that have been used for emotion recognition can be divided into two categories: prosodic and spectral. Prosodic features have been shown to provide key speaker emotional clues. Literature in SER reached important conclusions that can be used in AC applications [20, 21]. The major conclusion relies in the possibility to recognize emotion from a speech. Although there is some difficulty in recognizing certain kind of emotions like disgust, there are others that can be recognize with more accuracy like sadness and fear. Also, the pitch of a speech seems to be connected to the level of arousal. In addition affect discovery in speech has accuracy rates lower than facial expression for recognizing basic expressions.

The field of Affective Neuroscience tries to map the neural circuitry that occurs during an emotional experience [23, 24], the techniques used by neuroscientists include fMRI, EEG, among others. fMRI is based on the MRI technology. It is a non-invasive test that uses a strong magnetic field and radio waves to create detailed body images. It monitors the blood flow in the brain to detect activity areas. Therefore fMRI provides a map of which parts of the brain are active during an emotion or feeling. A study performed by Yang and Damasio [25] with brain damaged patients show that the emotional course was required for learning and it is essential to the decision making process.

Affective detection in a written text consists in determine the emotional or attitude context within the written language or transcripts of oral communication. The initial work on this matter aimed to understand how text could express an emotion or how text could generate different emotions. These studies began by find resemblance in how people of different cultures communicate [26, 27]. Osgood [26] used a Multidimensional scaling (MDS) procedure to create models of affective words based on words evaluations provide by diverse cultures. The dimensions considered in these studies were: evaluation, potency and activity. Evaluation refers to quantification of a word in relation with an event it portrays, if that event is pleasant or unpleasant. Potency refers to how the word is related with a level of intensity, strong words as opposed to weak words. Activity refers to a word as active or passive [22].

The personality research aims to study what distinguish one individual from another [28]. Personality research depends on quantifiable concrete data that can be used to comprehend what people are like. In several publish papers is acknowledged that personality traits as influence your performance in numerous areas in your life [29, 30]. Also the relationship between personality and learning is largely accepted [5, 30]. Leading with personality one of the models largely use is the Big Five Model [6, 29]. The Big Five dimensions are Openness, Conscientiousness, Extraversion, Agreeableness, and Neuroticism (OCEAN). Openness dimension include personalities that are open to experiences, like art, emotions, adventure, uncommon ideas and a wide range of experiences. Openness dimension returns an amount of intellectual curiosity, originality and an inclination for innovation and variety. Conscientiousness dimension it has a tendency to self-discipline, sense of duty, and aims for achievement. It is organized and dependable; he plans rather than acts spontaneous. Extraversion is mirrored as energy, positive emotions, confidence, sociability and the tendency to seek stimulation in the company of others, and talkativeness. Agreeableness has the tendency to be empathetic and accommodating rather than suspicious and antagonistic towards others. Neuroticism has the propensity to experience disagreeable feelings easily, such as anger, anxiety, depression, or vulnerability.

3 Adaptive System, Student Model and Personality

An Adaptive System (AS) builds a model of the objectives, preferences and knowledge of each user and uses it, dynamically, through the Domain Model and the Interaction Model, to adapt its contents, navigation and interface to the user needs [31]. In Educational Adaptive Systems, the emphasis is placed on students' knowledge in the domain application and learning style, in order to allow them to reach the learning objectives proposed in their training [31]. In generic Adaptive Systems (AS), the User Model allows changing several aspects of the system, in reply to certain characteristics (given or inferred) of the user [32]. These characteristics represent the knowledge and preferences that the system assumes that the user (individual, group of users or no human user) has. In Educational AS, the UM (or Student Model) has increased relevance: when the student reaches the objectives of the course, the system must be able to re-adapt, for example, to his knowledge [32]. A Student Model (SM) includes the Domain Dependent Data (DDD) and the Domain Independent Data (DID). The Domain Independent Data (DID) are composed of two elements: the Psychological Model and the Generic Model of the Student Profile, with an explicit figure [33].

4 Development

To define and develop of a new framework that could find out if emotions can influence the learning process and also the construction of a new learning platform that takes into account the student emotional state, personality traits and learning preferences, in order to adapt the course to the student's needs, can improve the student's learning results, a prototype was developed. The prototype, entitled Emotion Test, was developed to be used in a web environment. It was implemented to be an engaging learning environment with multimedia interactivity. Emotion Test prototype simulates the entire learning process, from the explanation of the subject, to exercises and assessment test. Through the entire process student's emotional state, personality traits and learning preferences are considered. Next figure shows the architecture proposed (Fig. 1).

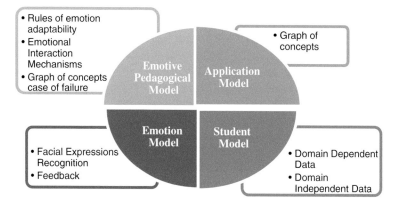

Fig. 1 Architecture

The architecture proposed for this prototype is composed of 4 main models: Student Model, Emotional Model, Application Model and Emotive Pedagogical Model. The student model represents student information and characteristics. This includes personal information (name, email, telephone, etc.), demographic data (gender, race, age, etc.), knowledge, deficiencies, learning styles, emotion profile, personality traits, etc. This information is useful to better adapt the prototype to the student. The emotion model gathers all the information of facial recognition software and feedback of the students. The application model is composed by a series of modules contain different subjects. The subject consists in a number steps that the student has to pass in order to complete the learning program. Usually each subject is composed by a diagnostic test in order to access the student level of knowledge. Followed by the subject content in which the subject is explained and followed by the subject exercises and final evaluation test. The last model is the emotive pedagogical model which is composed by three sub-models: the rules of emotion adaptability, the emotional interaction mechanisms and the graph of concepts in case of failure. The rules of emotion adaptability comprise the ways the subject content is presented. The subject content and subject exercises are presented according the student's learning style and personality. The emotional interaction mechanisms consist in the trigger of an emotion interaction, when is captured an emotion that need to be contradicted in order to facilitate the learning process. The emotions to be contradicted are: anger, sadness, confusion and disgust. The interaction can depend on the student's learning style and personality. Finally the graph of concepts in case of failure this indicates the steps to be taken when a student fails to surpass a subject. For this framework the prototype built allows students to consolidate knowledge, autonomously and with ongoing support through teaching methodologies, educational activities and emotional interactions. The prototype content, activities and interactions are defined by the teacher, but dynamically adapted and personalized according to the knowledge level, learning preferences, personality, skills and student learning pathway.

5 Prototype Evaluation and Data Analyses

The participants in this prototype evaluation were the 1st year students of Higher Education Establishment of Computer Engineering of Oporto of two courses: Informatics Engineering and Systems Engineering. The total number of students involved in these testes was 115 students with ages between 17 and 42 years old. This group of students was composed of 20 % female (n = 23) and 80 % male (n = 92) participants mainly from the districts of Oporto, Aveiro and Braga. To evaluate the prototype were conducted 2 pre-tests and a final test. For this paper only be showed the data gather for the final test. The student's participants in this prototype evaluation did not have any prior knowledge of the content of the subject approach by the prototype. For the test the students were group randomly in 2 groups as we can see in the table below. Group v1 tested the prototype with the emotional interaction and learning style with a low level of difficulty and Group v2 test the prototype without the emotional interaction and learning style with a low level of difficulty. In each group evaluation process was different: The Group v1 had to do Diagnostic test (in paper) to help grade the initial knowledge of the student. Followed by the test of the prototype with the emotional interaction and learning style. This comprehends the completism of subject module composed by a diagnostic test in order to access and update the student level of knowledge. Followed by the subject content in which the subject is explained and follow by the subject exercises and final test. After this test the students had to do a final test (in paper) to help grade the final knowledge of the student. The evaluation by this group ended with the answer of the Acceptability questioncr to determine the acceptability of the prototype. This questionnaire consists in first determine the acceptance of the prototype and second the degree of difficulty of use of each feature of the prototype. The Group v2 had to do diagnostic test (in paper) to help grade the initial knowledge of the student. Following the testing of the prototype and this time the students tested the prototype with same subject content as group v1 but without any emotional integration. After this test the students had to do a final test (in paper) to help grade the final knowledge of the student. The evaluation by this group also ended with the answer of the Acceptability questioner the same questionnaire given to group v1. To assess the learning preferences of each student, the VARK questionnaire was answer by the students. The analyses of the data showed that the preferred style was aural with 57 % of the students, followed by visual preference with 25 %, kinetic preference with 14 % and read/write with 4 %. In the test groups the distribution of participants according to their answer of the VARK questionnaire is showed in the following table. The aural preference is the preference with the highest percentage in group v1 and visual preference is highest in group v2. The read/write preference has the lowest percentage in all the groups. To determine the personality of the students the responses to the questionnaire TIPI were also analyzed. The data showed that 43 % of students fell into personality of neuroticism followed by Conscientiousness with 25 %, openness with 18 % and agreeableness and stability with 14 %. We had no students in extroversion personality. In the test groups the distribution of participants according to their answer of the TIPI questionnaire is showed in the following table. The Neuroticism personality was the personality with the highest percentage in all the groups and the agreeableness personality was the personality with the lowest percentage. From the data gather from the final teste it was concluded that the distributions are not normal after having applied the Kolmogorov-Smirnov test shown in the following Table 1.

Analyzing the results shown in Table 2 only group v2 for the diagnostic test has a p value with low bound but of true significance. All the other groups for the p values are not statistically significant. So we can conclude that the groups, with the exception of group v2 in the diagnostic test, do not have normal distributions. The analyses of the student's grades showed the following. The next table shows the descriptive statistics of the diagnostic test and final test across the groups.

As data does not have a normal distribution for the two groups they were compared using a non-parametric test Mann-Whitney. For group v1 for diagnostic

Table 1 Tests of normality

	Groups	Kolmogorov-Smirnov[a]		
		Statistic	df	P
Diagnostic test in paper	Group v1	0.229	14	0.046
	Group v2	0.184	14	0.200*
Final test in paper	Group v1	0.323	14	0.000
	Group v2	0.281	14	0.004

*This is a lower bound of the true significance
[a]Lilliefors significance correction

Table 2 Descriptive statistics

Groups		N	Minimum	Maximum	Mean (%)	Std. Deviation
Group v1	Diagnostic test in paper	14	0	100	45.7	40.3
	Final test in paper	14	60	100	85.7	12.2
Group v2	Diagnostic test in paper	14	0	80	37.1	29.2
	Final test in paper	14	0	100	61.4	33.7

test we have a mean of 45.7 % (SD = 40.3) and for the final test a mean of 85.7 % (SD = 12.2). For group v2 for diagnostic test we have a mean of 37.1 % (SD = 29.2) and for the final test a mean of 61.4 % (SD = 33.7). For the diagnostic test we have Mann–Whitney U = 83.0 and for a sample size of 14 students. For this analysis we found a P value of 0.479 which indicates that we don't have any statistical difference which is understandable because it was assumed that all students had more or less the same level of knowledge. For the final test we have Mann-Whitney U = 54.0 and for an equal sample size of the diagnostic test. For this analysis we found a P value of 0.029 in this case the differences observed are statistical different. Using the two-way ANOVA to we tried running a series of test to compare the means of the students by group and by learning preference and by group and personality. The objective of running these tests it is to see if learning preference, personality and emotional state had any influence on the outcome of the final test. For the first test, we tested by group and by learning preference however, no statistically significant differences in the results were obtained (P = 0.614). Therefore we cannot conclude that the learning preference in each group had any influence in the final test outcome. For the second test, we tested by group and by personality however, no statistically significant differences in the results were obtained (P = 0.988). Therefore we cannot conclude that the personality had any influence in the final test result. To prove this assumption we need a larger sample size. Another test was made regarding the emotional state of the student during this experiment. Our assumption was that the student in a negative state throughout the

Fig. 2 Estimated marginal means of the final test— emotional state

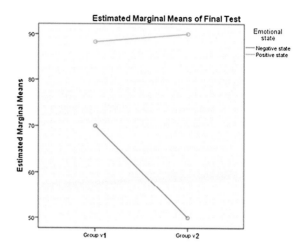

experiment had lower results. The differences found were statistically significant (P = 0.011). Consequently we can conclude that the emotional state had influence in the final test results of the students (Fig. 2).

6 Conclusions

During the course of this work it was attempted to find an answer to this question. The gathered data from the performed tests showed that there is a statistical difference between students' learning results while using two learning platforms: one learning platform that takes into account the student's emotional state and the other platform that does not have that in consideration. This gives an indication that by introducing the emotional component, the students' learning results can possibly be improved. In the development of this work and in the attempt to answer the central question one issue became apparent. This beneficial issue was the use of a new approach in user modelling process that uses learning and cognitive styles and student emotional state to adapt the user interface, learning content and context. This was observed to be very advantageous because the user interface, learning content and context was presented in a way that each individual student could best comprehend and associated with an emotional component enhancing their learning outcomes. Noting that this user modelling process was used in both learning platforms but only one platform used the emotional component.

Acknowledgements This work is supported by FEDER Funds through the "Programa Operacional Factores de Competitividade—COMPETE" program and by National Funds through FCT "Fundação para a Ciência e a Tecnologia" under the project: FCOMP-01-0124-FEDER-PEst-OE/EEI/UI0760/2014.

References

1. Shen, L., Wang, M., Shen, R.: Affective e-learning : using 'emotional' data to improve learning in pervasive learning environment related work and the pervasive e-learning platform. Educ. Technol. Soc. **12**, 176–189 (2009)
2. Kort, B., Reilly, R., Picard, R.W.: An affective model of interplay between emotions and learning: Reengineering educational pedagogy-building a learning companion. In: Proceedings IEEE International Conference on Advanced Learning Technologies, ICALT 2001, pp. 43–46 (2001)
3. Stafford, T.: What good are positive emotions? Psychologist **17**(6), 331 (2004)
4. Morgan, R., Baker, F.: A Comparison of VARK Analysis Studies and Recommendations for Teacher Educators. Ron Morgan and Fred Baker, University of South Alabama, Alabama (2012)
5. Ibrahimoglu, N., Unaldi, I., Samancioglu, M., Baglibel, M.: The relationship between personality traits and learning styles: a cluster analysis. Asian J. Manage. Sci. Educ. **2**(3) ISSN 2186-845X (2013)
6. Komarraju, M., Karau, S.J., Schmeck, R.R., Avdic, A.: The Big Five personality traits, learning styles, and academic achievement. Personal. Individ. Differ. **51**(4), 472–477 (2011)
7. Picard, R.W., Papert, S., Bender, W., Blumberg, B., Breazeal, C., Cavallo, D., Machover, T., Resnick, M., Roy, D., Strohecker, C.: Affective learning—a manifesto. BT Technol. J. **22**(4), 253–268 (2004)
8. Kleinginna, P.R., Kleinginna, A.M.: A categorized list of emotion definitions, with suggestions for a consensual definition. Motiv. Emot. **5**(4), 345–379 (1981)
9. Chibelushi, C., Bourel, F.: Facial expression recognition: a brief tutorial overview. In: Fisher R. (ed.) CVonline: On-Line Compendium of Computer Vision, vol. 9. School of Computing, Staffordshire University. http://www.dai.ed.ac.uk/CVonline (2003)
10. Foa, E.B.: Emotion in the Human Face. J. Behav. Ther. Exp. Psychiatry **4**(1), 87–88 (1973)
11. Fischer, R.: Automatic Facial Expression Analysis and Emotional Classification, Octber 2004
12. Ekman, P., Friesen, W.V.: Facial Action Coding System: A Technique for the Measurement of Facial Movement, vol. 12. Consulting Psychologists Press, Palo Alto (1978)
13. McDaniel, B., D'Mello, S., King, B., Chipman, P., Tapp, K., Graesser, A.: Facial features for affective state detection in learning environments. In: 29th Annual Meeting of the Cognitive Science Society (2007)
14. Pantic, M., Patras, I.: Dynamics of facial expression: recognition of facial actions and their temporal segments from face profile image sequences. IEEE Trans. Syst. Man Cybern. Part B Cybern. **36**(2), 433–449 (2006)
15. De Lemos, J., Reza Sadeghnia, G., Ólafsdóttir, Í., Jensen, O.: Measuring emotions using eye tracking. In: Proceedings of Measuring Behavior, vol. 2008, p. 226 (2008)
16. Partala, T., Surakka, V.: Pupil size variation as an indication of affective processing. Int. J. Hum Comput Stud. **59**(1–2), 185–198 (2003)
17. Dawson, M.E., Schell, A.M.: Startle modification: implications for neuroscience, cognitive science, and clinical science. Implications for Neuroscience, Cognitive Science, and Clinical Science, vol. xiv, 383 pp. Cambridge University Press, New York (1999)
18. Calvo, M.G., Lang, P.J.: Gaze patterns when looking at emotional pictures: motivationally biased attention. Motiv. Emot. **28**(3), 221–243 (2004)
19. Wu, S., Falk, T.H., Chan, W.-Y.: Automatic speech emotion recognition using modulation spectral features. Speech Commun. **53**(5), 768–785 (2011)
20. Russell, J.A., Bachorowski, J.-A., Fernandez-Dols, J.-M.: Facial and vocal expressions of emotion. Annu. Rev. Psychol. **54**, 329–349 (2003)
21. Scherer, G., Johnstone, T., Klasmeyer, G.: Vocal expression of emotion. In: Davidson, R.J., Scherer, K.R., Barrett, L.F. (eds.) Handbook of Affective Sciences, pp. 433–456. Oxford University Press, New York (2003)

22. Calvo, R.A., Member, S., Mello, S.D.: Affect detection : an interdisciplinary review of models, methods, and their application to learning environments. Rev. Lit. Arts Am. 1(1), 1–23 (2010)
23. Dalgleish, T., Dunn, B., Mobbs, D.: Affective neuroscience: past, present, and future. Emot. Rev. 1(4), 355–368 (2009)
24. Paradiso, S.: Affective Neuroscience: The Foundations of Human and Animal Emotions, vol. 159, no. 10. Oxford University Press, New York (2002)
25. Immordino-Yang, M.H., Damasio, A.: We feel, therefore we learn: the relevance of affective and social neuroscience to education. Mind Brain Educ. 1(1), 3–10 (2007)
26. Osgood, C.E., May, W.H., Miron, M.S.: Cross Cultural Universals of Affective Meaning. University of Illinois Press, Urbana (1975)
27. Lutz, C.: The Anthropology of emotions. Annu. Rev. Anthropol. 15(1), 405–436 (1986)
28. Santos, R., Sistema de Apoio à Argumentação em Grupo em Ambientes Inteligentes e Ubíquos considerando Aspectos Emocionais e de Personalidade (2010)
29. Kumar, K., Bakhshi, A., Rani, E.: Linking the Big Five personality domains to organizational citizenship behavior. Int. J. Psychol. Stud. 2, 73–82 (2009)
30. Diseth, Å.: Personality and approaches to learning as predictors of academic achievement. Eur. J. Personal. 17(2), 143–155 (2003)
31. Martins, A.C., Faria, L., De Carvalho, C.V.: User modeling in adaptive hypermedia educational systems. Educ. Technol. Soc. 11, 194–207 (2008)
32. Martins, C., Couto, P., Fernandes, M., Bastos, C., Lobo, C., Faria, L., Carrapatoso, E.: PCMAT—mathematics collaborative learning platform. Adv. Intell. Soft Comput. 89, 93–100 (2011)
33. Kobsa, A.: User modeling: recent work, prospects and hazards. In: Schneider-Hufschmidt, M., Kühme, T., Malinowski, U. (eds.) Adaptive User Interfaces: Principles and Practice. North-Holland, Amsterdam (1993)

A Reasoning Module for Distributed Clinical Decision Support Systems

Tiago Oliveira, Ken Satoh, Paulo Novais, José Neves, Pedro Leão
and Hiroshi Hosobe

Abstract One of the main challenges in distributed clinical decision support systems is to ensure that the flow of information is kept. The failure of one or more components should not bring down an entire system. Moreover, it should not impair any decision processes that are taking place in a functioning component. This work describes a decision module that is capable of managing states of incomplete information which result from the failure of communication between components or delays in making the information available. The framework is also capable of generating scenarios for situations in which there are information gaps. The proposal is described through an example about colon cancer staging.

1 Introduction

Computer-Interpretable Guidelines (CIGs) encode medical knowledge and process knowledge in step-by-step algorithms which, in CIG systems, are interpreted by execution engines [10]. They are machine-readable versions of Clinical Practice Guidelines (CPGs) CIG systems are a kind of distributed decision support system that draw information from other information systems, scattered across organizations. This information may be provided by: a human agent, such as a physician or a nurse,

T. Oliveira (✉) · P. Novais · J. Neves
Algoritmi Centre/Department of Informatics, University of Minho, Braga, Portugal
e-mail: toliveira@di.uminho.pt

K. Satoh
National Institute of Informatics, Sokendai University, Tokyo, Japan
e-mail: ksatoh@nii.ac.jp

P. Leão
ICVS/3B's—PT Government Associate Laboratory, Braga/guimarães, Portugal
e-mail: pedroleao@ecsaude.uminho.pt

H. Hosobe
Department of Digital Media, Hosei University, Tokyo, Japan
e-mail: hosobe@hosei.ac.jp

© Springer International Publishing Switzerland 2016
P. Novais et al. (eds.), *Intelligent Distributed Computing IX*,
Studies in Computational Intelligence 616,
DOI 10.1007/978-3-319-25017-5_36

interacting with the system; or extracted from other clinical information systems. However, if there is no knowledge about the state of a patient, it becomes impossible to respond with actions to medical events. This might be due to delays in medical tests or failures in the communication. One of the major issues in these careflow management systems is the occurrence of exceptions or deviations of what is established in the protocols, of which incomplete information is a case. This calls forth the need for higher functions in clinical decision support systems that go beyond the display of information and inference. This work focuses on the development of a Speculative Module for CIG systems that allows them to cope with incomplete information. The decision process treated in the module follows a structure based on a logical framework of Speculative Computation [5], which uses default constraints to advance the computation of decisions. As such, the main contributions of this work are the following: (i) a decision module for CIG systems capable of coping with missing information; (ii) a default generation method for clinical decision criteria; and (iii) a procedure for the generation of clinical scenarios.

2 Related Work

Existing CIG models are mainly task-network models, which means that they organize clinical procedures in networks of mutually dependent tasks. Models such as GLIF3 [2], PROforma [3], and SAGE [13] follow this representation paradigm, although with different foci. While GLIF3 is more focused on the procedural logic of CPGs, PROForma is focused on the argumentation of clinical decisions, and SAGE on the contextual elements of unfolding clinical processes. Yet, the execution engines for these CIG languages only provide straightforward reasoning and do not provide mechanisms to deal with uncertainty [10]. These mechanisms would allow health care professionals to devise scenarios, to anticipate the specific needs of patients, and to mobilize resources beforehand for the clinical tasks ahead. There are techniques derived from probability theory, such as Bayesian Probabilities [7], Certainty Factors [11], Dempster-Shafer Theory [6], and Fuzzy Sets [12], which provide a way to deal with different types of uncertainty and have been used to some extent in the medical domain. However, incomplete information, which is the object of study in this work, deviates from the type of process uncertainty treated by these techniques. In the setting presented herein, the process is established by the procedural logic of CPGs, and a decision module with a logic programming component and a machine learning component is employed to serve as an interface between guideline procedural knowledge and a data-driven default generation method.

3 Setting for Clinical Decision Support

3.1 Clinical Example

The clinical situation used as an example to demonstrate the Speculative Module is that of colon cancer staging and selection of adjuvant therapy. In most cases of colon cancer, surgery, or more specifically a colectomy, is performed. After surgery, it is necessary to assess the patient based on the most recent findings, which typically occurs in a group appointment of physicians. A decision is made based on the TNM staging system for the classification of malignant tumors, which consists of three criteria: T (tumor), which stands for the degree of tumor invasion into the wall of the colon; N (nodes), the number of metastases in regional lymph nodes; and M (metastasis) represents the detection of distant metastases in other organs such as the liver or the lungs. In Fig. 1, the clinical situation described above is represented according to the CompGuide model [9] for CPGs. It is an ontology defined in Ontology Web Language (OWL) following the task-network model. In Fig. 1 first there is a *Question* task which aims to obtain the values for the T, N and M parameters. This is a data entry point in the guideline for the retrieval of patient information. This *Question* task is linked to five other tasks through the *hasAlternativeTask* property, which means that these tasks should be executed alternatively to one another, according to the validation of specific trigger conditions, also represented in Fig. 1. The tasks are of the *Action* type for they recommend the execution of specific clinical procedures by health care professionals. Depending on the stage of the cancer, it is necessary to decide the best adjuvant therapy, which involves choosing between different observation and chemotherapy schemes. There are cases in which it is difficult to accurately assess the degree of tumor invasion. Furthermore, to know if there are metastases in regional lymph nodes, it is necessary to wait for a pathology report which, in turn, depends on the completion of laboratory tests. The same may hap-

Conditions	T (t)	N (n)	M (m)	Task	Recommendation
A	t0 *or* tis	--	--	Action 1	No adjuvant therapy, colonoscopy in 1 year
B	t1 *or* t2	n0	m0	Action 1	No adjuvant therapy, colonoscopy in 1 year
C	t3	n0	m0	Action 2	Clinical trial or observation or consider capecitabine or 5-FU/leucovorin
D	t4	n0	m0	Action 3	Capecitabine or 5-FU/leucovorin or FOLFOX or CapeOX
E	t1 *or* t2 *or* t3 *or* t4	n1 *or* n2	m0	Action 4	FOLFOX or CapeOX or FLOX
F	t1 *or* t2 *or* t3 *or* t4	n0 *or* n1 *or* n2	m1	Action 5	Colonoscopy, chest and abdominal/pelvic CT, platelets, chemistry profile

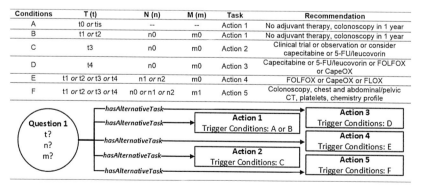

Fig. 1 Clinical setting for the example regarding colon cancer represented according to the CompGuide model

pen with the evaluation of distant metastases. For these reasons, there may be cases in which the values for T, N and M are unknown. The whole clinical situation was extracted from the Clinical Practice Guideline in Oncology for Colon Cancer from the National Comprehensive Cancer Network [1]. The medical content is greatly simplified for the sake of conveying the key features of the Speculative Module.

3.2 Structure of the CIG System

CIG systems have a similar architecture to that of the CompGuide system, represented in Fig. 2. The main component is the CIG Engine. It is responsible for executing CIGs encoded in the Knowledge Base, containing situations like the one described in Sect. 3.1, against patient information retrieved from distributed sources such as other clinical information systems. The system keeps a local repository of previous guideline executions, with patient information fed to the system in previous installments. The CIG Engine is responsible for providing the recommendations about the next clinical task. When there is a decision point, a Speculative Module is deployed to deal with cases of incomplete information. Another cause of incomplete information might be the failure of communication between the CIG system and the information sources. The setting is the following: (1) the CIG Engine will recommend the next task in the clinical careflow; (2) the transition from one task to another is only possible if the first is connected to the latter through the *hasAlternativeTask* property (just like in the example); (3) to move to one of the alternative tasks, the trigger conditions of such task must be met; (4) the information necessary to verify the trigger conditions will be acquired from an external information source, the oncology information system (*ois*) in this case; and (6) the system has a Speculative Module which uses default values to continue the execution and generate clinical scenarios in the event that there is no answer from the (*ois*).

4 Description of the Speculative Module

4.1 Generation of Defaults

The Generation of Defaults is a part of the Speculative Module and consists in a procedure that seeks to acquire the most likely values for the decision parameters, taking into account possible dependence relationships between them. Bayesian Networks (BNs), for their set of characteristics [14, 15], provide an ideal support for a default generation model. The Generation of Defaults is depicted in Fig. 2, inside the Speculative Module. It comprises five sequential stages. The first is the identification of askable atoms, in which the clinical parameters for the decision are identified and isolated. Next, the module retrieves relevant data about previous guideline exe-

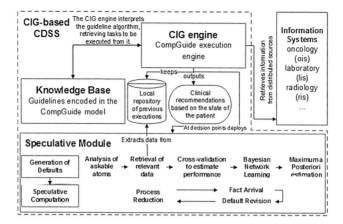

Fig. 2 CompGuide system architecture with its main components: *CIG Engine*, *Knowledge Base*, and *Speculative Module*

cutions regarding the isolated parameters from the local repository. In the following stage, six different BN learning algorithms are used in cross-validation. The purpose is to select the one with the lowest log likelihood loss (*logl*), which is a measure of the entropy exported by a model in order to keep its own entropy low [4]. So, low values are synonymous with a good fit between the model and the data. After the best performing algorithm is selected, a BN is generated. The last stage consists in a Maximum a Posteriori (*MAP*) estimation, which is a query to the BN that provides the most likely values for the parameters, given the evidence, along with a probability value [8]. It is possible to perform a MAP query with no evidence, which provides the most likely configuration for all the parameters in the network. Figure 3 shows the results of the procedure when applied to data of 515 patients who received treatment for colon cancer at the Hospital of Braga, Portugal. In Fig. 3a, it is possible to see that the algorithm that produces the lowest *logl* is the *iamb*. Figure 3b also shows the structures learned from the different algorithms. In the chosen structure (the one learned with the *iamb*) it was only possible to establish a dependence

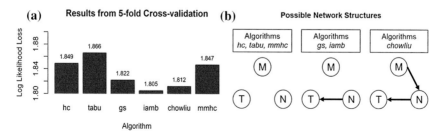

Fig. 3 **a** Results from 5-fold cross-validation of six structure learning algorithms; **b** Bayesian Network structures for the TNM parameters learned with the different algorithms

relationship between T and N. This is particularly useful for it indicates to the CIG
Engine that tumor invasion and local lymph nodes metastases might be correlated,
but distant metastases are not dependent on the other two parameters. After performing the MAP estimation on the network, the values obtained for T, N, and M were
respectively $T = t3, N = n2$, and $M = m1$, with $T, N, M_{MAP} \approx 0.4395$. These values are used as initial defaults in Speculative Computation. MAP queries will be
performed as information from the sources arrive in order to adjust the remaining
default values. The Generation of Defaults was developed using the *bnlearn* package
for R and the *inflib* Java library for inference in BNs from the SamIam project.

4.2 Speculative Computation

The Speculative Computation part of the Speculative Module is based on a logic
programming framework with constraint processing.

4.2.1 Formalization of the Clinical Example According to the Framework

A Framework for Speculative Computation in clinical decision support systems
which features disjunctive constraint processing is a tuple $\langle \Sigma, \mathcal{E}, \Delta, \mathcal{P} \rangle$. This formulation is based on the work presented in [5]. Σ is a finite set of constants. An
element in Σ is a system component identifier, an information source with which
the CIG system communicates. $\mathcal{E}$ is a set of predicates called external predicates,
representing the decision criteria, i.e., the clinical parameters in the trigger conditions. When Q is an atom with an external predicate and S is the identifier of a remote
information source, $Q@S$ is called an askable atom. Δ is the default answer set, consisting of set of default rules, obtained from the Generation of Defaults, called default
rules with respect to $Q@S$, of the following form: $Q@S \leftarrow C \parallel$, where $Q@S$ is an
askable atom and C is a set of constraints called default constraints for $Q@S$. As for
$\mathcal{P}$, it is a constraint logic program of the form: $H \leftarrow C \parallel B_1, B_2, \ldots, B_n.$, where H is
a positive ordinary literal called a head of rule R; C is a set of constraints; and each
$B_1, B_2, \ldots, B_n$ is an ordinary literal, or an askable literal. The situation described in
Sect. 3.1 regarding the staging of colon cancer can be represented according to the
framework. The predicate $nt(a, b)$ indicates that b is the task that follows a and is
used in the initial query in order to know what procedure should be performed next.
$alt(a, b)$ indicates that b is an alternative task linked to a. $tcv(b)$ means that the trigger
conditions for task b are validated. The complete representation is:

- $\Sigma = \{ois\}$
- $\mathcal{E} = \{t, n, m\}$
- Δ is the following set of rules: $t(T)@ois \leftarrow T \in \{t3\} \parallel .$ $n(N)@ois \leftarrow N \in \{n2\} \parallel .$ $m(M)@ois \leftarrow M \in \{m1\} \parallel .$
- $\mathcal{P}$ is the following set of rules:

$nt(X,F) \leftarrow \| alt(X,F), tcv(F).^{rule\ 1}$

$tcv(F) \leftarrow F \in \{action1\}, T \in \{tis, t0\} \| t(T)@ois.^{rule\ 2}$

$tcv(F) \leftarrow F \in \{action1\}, T \in \{t1, t2\}, N \in \{n0\}, M \in \{m0\} \| t(T)@ois, n(N)@ois, m(M)@ois.^{rule\ 3}$

$tcv(F) \leftarrow F \in \{action2\}, T \in \{t3\}, N \in \{n0\}, M \in \{m0\} \| t(T)@ois, n(N)@ois, m(M)@ois.^{rule\ 4}$

$tcv(F) \leftarrow F \in \{action3\}, T \in \{t4\}, N \in \{n0\}, M \in \{m0\} \| t(T)@ois, n(N)@ois, m(M)@ois.^{rule\ 5}$

$tcv(F) \leftarrow F \in \{action4\}, T \in \{t1, t2, t3, t4\}, N \in \{n1, n2\}, M \in \{m0\} \| t(T)@ois, n(N)@ois, m(M)@ois.^{rule\ 6}$

$tcv(F) \leftarrow F \in \{action5\}, T \in \{t1, t2, t3, t4\}, N \in \{n0, n1, n2\}, M \in \{m1\} \| t(T)@ois, n(N)@ois, m(M)@ois.^{rule\ 7}$

$alt(question1, F) \leftarrow F \in \{action1\} \|.$ $\quad alt(question1, F) \leftarrow F \in \{action2\} \|.$

$alt(question1, F) \leftarrow F \in \{action3\} \|.$ $\quad alt(question1, F) \leftarrow F \in \{action4\} \|.$

$alt(question1, F) \leftarrow F \in \{action5\} \|.$

Speculative Computation starts with a top goal. The notion of goal is central for it represents what is necessary to achieve in the computation, or, in other words, it is the outcome of the decision. In the framework, a goal has the form of $\leftarrow C \| B_1, \ldots, B_n$ where C is a set of constraints and each of $B_1, \ldots, B_n$ is either an askable atom or an atom that is unifiable with the head of a rule in $\mathscr{P}$. During the computation the framework keeps a set of beliefs about the askable atoms, i.e., the clinical parameters in the decision, in a current belief state (CBS). In the beginning, the CBS assumes the default rules in Δ and, afterwards, the top goal is reduced according to CBS and $\mathscr{P}$. Goals and the product of their reduction are kept in processes. They are structures that represent the different alternative computations in the framework. In this regard, there are two types of processes: active and suspended. Active processes are those whose constraints C are contained in, and therefore are consistent with, the CBS. They are regarded as the valid scenarios. Processes that do not fulfill this condition are suspended. In order to keep track of the different processes, active processes are kept in the active process set (APS), whereas suspended processes are kept in the suspended process set (SPS).

4.2.2 Phases of Speculative Computation

Speculative Computation includes: the *process reduction* phase, the *fact arrival* phase, and the *default revision* phase. Throughout these phases active processes are represented according to the tuple $\langle \leftarrow C \| GS, UD \rangle$, where C is a set of constraints, GS contains the goals in the form of literals to be proved, and UD is the set of used defaults. If an atom is reduced through a default, it is added to the UD of the process. Suspended processes have a similar structure, they are represented by a tuple $\langle SP, \leftarrow C \| GS, UD \rangle$, where SP is a set containing the atoms that were responsible for suspending the process and whose constraints are not consistent with the CBS.

The first phase is *process reduction*, which is depicted in Fig. 4a. The initial query to the system becomes the initial goal set in the first active process. The process is

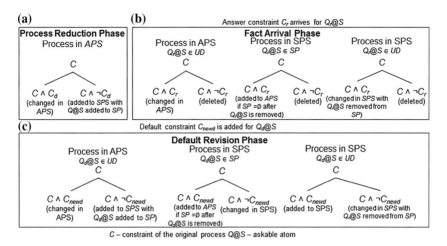

Fig. 4 Diagram of Speculative Computation. Each bifurcation represents two new processes with the displayed constraints

further reduced according to its goals. If an atom in the goal set is unifiable with a rule in $\mathscr{P}$ then that atom is replaced in GS with the body of the rule, and the constraints of the rule are added to the constraints of the process. This originates as many active processes as there are rules with which the atom in GS is unifiable. This ensures that different execution traces are explored, keeping all possible scenarios open. But, if the atom in the goal is an askable atom $Q@S$, with and attached constraint C, for which there is a default constraint C_d, the module verifies the consistency of $C \wedge C_d$ and $C \wedge \neg C_d$. If $C \wedge C_d$ is consistent, then a new active process is created from the initial process with this constraint, and $Q@S$ is added to UD. If the same happens with $C \wedge \neg C_d$, a suspended process is created as well, and $Q@S$ is added to SP. With this disjoint constraint processing, the objective is not to exclude any possible scenario and keep active only the processes that are consistent with the CBS. While this takes place, the real value of the atom is asked from the information source. These procedures are applied to active processes until they have an empty goal set, which means that there is nothing left to prove and they are valid scenarios according to the default constraints. *Fact arrival* occurs when a constraint $Q_r@S \leftarrow C_r \parallel$ is returned from an information source S for a question regarding an askable atom $Q_r@S$. After a step of process reduction is finished, the new constraint replaces the default in the CBS for the corresponding askable atom. It is then necessary to revise the processes in APS and SPS that have $Q_r@S$ either as a used default (in UD) or a suspended atom (in SP). The revision of constraints occurs as shown in Fig. 4b. Since the replies from the information sources are facts and, thus, regarded as definitive, there are execution traces that are deleted when they are inconsistent with the CBS. The objective with this phase is to remove scenarios that can no longer occur. The *default revision* phase consists in changing all the processes according to the new default rules provided by the method for the Generation of Defaults. The arrival of a new fact may change the remaining default rules. As such, a new MAP estimation is done using the BN with the available facts as evidence. For every changed default, *default revision* is performed. Processes in *APS* and *SPS* are revised in order to determine if their constraints are consistent with the new default constraint $Q_d@S \leftarrow C_{newd}$ ||. According to the position of the atom attached to the new default, i.e., if it is a suspended atom (in *SP*) or a used default (in *UD*), the existing processes may generate new active and suspended processes, but their execution trace is never erased, unlike what happens in the *fact arrival* phase. The revision of processes takes place as shown in Fig. 4c. This keeps the scenarios coherent and increasingly aligned with the newly arrived facts. It is a dynamic mechanism because it is triggered by changes in the default rules. These changes, in turn, are caused by the arrival of facts from the information sources. The treatment of processes in all of these three phases is implemented in Prolog.

4.3 Example and Discussion

According to the situation described in Sect. 3.1 and the formalization in Sect. 4.2.1, the question to pose to the Speculative Module would be $nt(question1, F)$, which seeks to determine which task should follow *Question1* in the management of the patient. The initial *CBS* assumes the values in Δ. Then, process reduction occurs with the unification of the goals in the *GS* with the head of rules and facts in $\mathscr{P}$. This will happen through rules 1 and 2. By the time process reduction reaches rules 3, 4, 5, 6, and 7, one will have five active processes representing the five possible alternative tasks. At this point, it becomes necessary to reduce askable atoms. For instance, if the system needs a value for the goal $t(T)@ois$, since there is only a default constraint for that goal in the *CBS*, that constraint is used in process reduction, yielding active processes which are consistent with the default and suspended processes which are not. As such, a process representing *action5* is split into two processes, an active process using the default constraint $t(T)@ois \leftarrow T \in \{t3\}$ ||, and a suspended process using the negation of the default constraint. Meanwhile, a question is sent to *ois* so as to know the real value of $t(T)@ois$. The procedure is similar whenever an askable atom is found. By continuing process reduction, it is possible to arrive at a state in which there is a process with an empty goal set, achieved purely by relying on default constraints. As such, the process is a tentative scenario. By outputting constraints C and used defaults UD, the system provides the task that will most likely follow *question1*. In this case $C = \{F \in \{action5\}, T \in \{t3\}, N \in \{n2\}, M \in \{m1\}\}$ and $UD = \{t(T)@ois, n(N)@ois, m(M)@ois\}$, which means that *action5* should be the next task, featuring a series of workup exams such as a colonoscopy, an abdominal/pelvic CT, and so forth. Based on this, the health care professional may start preparations. There might be cases in which more than one process with an empty goal set are produced. In such situations, both are presented as scenarios.

In *fact arrival*, answers from the *ois* trigger the revision of active and suspended processes. Assuming that the fact $(n(N)@ois \leftarrow N \in \{n0\} \parallel)$ arrives, it becomes the new rule for $n(N)@ois$ in the *CBS*, and all the processes consistent with this and the remaining *CBS* become active, whereas the others are deleted. Additionally, a new MAP estimation is submitted to the BN with the new fact as evidence. It produces the result $T, M_{MAP} = argmax_{T,M} P(T, M \mid N = n0) \approx 0.9126$ with $T = t1$ and $M = m1$. The default value for $t(T)@ois$ is the only one that changes. Through *default revision*, $t(T)@ois \leftarrow T \in \{t1\} \parallel$ replaces the old default in the *CBS* and all the processes are revised accordingly, but no execution trace is deleted. In the end of these steps there is an active process representing the most likely scenario given the configuration of the information, achieved through fact and default constraints. Outputting constraints and used defaults would give $C = \{\leftarrow F \in \{action5\}, T \in \{t1\}, N \in \{n0\}, M \in \{m1\}\}$ and $UD = \{t(T)@ois, m(M)@ois\}$. Curiously, the scenario recommends the same task, but achieved through a different route. In this case, the new default is included in the process and $n(N)@ois$ in no longer in UD because it is not a default anymore. As information arrives for the other askable atoms, these procedures are repeated. The Speculative Module provides a filter for rule-based clinical decision support systems, managing different information states.

5 Conclusions and Future Work

Speculative Computation provides tentative scenarios for the state of a patient and, based on them, the most appropriate clinical task. Speculative Computation is used as an interface to clinical algorithms, bringing to clinical decisions a set of mechanisms to reduce and revise scenarios according to facts and newly derived defaults. As such, they ensure a convergence of all the scenarios towards the real state of the patient. This kind of mechanism does not have a parallel in other CIG systems. The computation is dynamic in the sense that it accounts for changes in the original beliefs. As future work, it is necessary to incorporate the probability provided by the MAP estimation into Speculative Computation so as to provide a measure of the likelihood of the processes holding true.

AcknowledgementsThis work is part-funded by ERDF—European Regional Development Fund through the COMPETE Programme (operational programme for competitiveness) and by National Funds through the FCT—Fundação para a Ciência e a Tecnologia (Portuguese Foundation for Science and Technology) within project FCOMP-01-0124-FEDER-028980 and project scope UID/CEC/00319/2013. The work of Tiago Oliveira is supported by a FCT grant with the reference SFRH/BD/85291/ 2012.

References

1. Benson, A., Bekaii-Saab, T., Chan, E., Chen, Y.J., Choti, M., Cooper, H., Engstrom, P.: NCCN Clinical Practice Guideline in Oncology Colon Cancer. Technical Report, National Comprehensive Cancer Network (2009)
2. Boxwala, A.A., Peleg, M., Tu, S., Ogunyemi, O., Zeng, Q.T., Wang, D., Patel, V.L., Greenes, R.A., Shortliffe, E.H.: GLIF3: a representation format for sharable computer-interpretable clinical practice guidelines. J. Biomed. Inf. **37**(3), 147–161 (2004)
3. Fox, J., Ma, R.T.: Decision support for health care : the PROforma evidence base. Inf. Prim. Care **14**(1), 49–54 (2006)
4. Hastie, T., Tibshirani, R., Friedman, J., Hastie, T., Friedman, J., Tibshirani, R.: The Elements of Statistical Learning: Data Mining, Inference, and Prediction, 2nd edn. Springer-Verlag, New York (2009)
5. Hosobe, H., Satoh, K., Codognet, P.: Agent-Based speculative constraint processing. IEICE Trans. Inf. Syst. **E90-D**(9), 1354–1362 (2007)
6. Hua, Z., Gong, B., Xu, X.: A dsahp approach for multi-attribute decision making problem with incomplete information. Expert Syst. Appl. **34**(3), 2221–2227 (2008)
7. Kononenko, I.: Inductive and bayesian learning in medical diagnosis. Appl. Artif. Intell. **7**(4), 317–337 (1993)
8. Korb, K., Nicholson, A.: Bayesian Artifical Intelligence, 2nd edn. CRC Press, London (2003)
9. Oliveira, T., Novais, P., Neves, J.: Representation of clinical practice guideline components in owl. In: Trends in Practical Applications of Agents and Multiagent Systems, Advances in Intelligent Systems and Computing, vol. 221, pp. 77–85. Springer (2013)
10. Peleg, M.: Computer-Interpretable clinical guidelines: a methodological review. J. Biomed. Inf. **46**(4), 744–763 (2013)
11. Shortliffe, E.H., Davis, R., Axline, S.G., Buchanan, B.G., Green, C., Cohen, S.N.: Computer-based consultations in clinical therapeutics: explanation and rule acquisition capabilities of the mycin system. Comput. Biomed. Res. **8**(4), 303–320 (1975)
12. Straszecka, E.: Combining uncertainty and imprecision in models of medical diagnosis. Inf. Sci. **176**(20), 3026–3059 (2006)
13. Tu, S.W., Campbell, J.R., Glasgow, J.: Nyman, M.a., McClure, R., McClay, J., Parker, C., Hrabak, K.M., Berg, D., Weida, T., Mansfield, J.G., Musen, M.a., Abarbanel, R.M.: The SAGE guideline model: achievements and overview. J. Am. Med. Inf. Assoc. JAMIA **14**(5), 589–98 (2007)
14. Van der Heijden, M., Lucas, P.J.F.: Describing disease processes using a probabilistic logic of qualitative time. Artif. Intell. Med. **59**(3), 143–155 (2013)
15. Visscher, S., Lucas, P.J.F., Schurink, C.A.M., Bonten, M.J.M.: Modelling treatment effects in a clinical Bayesian network using boolean threshold functions. Artif. Intell. Med. **46**(3), 251–266 (2009)

Part X
2nd Workshop on Cyber Security and Resilience of Large-Scale Systems (WSRL 2015)

2nd Workshop on Cyber Security and Resilience of Large-Scale Systems

Massimo Ficco and Salvatore D'Antonio

Abstract The *2nd Workshop on Cyber Security and Resilience of Large-Scale Systems*, *WSRL 2015*, features contributions in the topics of resilience, dependability, and security of distributed systems.

Control and intelligent applications are a vital part of modern large-scale systems in various application areas, such as energy generation and distribution, traffic control and critical infrastructures. Centralized control is being replaced by distributed and more open control systems that possess increasing levels of autonomy. Communication networks and intelligent systems are at the core of these developments. Such systems must be able to recover from failures and intrusions in order to preserve their functions. Therefore, achieving resilience and security in today's complex, interconnected, and interdependent systems requires an integrated engineering approach to address resilience issues affecting both cyber and physical systems.

M. Ficco
Second University of Naples, Caserta, Italy
e-mail: massimo.ficco@unina2.it

S. D'Antonio (✉)
University of Naples Parthenope, Naples, Italy
e-mail: salvatore.dantonio@uniparthenope.it

© Springer International Publishing Switzerland 2016
P. Novais et al. (eds.), *Intelligent Distributed Computing IX*,
Studies in Computational Intelligence 616,
DOI 10.1007/978-3-319-25017-5_37

401

A Semantic Driven Approach for Consistency Verification Between Requirements and FMEA

Gabriella Gigante, Francesco Gargiulo, Massimo Ficco and Domenico Pascarella

Abstract Consistency within the system life cycle is difficult to guarantee, due to the cross of different skills and requirements, often expressed by means of different languages. In particular, in safety-critical systems consistency between software requirements and safety analysis requires checks to guarantee that safety engineer needs are feasible and implemented by the system. Failure Mode and Effects Analysis (FMEA) is a systematic technique to analyze the failure modes of components, evaluating their impact and their mitigation actions, which are procedures to be implemented by operators or by the system itself (usually by the software). Although the actual efforts to centralize system information in a structured way, safety analysis is not tied in a structured manner to other systems, in particular to software. This paper proposes an automatic approach to check consistency between FMEA and software requirements with a bit effort of formalization. The approach models FMEA and software requirements with Resource Description Framework (RDF) triplets and checks their consistency on the basis of consistency rules.

Keywords Requirements engineering · Requirements verification · Consistency · RDF · Semantic distances · Ontologies

G. Gigante (✉) · F. Gargiulo · D. Pascarella
CIRA (Italian Aerospace Research Centre), Capua, Italy
e-mail: g.gigante@cira.it

F. Gargiulo
e-mail: f.gargiulo@cira.it

D. Pascarella
e-mail: d.pascarella@cira.it

M. Ficco
Department of Industrial and Information Engineering,
Second University of Naples, Aversa, Italy
e-mail: massimo.ficco@unina2.it

© Springer International Publishing Switzerland 2016
P. Novais et al. (eds.), *Intelligent Distributed Computing IX*,
Studies in Computational Intelligence 616,
DOI 10.1007/978-3-319-25017-5_38

403

1 Introduction

Consistency and completeness represent some of the main quality factors to design a "dependable system": inconsistent work-products imply system failures and unexpected behavior. In evolving software systems, consistency seems hard to maintain and to verify [2]. In safety-critical systems, consistency is hard to guarantee at process level due to the different skills to be provided.

Model-based development allows to define the system at different levels of abstraction, improving the consistency, the correctness and the completeness of its functional aspects. On the other hand, non-functional aspects such reliability and safety are more difficult to integrate. Safety standards require analysis during the system development [9–11], which provides a further view of the system by means of different formalisms. These dimensions of heterogeneity can easy lead to inconsistencies. In this context, providing evidence of consistency among the different work-products is challenging [13].

System safety analysis often are not linked to system software, which is usually in charge of recovery actions. These are identified in Failure Mode and Effects Analysis (FMEA) and define further software safety functions. Thus, software requirements model has to be consistent and complete against the FMEA model.

The idea discussed in this paper investigates a simple approach to verify inconsistencies between requirements and FMEA, both expressed in natural language. It proposes to model both requirements and FMEA by means of RDF (Resource Description Framework) triplets and to check consistency by means of empiric rules on similarity measures between inconsistent triplets [14].

2 Background

2.1 Consistency Concept

Engineering communities have built a conceptual framework for inconsistencies in system models, which proposes inconsistency management as a process. In literature, two general frameworks have been proposed: one by Finkelstein and one by Nuseibeh [5]. The activities of the frameworks are: detection of overlaps; detection, diagnosis, handling and tracking of inconsistencies; specification and application of a management policy for inconsistencies [6, 14]. Such activities have been studied at different levels. At process level, consistency is checked to hold between different models: it is horizontal if the models to check have the same abstraction level, otherwise it is vertical. At single level, consistency is checked within an artifact and is referred as internal or evolution consistency [4].

Any idea of consistency starts from the intuitive concept of contradiction: it denotes any situation in which a relationship $\Re$, that should hold between models or elements of a model, is found not to hold. Reference [3] defines inconsistency as "*any situation in which two descriptions do not obey some relationship that should hold between them*". Although such definition gives rise to a large debate, it remains a reference starting point. In this way, the problem of detection can be completely addressed by a robust set of consistency rules and by a good algorithm, that shall browse the model and check the rules violation. Consistency rules can be simple when they refer to notations, development and local contingencies, wherein contradiction is a direct refutation of previously stated presented information. Moreover, information within the software models can be refuted in an indirect manner. A given set of facts could establish a potential situation that would contradict other facts within the models. Therefore, establishing consistency within a software model and between different system models is also a semantic task.

2.2 FMEA

FMEA is one of the first systematic techniques for dependability analysis, but it is still very popular throughout the safety-critical domains. It is a single point of failure analysis for safety-critical systems [12]. It examines the consequences of potential failures on the functionality of a system. Typically, FMEA is practiced on physical systems and the considered failure modes are the failures of physical components. More recently, different kinds of FMEAs are used in many domains to analyze general safety-critical systems (also software or processes) and can either be qualitative or quantitative [13]. FMEA goes through the following steps: identification of the system boundaries; identification of each function and associated component; identification of the potential failures mode for each function/component; identification of the impact of each failure at component level and at system level; allocation of the probability of each failure (only for quantitative evaluations). Finally, in the case of quantitative evaluations an additional step consists in allocating the probability of a failure for the associated component.

2.3 Internal and External Inconsistencies of FMEA

FMEA uses natural language to describe failure modes, failure effects and mitigation actions. It usually contains hundreds of lines, filled by a team of engineers that can be different from the development team, especially from the software team. In this sense, the FMEA internal and external consistency cannot be easily guaranteed. Failure modes should be the same type for each function and for each class of components. They should be described by means of the same terms, otherwise the evaluation of occurrence probabilities is difficult. Furthermore, if a new failure mode is discovered during the design, it should be added, resulting in a time consuming and error prone operation. Failure effects should be consistent with the corresponding failure modes. But if they are described in different ways, such check is difficult.

The choice of a seamless modeling framework could be the proper solution to such problems. Recent researches highlight difficulties to conceal system functional and not functional properties. On the one hand, research effort focuses on proposing unified approaches to system modeling by defining a comprehensive formalism. The drawback of this approach is the difficulty to put into practices, as the development team has to learn a new language [15]. On the other hand, research proposes a multi-formalism approach, which models each aspect of the system using the proper formalism and combines the different models for the overall system model by means of powerful operators. The main problems are the operators and the consistence between different formalisms.

Proposals of knowledge-based approaches, such as taxonomy or ontologies to automatically define FMEA, are the scope of recent studies. Such techniques could enforce the internal consistence, but the external consistence is still not covered. In this paper, the external consistency between FMEA actions demanded to software and software requirements is checked, being both specified in natural language.

3 Consistency Verification Framework

3.1 The Approach

The proposed approach adopts RDF to model each system artifact written in natural language. It evaluates semantic distances between RDF triplets by using similarity measures on the basis of referring ontologies. The semantic distances are checked according to thresholds inferred from a previous study, proposing a high level classification of inconsistency in order to derive general consistency rules with a confident level of possible inconsistencies coverage [14]. Finally, it experiments the defined rules to check inconsistency between a part of system FMEA and a set of software requirements with known inconsistencies.

3.1.1 The RDF Triplet Extraction

The use of RDF triplets allows to move to a more structured representation of the artifacts. This representation is not semantically equivalent in strict sense, but the key concepts are captured. Before the extraction, some well-known NLP (Natural Language Process) tasks would be executed, such as: lemmatization, NER (Named Entity Recognition) both for common entities (i.e., people, places and organization) and domain specific entities; compound words, abbreviations or acronyms detection; the word sense disambiguation, etc. These tasks often rely on one or more general purpose or domain specific ontologies. The *subj*, *pred* and *obj* are not plain words but they represent concepts of an ontology. In our approach triplets are identified according to the following rules:

- the passive form is translated into the active form;
- the complement of specification related to the subject is treated as a single entity, with the subject that must be present in the domain ontology;
- the complement of specification related to the object is treated as a single entity, with the object that must be present in the domain ontology;
- the complements of term, of agent, of half, of time, of motion are translated defining a triplet for each of them and associating to each verb the preposition they imply (for example, the predicate and the term complement are translated in *predicate_TO, term complement*);
- if the requirement expresses a conditioned action, which is an action that should be performed only if a condition is true, the triplets are at least 3: the first asserting the action with the object (main) A, the second expressing the precondition B, the third <B, precondition, A> expressing that A is executed only if B is true;
- if the requirement expresses a constrained action, that is an action that should be performed only if a constraint is true, the triplets are at least 3: the first asserting the action with the object (main) A, the second expressing the constraint B, the third < B, constraint, A> expressing that A is executed only if B is true;
- FMEA mitigation actions are translated as conditioned actions, where precondition triplets are built up with the failing component as subject, the verb *"fails"* as predicate, the conjunction of failure mode and the phase to which FMEA refers as object.

In the above definitions, we use the terms precondition and constraint pointing out two different concepts. Precondition is a requirement that depends on the specific system, mission or operation, and it is extracted from the artifacts. Constraint expresses a necessary condition due to physical issues (not depending on the application) and it is provided by domain knowledge.

The extraction of precondition triplets has been implemented manually.

3.1.2 The Adopted Similarity Measure

First of all, it is necessary to identify a referring ontology to calculate similarity measure. We adopted Wordnet 2.1 [7, 8] and some libraries of Wordnet 3. We model some general main concepts of the domain under study by RDF triplets in a text file. We choose the edge-based Wu and Palmer [1] similarity measure:

$$WP(x, y) = \frac{2p(lcs)}{p(x) + p(y) + 2p(lcs)}, \tag{1}$$

where *lcs* is the last common subsumer of x and y and $p(x)$ and $p(y)$ are the levels of x and y in the wordnet tree. To overcome the problem of similarity measure calculated on non-homogenous ontologies, we adopt the same technique proposed in [14], extending the upper ontology by simply "attaching" the domain specific concepts to the common root concept. For example, to calculate the distance between ground software (concept of the domain ontology) and system, Eq. (1) becomes:

$$WP(\text{"ground software"}, \text{"system"})$$
$$= \frac{2p(lcs)}{p(\text{"ground software"}) + 1 + p(\text{"system"}) + 2p(lcs)} \qquad (2)$$

3.1.3 The Inconsistencies Rules

FMEA and software requirements are consistent if the following conditions hold: every FMEA action assigned to software is a software function implemented under a precise precondition, exactly the same of FMEA triplet precondition; no software function contradicts or obstacles any FMEA action, that is, software requirements are internally consistent. Such considerations led to the adoption of a subset of the inconsistency rules proposed in previous paper [14], mainly related to conflicting actions or conflicting preconditions.

Let us assume the RDF triplet $t_i = <s_i, p_i, o_i>$ and $T_i = <t_i>$ the set of triplets under study. To detect inconsistent triplets in the identified set T_i, we consider the following rules. Two triplets t_i and t_j are considered potentially inconsistent if:

Rule 1   1. *have the same subject or included in the t_i subject s_i;*
        2. *have the same object or included in the t_i object o_i;*
        3. *have the t_j predicate, p_j, equal to not t_i predicate, p_i;*
Rule 2   1. *have the same subject or included in the t_i subject s_i;*
        2. *have the same object or included the t_i object o_i;*
        3. *have the t_j predicate, p_j, equal to the "opposite" of the t_i predicate, p_i;*
Rule 3   1. *express an interaction;*
        2. *and have different objects;*
Rule 4   1. *t_i expresses a constraint;*
        2. *t_j object is the subject of t_i;*
Rule 5   1. *have a different number of preconditions or;*
        2. *every precodition of t_i is equivalent to at least one precondition of t_j and viceversa.*

Two triplets t_i and t_j define an interaction if they have:

Rule 3.1   1. *the t_i object is the t_j subject;*
          2. *the t_j subject is the t_i object.*

A triplet t_i expresses a constraint or a precondition of another triplet t_k if:

Rule 4.1   1. *t_k occurs only if t_i occurs.*

Furthermore, two triplets t_i and t_j are considered potentially equivalent if they:

Rule 6   1. *have the same subject or included in the t_i subject s_i;*
        2. *have the same object or included in the t_i object o_i;*

   3. *have the same predicate, p_j, equal or included in t_i predicate, p_i, or have respectively predicates "constraint" and "precondition".*

Rules from 1 to 4 allow to verify internal consistency of software requirements triplets. Rule 5 allows to identify inconsistencies between conditioned triplets to verify both internal inconsistencies in software requirements and inconsistencies with FMEA. Rule 6 allows to verify the presence of FMEA actions among software requirements functions.

3.2 The Framework

Figure 1 shows a schema of the proposed framework implemented in Java. The green boxes represent documents, the blue boxes represent the software functions, the orange boxes represent the input rules according to which software functions check inconsistencies. Input documents are pre-processed according to some NLP basic rules. Then, they are processed by the software function Triplet Extractor. It transforms the input documents into a set of triplets reported in text files: respectively, File set *Ri* related to software requirements triplets and File set *Fi* related to FMEA triplets. Input documents must have predefined formats. FMEA file is processed by

Fig. 1 The software framework

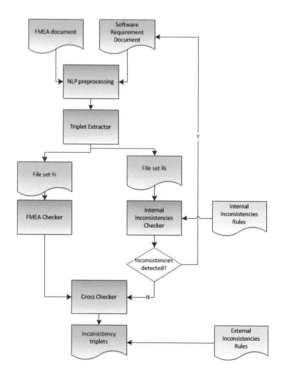

FMEA checker. It verifies the completeness of each row and groups the equivalent actions, in order to have the complete set of preconditions for it. Requirements triplets are checked for internal consistency by the Internal Consistencies Checker function according to the rules identified in Sect. 3.1. In the case of possible inconsistencies, the file is verified and manually updated, and the processing starts again until any internal inconsistency is removed. At the end, the two text files are given in input to the Cross checker function, which implements the rules related to external consistency. In the specific case study, it checks if each triplet in FMEA is equivalent at least to one triplet in the requirements set. In the negative case, it highlights a possible inconsistency. In the positive case, it checks if preconditions are equivalent to the current triplets. In the negative case, it highlights possible inconsistencies. To detect contradictory or equivalent triplets, the software function searchs the relative terms in the ontologies and evaluates semantic distances by means of Wu and Palmer semantic metric. To detect if a term x is equivalent to a term y, it identifies the set of all synonymous of y, verifies that x does not belong to it and evaluates the medium of distances between y and each synonym. If the distance between x and y is between 0 and such medium, x can be considered equivalent to y. To detect if a term x is opposite to a term y, it identifies the set of all antimony relations of y, verifies that x does not belong to it and evaluates the mean of distances between x and each word of such set. If the distance between x and y is major of the evaluated distance, x can be considered opposite to y as well as is more "semantically related" to the opposite terms.

The combination of such basic algorithms with the inconsistency rules provides the underline logic of the verification framework.

3.3 Experimental Results

In our case study, we are interested to detect inconsistencies between a part of system FMEA and a set of fifty software requirements. FMEA is relative to an unmanned space vehicle and requirements are related to its on-board data handling software. In such set, at least ten sentences express well-known inconsistencies. Requirements describe some software functions: the mission organization in different phases from pre-launch to vehicle landing; the exception reporting to the ground segment; the issuing to ground segment of actual collected data related to vehicle health status; the safety constraint, according to which ground operator shall always know the position of the vehicle; the typical reaction to an on board over temperature problem. Hereafter, the set of inconsistent software requirements related to the on board software:

R1 it shall not never power off equipments until deceleration phase;
R3 it shall disable reporting in preflight;
R4 it shall activate the flight termination system only during flight or deceleration phase in the case of RF power amplifier failure;

System				
Subsystem			Failure Mode and Effects Analysis	
Phase	Pre flight			
Item / Function	Potential Failure Mode(s)	Potential Effect(s) of Failure	Potential Cause(s)/ Mechanism(s) of Failure	Recommended Action(s) to Software
RF power amplifier	no function	Loss of mission	Loss of physical conection between amplifier and power amplifier	
		Loss of mission	Temperature overcomes 80°	Power off the power amplifier
	Intermittent function	Degraded mission	Temperature overcomes 80°	Power off the power amplifier
GPS	No function	Mission degradation due to loss of accurate position evaluation	GPS phisical damage	Reports the problem to ground
			Phisical connection with GPS	Reports the problem to ground
Flight Data recorder	No function	Post flight analysis can't be done	Loss of physical connetion between on baord computer and FDR	
		Post flight analysis can't be done	Mass memory full	Disable on board software storing function
	Untimely function	Post flight analysis can't be accurate	on board software tasking not working	Disable on board software storing function
	erreneous function	Post flight analysis can't be done	Mass memory damaged or on board software not working properly	
bus Spacewire	No function	Loss of mission	loss of physical connetction	Activate the flight termination system
Software scheduling	Intermittent function	Degraded mission		Reboot the software

Fig. 2 Little part of FMEA with known inconsistencies

R5 it shall reboot itself if it does not receive data from equipment.

The part of FMEA under study with known inconsistencies with the above requirements is reported in Fig. 2. The algorithm processed at least 50 requirements triplets and 50 FMEA rows. Some examples of known inconsistent triplets between requirements and FMEA are reported in Fig. 3. R1 and F1, and R3 and F3 are in conflict

	Requirement		FMEA Row
R1	R1.1<OBSW, not power_off, equipments> R1.2<OBSW, power_off_IN, deceleration>	F1	OBSW shall power off RF power amplifier if temperature overcomes 80° F1.1<OBSW, power-off, RF power amplifier> F1.2<OBSW, power-off_IN, preflight> F1.3 <RF power amplifier, overcomes, 70°> F1.4 <F1.3, pre-condition, F1.1>
	Any corresponding requirement	F2	OBSW shall disable storing in the case of memory full F2.1<OBSW, disable,storing> F2.2< memory,is,full> F2.3<OBSW_disable_IN, preflight> F2.4 <F1.2, pre-condition,F1.1>
R3	T11<OBSW,disable,report> T12<OBSW, disable_IN, preflight >	F3	OBSW shall report errors to ground in the case of GPS malfunction F3.1<OBSW, report, error> F3.2<OBSW, report_to, ground> F3.3<OBSW, report_IN, preflight> F3.4 <gps, failure, true> F3.5 <F3.4, pre-condition, F3.1>
R4	T4.1<OBSW, activate, FTS> T4.2<OBSW, activate_IN, flight > T4.3<OBSW, activate_IN, deceleration> T4.4 <RF power amplifier, failure, true> T4.5 <T4.4, pre-condition, T4.1>	F4	OBSW activate the flight termination system F4.1<OBSW, activate, FTS> F4.2<OBSW, activate_IN, preflight> F4.3 <RF power amplifier, failure, true> F4.4 <spacewire, failure, true> F4.5 <F2.4, or, F2.3> F4.6 <F2.5, pre-condition, F2.1>
R5	R5.1<OBSW, reboot, null> R5.2<OBSW, not receive, data> R5.3<OBSW , not receive_FROM, equipments> R5.4 <R5.2, pre-condition, R5.1>	F5	OBSW shall reboot itself if tasks exceeds the assigned time F5.1<OBSW, reboot, null> F5.2<task, exceed, time> F5.3 <F5.2, pre-condition, F5.1>

Fig. 3 Found inconsistencies between FMEA rows and requirement

because the action required by FMEA is in contrast with that provided by requirements. F2 is inconsistent because the action required by FMEA is not present in requirements. R4 and F4 are in conflict because they have different number of preconditions. R5 and F5 are in conflict because they have different preconditions.

The program has detected all the known inconsistencies with 20 % of false positives. We think that this is mainly due to the fact that we do not use a precise word sense but we consider all word senses of each word. This could be avoided by running first the word sense disambiguation that outputs the index of the intended word sense for each element of the triplet. This index can allow to calculate distances among Wordnet synsets belonging to the given word sense. We do not think results can be influenced by the chosen metric. Obtained results encourage to go on verifying the framework. Feedback from a more exhaustive testing should help to understand the coverage level of inconsistencies in order to improve the RDF extractor model, which should also take into account the software procedures formalization, usually addressed by FMEA instead of single mitigation actions.

4 Conclusion and Future Work

In this paper, we use RDF to model software and safety artifacts like FMEA. If the set of triplets represents a single software artifact, the presented algorithm will check its internal consistency and the consistency with FMEA triplets according to identified rules. RDF allows to check consistency from a semantic point of view in a simple manner. We start our research from a simple case study: the verification of known inconsistencies between a set of 50 requirements written in English and FMEA rows. Results encourage to further investigate such approach. Future work shall include the refinement of the framework, and the extension of the framework itself, by automating all the steps of the analysis, as well as proposing a well-defined domain ontology modeling the PUS standard for space data handling software.

Acknowledgments This work has been partially supported by EU with the project CRYSTAL (SP1-JTI-ARTEMIS-2012-AIPP1-332830).

References

1. Wu, Z., Palmer, M.: Verbs semantics and lexical selection. In: Proceedings of the 32nd annual meeting on Association for Computational Linguistics, pp. 133–138 (1998)
2. Nuseibeh, B., Russo, A.: Completeness in formal specification language design for process-control systems. In Proceedings of the 3rd Workshop on Formal Methods in Software Practice, pp. 75–87 (2000)
3. Nuseibeh, B., Easterbrook, S., Russo, A.: Leveraging Inconsistency in Software development. IEEE Comput. **33**(4), 24–29 (2000)

4. Mens, T., Van Der Straeten, R., Simmonds, J.: A Framework for Managing Consistency of Evolving UML Models. http://citeseerx.ist.psu.edu/viewdoc/summary?doi=10.1.1.130.9786 (2005)
5. Kroha, P., Gayo, L.: Using semantic web technology in requirements specifications. ChemnitzerInformatik-Berichte CSR-08-02, ISSN 0947-5125, TU Chemnitz (2008)
6. Ficco, M., Daidone, A., Coppolino, L., Romano, L., Bondavalli, A.: An event correlation approach for fault diagnosis in SCADA infrastructures. In: Proceedings of the 13th European Workshop on Dependable Computing, May 2011, pp. 15–20 (2011)
7. Wordnet search 3.1. http://wordnet.princeton.edu/
8. Liu, X.Y., Zhou, Y.M., Zheng, R.S.: Measuring semantic similarity in WordNet. In: Proceedings of the IEEE International Conference on Machine Learning and Cybernetics, vol. 6, August 2007, pp. 3431–3435 (2007)
9. ISO, ISO 26262 Road vehicles Functional Safety, Part 1–10 (2011)
10. ECSS-E-40C, Safety Space Product Assurance, ECSS Secretariat ESA-ESTEC Re-quirements & Standards Division Noordwijk, The Netherlands, 6 March 2009
11. Zazzaro, G., Gigante, G., Zaccariello, E., Ficco, M., Di Martino, B.: Supporting development of certified aeronautical components by applying text analysis techniques. In: Proceedings of the 8th International Conference on Complex, Intelligent and Software Intensive Systems, pp. 602–607 (2014)
12. Ficco, M., Avolio, G., Battaglia, L., Manetti, V.: Hybrid simulation of distributed large-scale critical infrastructures. In: Proceedings of the International Conference on Intelligent Networking and Collaborative Systems, September 2014, pp. 616–621 (2014)
13. Höfig, A., Zeller, M., Grunske, L.: MetaFMEA-A framework for reusable FMEAs. In: Proceedings of the 4th International Symposium on Model-Based Safety and Assessment, pp. 110–122 (2014)
14. Gigante, G., Gargiulo, F., Ficco, M.: A semantic driven approach for requirements consistency verification. Intell. Distrib. Comput. VIII **570**, 427–436 (2015)
15. Gribaudo, M., Iacono, M.: An introduction to multiformalism modeling. In: Theory and Application of Multi-Formalism Modeling, pp. 314–329 (2013)

An Analytical Approach for Optimal Resilience Management in Future ATM Systems

Domenico Pascarella, Francesco Gargiulo, Angela Errico and Edoardo Filippone

Abstract The air transportation system is a large-scale socio-technical system and its modelling approaches emphasize the sociological dimension due to the increasing importance of collaborative decision-making processes in the future Air Traffic Management (ATM). Resilience is assuming an increasing importance within ATM, but it is difficult or even impossible to establish the resilience role in realizing the targeted performance levels of an air traffic system. This paper proposes a systematic methodology for resilience management in ATM. It introduces an analytical definition of a resilience metric for an ATM system and formally states the resilience management problem as an optimization problem. Moreover, it describes a strategy for the problem solution and provides some preliminary results in order to quantitatively prove the validity of the methodology.

Keywords Resilience engineering · Resilience management problem · Resilience metric · Atm systems · Key performance indicators · Key performance areas

1 Introduction

Resilience is assuming an ever increasing importance within the future Air Traffic Management (ATM). The increasing traffic level, the typology of airspace users and a stronger interconnection among world areas can affect the ATM system performances in presence of unexpected events, not only from a safety perspec-

D. Pascarella (✉) · F. Gargiulo · A. Errico · E. Filippone
CIRA (Italian Aerospace Research Centre), Capua, Italy
e-mail: d.pascarella@cira.it

F. Gargiulo
e-mail: f.gargiulo@cira.it

A. Errico
e-mail: a.errico@cira.it

E. Filippone
e-mail: e.filippone@cira.it

© Springer International Publishing Switzerland 2016
P. Novais et al. (eds.), *Intelligent Distributed Computing IX*,
Studies in Computational Intelligence 616,
DOI 10.1007/978-3-319-25017-5_39

tive, but from a more comprehensive point of view. This paper provides a definition of resilience for the ATM framework, by taking into account several performance aspects and by implementing an innovative approach to support authority sharing, as described within the on-going project SESAR (Single European Sky ATM Research) JU (Joint Undertaking) E2.21 SAFECORAM (Sharing of Authority in Failure/Emergency Condition for Resilience of Air traffic Management).

The project aims to:

- study, analyze and define a classification approach to non-nominal, abnormal and emergency conditions in a highly automated ATM environment;
- identify a model to evaluate performance degradation on the base of classified failures and emergency conditions;
- develop a concept for tasks allocation and authority sharing between humans and systems, in those relevant degraded modes;
- build a software simulation environment to validate the concept with reference to a meaningful highly automated ATM scenario.

The proposed concept will be based on an evaluation of residual system performance and on a re-allocation strategy according to an assessment of which is the most performing element (human or system).

2 Background

Socio-technical systems are *systems that involve a complex interaction between humans, machines and the environmental aspects of the work system* [1]. Large-scale socio-technical systems (LSSTS) are a class of socio-technical systems that span technical installations embedded in large-scale social networks [2]. The air transportation system is a (large-scale) socio-technical system that is constantly influenced by internal and external events [3, 7], i.e., it is an open LSSTS. Therefore, it is an intractable system [5].

In the following, we firstly introduce some relevant terms for the ATM resilience framework and then we describe resilience engineering in air transportation.

2.1 Relevant ATM Terms

The following definitions are partly taken from references [9, 10].

An ATM system is performance-oriented due to the economic interests of its stakeholders. The resilience framework shall address the ATM ability to reduce the magnitude and the duration of the deviations from targeted performance levels. As a consequence, a set of Key Performance Indicators (KPIs) shall have to be rigorously established in order to include all the relevant performance dimensions. According to ICAO (International Civil Aviation Organization) [11], KPIs are quantitative indicators of current/past performance, expected future performance, as well as actual progress in achieving performance. KPIs are also grouped into Key Performance Areas (KPAs), which are *a way of categorizing performance subjects related to high-level ambitions and expectations* [11]. ICAO has defined eleven KPAs: safety, security, environmental impact, cost effectiveness, capacity, flight efficiency, flexibility, predictability, access and equity, participation and collaboration, interoperability.

The definition of state of an ATM system takes into account its performance indicators. The current state of an ATM system is defined by the current values of its KPIs, whereas the reference state of an ATM system is the specified set of values of its KPIs. A disturbance is a phenomenon which may cause a stress in a system. A stress is the state of a system caused by a disturbance which differs from the reference state. It is characterized by a deviation from the reference condition. A perturbation is the response of a system to the possible or current significant changes of the state caused by a disturbance. Perturbation aims at preventing the state changes and/or at minimizing the deviation from the reference state.

2.2 Resilience Definition for ATM

Considering that ATM is an open system, its operation is constantly perturbed by disturbances, which may interact with each other, potentially creating a cascade of adverse events. These events may pass without any discomfort for passengers, may result in a small passenger discomfort or may arise a discomfort that is out of any proportion [3, 7]. In the latter case, there are two categories of events: catastrophic accidents and events that push the ATM state away from its point of operation. These events are rare and exceptional, but they have large impacts. Hence, they have triggered several studies that have led to an ultra-safe ATM, but with a conflicting safety in respect to capacity, economy and environment requirements. Moreover, it is difficult or even impossible to establish the resilience role in realizing the ATM safety levels: currently, there is only a qualitative understanding of ATM resilience and no quantitative results exist. Thus, we are not able to assess whether an ATM system design is more or less resilient then another one [3, 7].

EUROCONTROL defines resilience as *the intrinsic ability of a system to adjust its functioning prior to, during, or following changes and disturbances, so that it can sustain required operations under both expected and unexpected conditions* [5]. This definition of resilience is apparently similar to robustness. Actually, resilience and robustness are complementary properties [9, 10]. They are both related to the system reaction that is triggered by a disturbance. Robustness of an ATM system is the ability of the system to experience no stress since a disturbance had occurred. On the other hand, resilience of an ATM system is the ability of the system to respond on disturbance within a time horizon by transient perturbation.

Currently, only an indirect measurement of ATM resilience has been defined [9, 10], whose concept is originated in material testing. On the contrary, performance-based methods measure the consequences of disruptions and the impact that system attributes have on mitigating those consequences, but they are not available for ATM systems. Alternative approaches related to hybrid simulation are used in [6].

3 ATM Resilience Management in SAFECORAM

The evolution of the resilience engineering concept follows three basic aspects: what is intended for resilience, how it can be measured and according to which modeling paradigm the ATM system can verify resilience measurements. In order to address all these aspects, the SAFECORAM approach mainly refers to the ability of the ATM system to "optimally" recover its global performance level, which is defined in terms of the system behavior with respect to specific KPAs and their related KPIs [11, 13].

In the following, the adopted performance framework, the proposed methodology and the experimental results are described.

3.1 Performance Framework

If a disturbance occurs, the ATM system can no longer perform its nominal conditions. The system can absorb disturbance and no change of tasks and authority reallocation is required. But if this robustness intrinsic property is no longer able to maintain the performance of the system, mitigation actions can be applied.

SAFECORAM project expresses the resilience as the level of the residual ATM system global performance as resulting from tasks reallocation due to a failure. This concept is graphically emphasized in Fig. 1, where the yellow area represents the nominal ATM system performance for a certain scenario under analysis. Once a failure occurs, different KPIs values will be associated to the system performance for each different mitigation action. In this way, every mitigation action will entail a different residual global performance, as represented by the red area in Fig. 1.

The proposed framework focuses on ATM in long-term vision (2050 perspective), based on the analysis of applicable documentation as resulting from the activities carried out in Europe and worldwide in ATM (SESAR concept of operations [12], ICAO [11] etc.). Starting from the 2050 scenario and a list of potential hazards [14], twelve study reference scenarios have been developed [4, 8] and validated as test cases for the proposed methodology of ATM resilience management.

Fig. 1 Residual ATM
system global performance

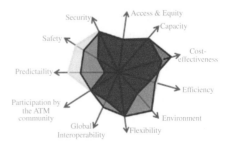

3.2 SAFECORAM Methodology for Resilience Management

In the remainder of this paper, we will refer with the term flow to the flows listed both in main (nominal) flow and alternative flows sections in each scenario.

The resilience metric of an ATM system should be expressed as a function of its performances. As a consequence, the statement of the resilience management problem within SAFECORAM project shall address a performance-based metric for resilience and shall take into account the KPAs and the KPIs that are prescribed by the SESAR performance framework. Among all KPAs selected from the SESAR performance framework, the following KPAs are considered within SAFECORAM project: safety; efficiency; capacity; environment.

In order to describe a general approach, suppose that there are n KPAs, named $\{A_1, \dots, A_n\}$, and suppose that the kth KPA is related to a set of KPIs, named $\left\{ KPI_1^{(A_k)}, \dots, KPI_{m_k}^{(A_k)} \right\}$, where m_k is the number of KPIs that are associated to A_k. We group all the KPIs into the following set

$$\Theta = \left\{ KPI_1^{(A_1)}, \dots, KPI_{m_1}^{(A_1)}, \dots, KPI_1^{(A_n)}, \dots, KPI_{m_n}^{(A_n)} \right\} \tag{1}$$

and we denote with $m = m_1 + \cdots + m_n$ the total number of KPIs.

Suppose also that S is a flow and $\{C_1, \dots, C_a\}$ are the actors involved in S. Each C_i can execute a set of tasks $T(C_i) = \{T_{i,1}, \dots, T_{i,h}\}$. From the performance point of view, each task $T_{i,j}$ may be also associated to a tuple $\left(k_{i,j}^{(1)}, \dots, k_{i,j}^{(m)} \right)$, wherein $k_{i,j}^{(t)}$ represents the "contribution" of (the execution of) $T_{i,j}$ in the evaluation of the tth KPI in Θ.

Here, we define the flow state (both nominal and non-nominal) as the set of values of its KPIs, that is, the state of a flow $F(S)$ is the following tuple

$$F(S) = \left\langle KPI_1^{(A_1)}, \dots, KPI_{m_1}^{(A_1)}, \dots, KPI_1^{(A_n)}, \dots, KPI_{m_n}^{(A_n)} \right\rangle \tag{2}$$

Note that, by means of the KPI definition, it is possible to define an order relation within a KPI associated to different flows of the same scenario and it is always possible to decide which is the "best" value of a certain KPI. In particular, we are interested in specifying an order relation amongst the whole states of the flows of a same scenario in order to establish if a flow S_p is better or worse than a flow S_q with respect to their states, i.e., their key performances. Therefore, we aim to identify an order relation between the following tuples

$$F(S_p) = \left\langle KPI_1^{(A_1)}(S_p), \dots, KPI_{m_1}^{(A_1)}(S_p), \dots, KPI_1^{(A_n)}(S_p), \dots, KPI_{m_n}^{(A_n)}(S_p) \right\rangle$$
$$F(S_q) = \left\langle KPI_1^{(A_1)}(S_q), \dots, KPI_{m_1}^{(A_1)}(S_q), \dots, KPI_1^{(A_n)}(S_q), \dots, KPI_{m_n}^{(A_n)}(S_q) \right\rangle \tag{3}$$

Clearly, some "conflicts" may arise because the flow S_p may be better than S_p with respect a KPI and, on the other hand, S_q may be preferable than S_q with respect to another KPI. These conflicts may be handled only by means of a global distance index for the whole state function $F(S)$. In order to establish a relation order amongst the flows of a scenario, we should define a similarity index for them, namely, a metric that quantifies their differences. For this reason, we introduce a flow distance function d. If Ω is the set of all the flows of the same scenario, a function $d : \Omega \longrightarrow \mathbb{R}$ is a flow distance if the following properties hold

$$
\begin{aligned}
d\left(S_p, S_p\right) &= 0, \quad \forall S_p \in \Omega \\
d\left(S_p, S_q\right) &= d\left(S_q, S_p\right), \quad \forall S_p, S_q \in \Omega \\
d\left(S_p, S_r\right) &\le d\left(S_p, S_q\right) + d\left(S_q, S_r\right), \quad \forall S_p, S_q, S_r \in \Omega
\end{aligned} \tag{4}
$$

A flow distance represents a quantitative measure of the similarity between two flows of a scenario. Obviously, the flow distance should be related to the flow states for our purposes. Indeed, S_p and S_q are similar and $d\left(S_p, S_q\right)$ is low if their states $F\left(S_p\right)$ and $F\left(S_p\right)$ (i.e., their global ATM performances) are close.

Based on the previous considerations, we define the resilience metric in the following way. Let S_0 be the nominal flow of a scenario $\mathbb{S}$, that is, the main flow of $\mathbb{S}$. Let S_i be an alternative flow of the same scenario $\mathbb{S}$ of S_0. The SAFECORAM resilience loss metric $RL_{\mathbb{S}}\left(S_i\right)$ in the scenario $\mathbb{S}$ is

$$
RL_{\mathbb{S}}\left(S_i\right) = d\left(S_0, S_i\right) \tag{5}
$$

This metric is a function of the selected scenario $\mathbb{S}$ (and its nominal flow) and of the triggered alternative flow S_i. It is a resilience loss metric because the more similar are the performed alternative flow S_i and the nominal flow S_0, the lower is $RL_{\mathbb{S}}\left(S_i\right)$. In this way, the proposed metric confirms that the ATM system is more resilient if the chosen alternative flow is more similar to the nominal flow, i.e., if their states (and so their global performances) are closer.

Several characterizations of the metric (5) may be provided according to the flow distance index. For example, suppose we have four KPAs, that there is a KPI for every KPA and that $KPI^{(A_i)} \in [0, 1], \forall i \in \{1, 2, 3, 4\}$. If S is a flow, we denote with $R(S)$ the area of the quadrangles with vertices $KPI_1(S)$, $KPI_2(S)$, $KPI_3(S)$ and $KPI_4(S)$. It may be seen as the "global performance area" of S. In this case, an intuitive definition of the flow distance between the flows S_1 and S_2 is

$$
d\left(S_1, S_2\right) = \left| R\left(S_1\right) - R\left(S_2\right) \right| \tag{6}
$$

Hence, according to (6), two flows are similar if they entail similar global performance area. This metric is also named area distance.

Another scalar real-valued formulation for the flow distance between the flows S_1 and S_2 of $\mathbb{S}$ is the difference distance, that has the following expression

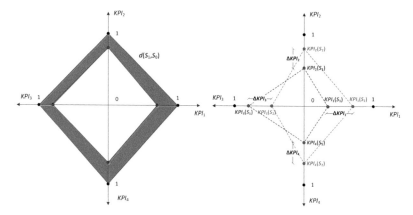

Fig. 2 Example of area distance (*left*) and difference distance (*right*) for the SAFECORAM resilience loss metric

$$d\left(S_1, S_2\right) = a_{1,1}\left|KPI_1^{(A_1)}\left(S_1\right) - KPI_1^{(A_1)}\left(S_2\right)\right| + \cdots + a_{1,m_1}\left|KPI_{m_1}^{(A_1)}\left(S_1\right) - KPI_{m_1}^{(A_1)}\left(S_2\right)\right| +$$
$$+ \cdots + a_{n,1}\left|KPI_1^{(A_n)}\left(S_1\right) - KPI_1^{(A_n)}\left(S_2\right)\right| + \cdots + a_{n,m_n}\left|KPI_{m_n}^{(A_n)}\left(S_1\right) - KPI_{m_n}^{(A_n)}\left(S_2\right)\right|$$
$$(7)$$

Figure 2 shows an example of the metrics (6) and (7).

Let δ be a disturbance in the nominal flow S_0 of $\mathbb{S}$ and let $\Gamma^{(\mathbb{S},\delta)} = \left\{S_1, \ldots, S_k\right\}$ be the set of alternative flows that may be executed in order to reach the same terminal condition of S_0. Here, we assume that δ is unique in $\mathbb{S}$ and $\Gamma^{(\mathbb{S},\delta)} = \Gamma^{(\mathbb{S})}$. This set can be modelled as a DAG (Directed Acyclic Graph) $G = \langle V, E \rangle$, where V is the set of vertices and E is the set of edges. Every vertex $v \in V$ corresponds to a single task $T_{i,j}$. An edge $(u, v) \in E$—with $u, v \in V$—states that the task u shall be executed before the task v starts. Then, it represents a precedence relation. We also assume that the there are a starting vertex v_{start} and an ending vertex v_{end}. The former conventionally represents a null task and also depicts the triggering condition (the disturbance) of $\Gamma^{(\mathbb{S})}$; the latter represents the terminal condition of $\mathbb{S}$. Thereby, an alternative flow $S_l \in \Gamma^{(\mathbb{S})}$ is a route from v_{start} to v_{end}. Figure 3 shows the structure of this resilience-based DAG. Every edge is labelled with the tuple $\left(k_{i,j}^{(1)}, \ldots, k_{i,j}^{(m)}\right)$, which represents the contribution of the destination vertex $T_{i,j}$ in the KPIs evaluation.

Given a flow distance function $d\left(\cdot\right)$, we define the resilience management problem as the following optimization problem

$$S_{opt} = \underset{S_l \in \Gamma^{(\mathbb{S})}}{\arg\min} RL_{\mathbb{S}}\left(S_l\right) = \underset{S_l \in \Gamma^{(\mathbb{S})}}{\arg\min} d\left(S_0, S_l\right) \qquad (8)$$

This problem consists in scheduling the best alternative flow S_{opt} in the scenario $\mathbb{S}$ as the alternative flow in $\mathbb{S}$ that has the minimum resilience loss (i.e., the flow distance) with respect to the nominal flow S_0.

Fig. 3 Directed acyclic
graph for the set of
alternative flows of a
scenario

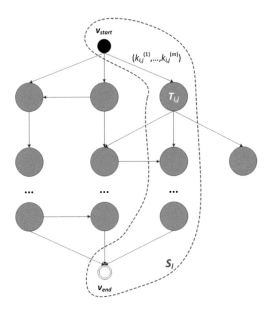

3.3 Experimental Results

A software demonstrator has been implemented in order to prove the feasibility and
the validity of the SAFECORAM resilience management process. It is in charge
of building the equivalent DAG of a scenario and of processing all the associated
alternative flows by means of a depth-first search strategy.

The following scenario has been used as an application example of the proposed
methodology in this paper.

Uplink loss during en-route phase

This scenario shows a commercial vehicle flying from Naples airport (LIRN) to London
Gatwick airport (EGKK), in normal traffic level and bad weather over the destination airport. The
on-board systems are continuously linked to the System Wide Information Management (SWIM)
and to local meteorological data in order to improve the capability of the automatic 4D Flight
Management System (4DFMS) to predict the trajectory and to accurately track the assigned 4D
contract. During the en-route phase, an uplink loss occurs that prevents the reception in uplink
of a new 4D contract elaborated by the ground segment. This failure stimulates many alternative flows with respect to the main flow. For instance, the Controller may perform the following
actions: activate a procedure in order to assure the safe flight of the vehicles surrounding the
affected aircraft; confirm the validity of the latest 4D contract assigned to the affected vehicle in
compliance with the traffic picture of its sector of responsibility; communicate by voice the new
contract data to be inputted by the Flight Crew to the 4D FMS, in order to divert the aircraft to
another airport.

The KPAs taken into account are efficiency, environment and capacity. The related KPIs are:

- K1-efficiency: fuel burn (kg/min per movement);
- K2-efficiency delay (minutes);
- K3-environment: emission of pollutant (kg of CO_2);
- K6-capacity: delay-predictivity (movements per one runway per hour);

This scenario involves 6 aircrafts. The adopted assumptions are the following:

1. movements: 15 movements per hour;
2. fuel burn: the average value of 65 kg per minute is used for all aircrafts;
3. duration: the time including TMA and Landing phases is 20 min;
4. emission of pollutant: $c_{CO_2} = 3.149$ kgs/kg of fuel burnt.

Both the area distance metric (6) and the difference distance metric (7) have been used as a resilience loss metric for the experimental phase. Even if coefficients may be assigned to each KPI in order to determine the importance of a KPI with respect to the others, it is assumed that all KPIs have the same weight (i.e., all coefficients are equal to 1). Besides, the KPIs values of the alternative flows are normalized with respect to the KPIs values of the nominal flow. As regards the KPI associated to Delay (K2-efficiency), the reported values refer to the whole duration of the flow instead of the delay.

The reference scenario contains 9 alternative flows. Tables 1 and 2 respectively report the KPIs values (absolute and normalized) of every flow and the distance values from the nominal flow. Moreover, the optimal flow and its related values are highlighted in bold.

Table 1 KPIs values of the nominal and alternative flows of the reference scenario

Flow	Fuel Burn		Pollution		Duration		Capacity	
	Absolute	Norm.	Absolute	Norm.	Absolute	Norm.	Absolute	Norm.
Nominal	1560.0	1	4093.7	1	20.0	1	5.0	1
1	2236.0	1.43	5867.6	1.43	52.0	2.6	4.5	0.9
2	**1846.0**	**1.18**	**4844.2**	**1.18**	**22.0**	**1.1**	**5.75**	**1.15**
3	1898.0	1.2166	4980.6	1.2166	26.0	1.3	5.5833	1.1166
4	1937.0	1.24166	5083.0	1.24166	29.0	1.45	5.45833	1.09166
5	2236.0	1.4333	5867.6	1.4333	52.0	2.6	4.5	0.9
6	1872.0	1.2	4912.4	1.199	24.0	1.2	5.66	1.133
7	2262.0	1.45	5935.8	1.45	54.0	2.7	4.4166	0.88333
8	1872.0	1.2	4912.4	1.199	24.0	1.2	5.66	1.133
9	2262.0	1.45	5935.8	1.45	54.0	2.7	4.4166	0.88333

Table 2 Alternative flows distances from the nominal flow of the reference scenario

Flow	Area distance	Difference distance
1	2.70554	2.56666
2	**0.66388**	**0.66388**
3	0.93611	0.84999
4	1.140277	1.02499
5	2.7055	2.5666
6	0.7999	0.7333
7	2.84166	2.71666
8	0.799	0.733
9	2.84166	2.71666

4 Conclusion and Future Work

This paper reports the design of a methodology to cope with the resilience management problem of the future ATM system. It defines the resilience metric for the reference system and formally states the resilience management problem as an optimization problem of tasks reallocation. The adopted strategy follows the scenario-based approach of the SAFECORAM project. Furthermore, the paper illustrates the achieved experimental results.

Future work will involve the development of a software simulation environment to validate the proposed methodology with reference to a meaningful highly automated ATM scenario. Moreover, the reaction time of the system for the resilience DAG building and exploration will be thoroughly analyzed in case of highly complex scenarios.

Acknowledgments This work is co-financed by EUROCONTROL acting on behalf of the SESAR Joint Undertaking and the European Union as part of Work Package E in the SESAR Programme. Opinions expressed in this work reflect the authors' views only and EUROCONTROL and/or the SJU shall not be considered liable for them or for any use that may be made of the information contained herein.

References

1. Baxter, G., Sommerville, I.: Socio-technical systems: From design methods to system engineering. Int. J. Hum. Comput. Stud. pp. 1-33 (2008)
2. Bijker, W., Hughes, T., Pinch, T.: The Social construction of technological systems: new directions in the sociology and history of technology, edited by MIT Press (1987)
3. CW Members: ComplexWorld Position Paper. Deliverable D23.2, SESAR Project Complex-World (E-01.01) (2012)
4. Di Vito, V., Torrano, G., Errico, A., and Filippone, E. (2014). Study Reference Scenarios. Deliverable D1.2, Edition 01.00.00, SAFECORAM Project (E.02.21)
5. EUROCONTROL: A White Paper on Resilience Engineering for ATM (2009)

6. Ficco, M., Avolio, G., Palmieri, F., Castiglione, A.: An HLA-based framework for simulation of large-scale critical systems (2015). doi:10.1002/cpe.3472
7. Francis, R.: Analysis of Resilience in Manmade and Natural Systems. Deliverable D1.1, FP7 Project Resilience 2050 (2013)
8. Gargiulo, F., Pascarella, D., Errico, A., Di Vito, V., Filippone, E.: Resilience Management Problem in ATM Systems as a Shortest Path Problem, SESAR Innovation Days (2014)
9. Gluchshenko, O., Foerster, P.: Performanced based approach to investigate resilience and robustness of an ATM system. In: Proceedings of Tenth USA/Europe Air Traffic Management Research and Development Seminar (ATM2013), Chicago (2013)
10. Gluchshenko, O.: Definitions of Disturbance, Resilience and Robustness in ATM Context. DLR Report IB 112-2012/28, release 0.07 (2012). http://elib.dlr.de/79571/
11. ICAO: Manual on Global Performance of the Air Navigation System. Doc 9883 (2009)
12. SESAR: The Concept of Operations at a glance. SESAR Definition Phase (2008)
13. SESAR: The Performance Target. SESAR Definition Phase, Deliverable 2 (2006)
14. Stroeve, S., Everdij, M., Blom, H.: Hazards in ATM: model constructs, coverage and human responses. Deliverable D1.2, Edition 1.0, MAREA Project (E.02.10) (2011)

Enabling Convergence of Physical and Logical Security Through Intelligent Event Correlation

Gianfranco Cerullo, Luigi Coppolino, Salvatore D'Antonio,
Valerio Formicola, Gaetano Papale and Bruno Ragucci

Abstract Until now, in most organizations, physical access systems and logical security systems have operated as two independent elements, and have been managed by completely separate departments. The lack of interoperability between the two sectors often resulted in a security hole of the overall infrastructure. An attacker who has physical access can not only steal a PC or confidential data, but can also compromise network security. Therefore, a combination of physical and logical security definitively allows for a more effective protection of the organization. In this work we present a correlation system which aims at bringing a significant advancement in the convergence of physical and logical security technologies. By "convergence" we mean effective cooperation (i.e. a coordinated and results-oriented effort to work together) among previously disjointed functions. The holistic approach and enhanced awareness technology of our solution allows dependable (i.e. accurate, timely, and trustworthy) detection and diagnosis of attacks. This ultimately results in the achievement of two goals of paramount importance, and precisely guaranteeing the protection of citizens and assets, and improving the perception of security by citizens. The effectiveness of the proposed solution is demonstrated in a scenario that deals with the protection of a real Critical Infrastructure. Three misuse cases have been implemented in a simulation environment in order to show how the correlation system allows for the detection of different attack patterns.

G. Cerullo (✉) · L. Coppolino · S. D'Antonio · V. Formicola · G. Papale · B. Ragucci
University of Naples "Parthenope", Napoli, Italy
e-mail: gianfranco.cerullo@uniparthenope.it

L. Coppolino
e-mail: luigi.coppolino@uniparthenope.it

S. D'Antonio
e-mail: salvatore.dantonio@uniparthenope.it

V. Formicola
e-mail: valerio.formicola@uniparthenope.it

G. Papale
e-mail: gaetano.papale@uniparthenope.it

B. Ragucci
e-mail: bruno.ragucci@uniparthenope.it

© Springer International Publishing Switzerland 2016
P. Novais et al. (eds.), *Intelligent Distributed Computing IX*,
Studies in Computational Intelligence 616,
DOI 10.1007/978-3-319-25017-5_40

427

1 Introduction

Technologies for implementing security services in the physical and in the logical domain are both stable and mature, but they have been developed independently of each other. Some of them have recently merged, but real convergence of physical and logical security technologies is still a faraway target. By "convergence" we mean: effective cooperation (i.e. a coordinated and results-oriented effort to work together) among previously disjointed functions. In the recent years, some achievements have been made, but much is yet to be done. As an example, Security Operation Centers (SOC) technology has improved significantly, but SOC solutions have typically been developed using vertical approaches, i.e. based on custom specific needs. In this paper we focus on SOC technology as key tool for increasing security of critical infrastructures through the convergence of physical and logical security. Specifically, a SOC aims to effectively detect and diagnose cyber-attacks and, in order to do so, it collects and analyses activity reports (e.g. system logs, notification and alert messages, traps, etc.)—also known as "events"—provided automatically by electronic and computer systems or manually by the personnel operating on the infrastructure. In order to be effective, the analysis performed by a SOC must be dependable, i.e. accurate, timely and trustworthy.

- Accurate—The detection rate will be high (i.e. a very high percentage—higher than what is currently achieved by state-of-the-art products—of real attacks will be detected) and the false positive rate will be low (i.e. a very low percentage—lower than what is currently achieved by state-of-the-art products—of real attacks will go undetected). It is worth emphasizing that in contexts such as highly available systems and applications (e.g. Critical Infrastructures) and crowded places (e.g. a stadium or an airport), false alarms can be as dangerous and harmful as false negatives [1]. Accuracy will be achieved by performing sophisticated correlation of the multitude of diverse events which will be collected in the two domains (namely: logical and physical). Evidence is demonstrating that this approach is effective [2–4].
- Timely—The aforementioned sophisticated correlation will be done in near real-time. This is a challenging task, since the amount of data that the system will have to process is massive and highly heterogeneous (both from the format and from the semantics point of view).
- Trustworthy—A largely overlooked issue in the design and development of security products is "who defends the defender" [5–7]. This means that the SOC platform has to be designed and implemented using fault- and intrusion-tolerant techniques. The platform will thus be resilient to fault and attacks, i.e. it will be able to perform its tasks correctly even in the presence of faults and/or if it will be under attack.

Typical systems that provide SOC with data are physical access control systems with real-time data processing features, service monitoring systems, infrastructure performance monitors, logical security systems. A number of facilities is also available to enforce controls for safeguarding the operation of the system, as well as for protecting the surrounding environment. In this paper we illustrate how a SOC can be used to detect attacks that are perpetrated by company employees and are usually referred to as "internal" attacks. In particular, the SOC is required to understand whether an outage is due to a misconfiguration caused by a legitimate maintenance operation or it is the effect of a malicious attack. In addition to information about regular operations, a SOC analyzes the information contained in the maintenance reports in order to distinguish the following cases:

- Events representing planned maintenance operations;
- Events representing failures of specific system components due to non-malicious faults;
- Events representing failures of specific system components due to malicious faults (attacks).

The paper is structured as follows. Section 2 presents the general architecture of a Security Operation Center. Section 3 illustrates the proposed correlation system, which allows to improve the capability of a Security Operation Center to combine physical and logical security technologies, thus achieving a higher attack detection performance. In Sect. 4 the correlation system is validated in three misuse cases where the attack strategy consists in exploiting both physical and logical security vulnerabilities. Finally, Sect. 5 provides some concluding remarks.

2 Architecture of a Security Operation Center

A Security Operation Center monitors and manages several types of security events to perform Real Time Device Monitoring (RTDM), Network Fault Management, Security Incident Management, Policy Management and Enforcement, Vulnerability Assessment and Policy Compliance Verification.

The following subsections present the tasks performed by a SOC that contribute to the implementation of the security scenarios addressed by this paper.

2.1 Real Time Device Monitoring

Real Time Device Monitoring is a continuous activity for real time monitoring of security-related events. It manages events generated by network devices (routers, switches), security devices (firewalls, IDS/IPS, antivirus, etc.), servers, and applications (e.g., web servers, application servers, proxies, etc.), physical access control systems (badges, intrusion detection systems, door and window alarms, etc.). RTDM systems are used for accurate analysis of alarms or events generated by monitored devices and initialization of corrective actions for alarms that exceed specific security thresholds; implementation of appropriate alerting mechanisms in accordance with defined procedures and escalation strategies; definition of new correlation rules to identify new threats; and tuning of existing correlation rules to avoid/reduce false positives. Security Information and Event Management (SIEM) systems are largely used to perform Real Time Device Monitoring activities.

A SIEM system is responsible for collection and correlation of all the events coming from the operational domain context and from the corporate areas.

2.2 Video Surveillance

In remote sites, the exterior zones of critical infrastructures are exposed to (sometime) extreme weather conditions and the use of video surveillance is critical and complex. In this case video surveillance is the complement to burglar alarms, and is used to minimize false alarms of physical violations. Indeed, borders have peculiarity for which intrusion can occur and must be accepted at a certain degree. Since guards are not everywhere, an attacker could tag along to a car in transit hiding himself from view. In these circumstances, video surveillance is a support for forensic and investigations, not being possible a continuous view and detection over all the facilities disseminated in the sites.

2.3 Physical Access Control

Physical Access Control systems must provide identification, authentication and authorization of people entering and exiting each zone of the infrastructure. Authentication can be single or double factor based. The single factor authentication—typically badge control—is less strong, but is the most commonly implemented, because there is no need for acquiring additional (e.g. biometric) parameters from the user. Also, in some countries there are laws limiting the usage of biometric data for physical security. One of the most important requirements for a physical access control is the analysis of authentication attempts. Indeed, physical access attempts must be recorded in order to discover when suspicious thresholds are exceeded. In order to properly supervise physical alarms, the SOC must be correctly tuned through severity or priority values. Indeed, in some operational contexts many physical access events are not so relevant, i.e. must be associated to low priority warnings. In other contexts, such as access to very critical rooms, attempts must trigger highest priority alerts.

Fig. 1   Main functional blocks of a real-time intelligent event correlation system

3 Correlation System

The main contribution of this paper is the development and validation of a correlation engine that has been implemented in order to enhance the capability of a Security Operation Center to protect a critical infrastructure from sophisticated attacks involving both the physical and logical domain. Real time correlation allows to combine huge amounts of micro-data generated by heterogeneous information sources, and obtain semantically richer macro description of faults in real time. This process is a key building block for a Situation Aware Security Operation Center since it enables timely and accurate detection and diagnosis of (both maliciously induced and not) faults on complex critical systems. The real time correlation process involves three main tasks, which are: collection and preprocessing of events at the edge of the SOC framework and in proximity of the data sources; distribution of these events from the edge to the core processing systems; data processing, i.e. correlation of information belonging to multiple layers of the infrastructure; semantic fusion of the information and final generation of ranked evidence. This processing chain is represented in the Fig. 1.

3.1 Data Collection

The collection task is in charge of gathering data generated from heterogeneous information sources, and to output these messages in a format which is processable by the centralized correlation engine. The main sub-tasks performed by the collection system can be summered in: data gathering, i.e. collection of data based on different transfer protocols; message format parsing, i.e. tokenization of fields from variously structured messages; message filtering, i.e. dropping of irrelevant messages; message pre-aggregation, i.e. coalescence of similar entries; format normalization, i.e. generation of fields in a standard format; forwarding, i.e. propagation of the events to the core processing. The parsing step extracts tokens from streams of events represented in syntactically and semantically heterogeneous data formats. In order to identify the input format, an "Event Id" is typically configured on the parsing system and is associated with the specific information source. For instance, IP address of the source, data transfer protocol used, collection port, session-id and tags can be used to identify the input format. Filtering of events is typically based on Regular Expressions (RegEx), so that it is possible to associate the specific filter to the required class of events. Filters wait for new Parsed Events to operate, or can be optimized to work during the parsing process. In the latter case, events matching the dropping RegEx rule are discarded and the parsing consumes the next message. Pre-aggregation step performs a first level of aggregation when similar entries are coalesced into a single message. In this case semantic reasoning is still required and is delegated to the core processing, as we see below.

3.2 Data Distribution

The message forwarding model is demanded to the requirements of the processing systems, i.e. how many processors will take charge of evaluating and processing the collected events. An effective solution comes from the Publish/Subscribe mechanism, which enables the publisher (i.e. the data collector in this case) to publish a single message that can be consumed by multiple subscribers (i.e. the correlation system and others): the Publisher only takes care of publishing its message on a "topic" hosted by some broker; the latter provides the message to the consumers subscribed to that topic. Finally, this messaging model supports asynchronous communication, and improves the scalability of the system. The messaging system must ensure reliability by guaranteeing the delivery of messages, through a trade-off between throughput and reliability. It is possible to introduce different levels of reliability, which assign to messages a greater or lesser relevance. Furthermore the messaging systems are usually supported by persistent storage systems that preserve messages and protect them from attacks aiming at violating data confidentiality and integrity. Another aspect is the robustness of the system. Actually publishers, subscribers and network can have failures, and redundancy can mitigate this issue. In addition to the common message brokering systems, such as Java Message Systems (JMS), a very effective technology is provided by Apache Kafka, that combines the model of messaging system with log aggregators. Also, it implements scalable and distributed processing of queues and topics.

3.3 Data Processing

Data processing task concerns with the centralized correlation of events coming from the distributed collectors at the edge. The centralized processing enables global view of the infrastructure state, and can take advantage of high computing resources available at the core systems. The correlation process aims at finding a relation among the data fields composing the normalized events, and eventually at producing an aggregated message. The aggregated output contains fields extracted from the input events and metrics obtained by combining each evidence. Whatever the metrics and the data fusion model—one will be discussed in the next section—, the aggregated message outputs the security risk level of aggregated events. In order to have effective situational awareness of the system state, it is of paramount importance refining the huge amount of information obtained from the monitoring systems. This ultimately means correlating both the information coming from the system operations (e.g. system logs, security incidents) with information related to the operational context (e.g. maintenance plans, physical events). One of the most widely and effective solutions to correlate streams of events is Complex Event Processing (CEP). We use CEP for defining relations among the events, i.e. to describe the correlation model (e.g. a simple matching rule or more sophisticated inter-event analysis), for combining metrics useful to rank the alerts, and for defining the structure of the output alert message. An example of CEP is EsperTech Esper [19]. The computational load of this processing is distributed by means of high performance and dependable computing technologies. An effective solution to merge the semantics of Esper with distributed, scalable and fault-tolerant processing is Apache Storm [18].

3.4 Data Fusion

Data Fusion [8] is the process of combining information from a number of different sources to provide a robust and complete description of an environment or process of interest. Data Fusion process is applied where a large amounts of data must be combined and fused to obtain information of appropriate quality and integrity on which decisions can be made. In any data fusion problem, there is an environment, process or quantity whose true value, situation or state is unknown. The sources provide some parameters, imperfect and incomplete knowledge, that are processed and then transformed in decisions, that provides effective support for human or automated decision making. Data fusion is the process of combining data to refine estimates and predictions of the state that is observed. Joint Directors of Laboratories (JDL) defines data fusion as a "multilevel, multifaceted process handling the automatic detection, association, correlation, estimation, and combination of data and information from several sources". The proposed correlation engine exploits the features provided by the Dempster-Shafer Theory that allows to combine multiple pieces of evidence for detecting an ongoing attack.

4 Misuse Cases

The Correlation Engine (CE) aims to correlate relevant information from the physical and the electronic domain in an effective way to ensure the security and the detection of potential attacks and threats. The information, coming from a huge amount of sources, is aggregated in real-time fashion using correlation rules. A correlation rule aggregates symptoms based on a set of parameters, such as the attack type, the target component and the temporal proximity. Alerts are not generated as results of all the monitored symptoms, but only when the correlation among such symptoms

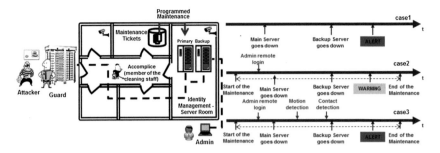

Fig. 2 Attack scenario storyboard

indicates a potential attack, thus reducing the number of false positives and improving the detection capability of the overall system. To explain the functioning of our solution a storyboard with three possible cases of attack is presented. The actors involved in the selected misuse cases are the Maintenance Scheduling Programming (MSP), the Network Management System (MS) and the Videosurveillance System (VS). An accomplice of an attacker wants to take advantage of a scheduled maintenance service on the server which manages the identity of users, named Primary Server, in order to allow the attacker to enter the building. When the Primary Server is down a Backup Server replaces it. The accomplice enters the Identity Management Server Room and opens the rack, in which the servers are placed. After that, he unplugs the Ethernet cord of the Backup Server, so that the system for identity management fails down, and the attacker can enter the building with a fake badge, as shown in Fig. 2.

- **Case 1**

  In this case we consider only the MS. The Primary Server goes down and the Backup Server replaces it. After a while also the Backup Server goes down and the CE is aware of the anomalous events occurring within the infrastructure and raises an alert. The CE observing an outage of the Primary Server and a non-operation of the Backup Server considers this situations as a possible symptom of an attack since the identity management system is out of service.

- **Case 2**

  The actors involved in this scenario are MSP and MS. The first provides a maintenance ticket in which it warns that the Primary Server stops the services offered, due to a maintenance job performed by the system administrator. The administrator logs in remotely and the Primary Server stops working. At the same time the Backup Server replaces it, afterwards also the Backup Server goes down. The CE correlates and aggregates the information provided by the two sources and raises a warning because it considers this event as a possible attack. It is aware that Primary Server is down for a maintenance operation. As soon as the Backup Server goes down, the CE detects this situation as a malfunction or an ongoing attack.

- **Case 3**

    MSP, MS and VS are the actors involved in this misuse case. The maintenance ticket is sent to the CE and the administrator logs in remotely on the Primary Server, that goes down. Now an accomplice of the attacker enters the Identity Management Server Room, the VS detects his presence and sends a motion detection log to the CE. The CE considers the aggregated event as a normal situation because the maintenance ticket warns that the Primary Server will be inactive for a certain time window and the motion detection does not indicate an imminent attack. After this the accomplice opens the rack, in which the Primary and Backup Server are placed. At the same time the VS detects the contact and sends a log to the CE that raises a warning since it interprets the information regarding the contact with the rack as relevant for safety purposes. After few seconds the accomplice unplugs the Ethernet cord of the Backup Server, which goes down. The CE aggregates the information in a single event that is processed and through the correlation rule returns the evidence of an attack. Then an alarm is raised.

5 Related Work

The use of correlation techniques for attack and intrusion detection has been largely explored in literature. Some relevant papers dealing with this research topic are presented below. In [8] the authors propose a Simple Event Correlator relying on a rule-based correlation approach, that is used to detect and filter out relevant symptoms useful for fault diagnosis in a Supervisory Control and Data Acquisition (SCADA) infrastructure. [9] presents a Generic Intrusion Detection and Diagnosis System for detection and diagnosis of complex attack patterns in large scale Critical Infrastructures. In [10] a comprehensive analysis of the cyber-security issues concerning Smart Grids, specifically network vulnerabilities, attack countermeasures, secure communication protocols and architectures, is performed. In [11] the authors present an Intrusion Detection System (IDS) for correlating attack symptoms from diverse information sources. The presented IDS relies on an ontology to drive the correlation task and is implemented as a distributed and highly scalable system. In [12] the authors identify limits of the current SIEM systems and propose a framework to enhance services for data treatment. They also provide prototypal deployment of a case study consisting in securing a dam monitoring and control system. In [13], the authors provide a performance comparison of the most popular open source rule based correlation engines. A distributed event correlation system which performs security event detection is presented in [14]. The system detects several misuse cases, with a low false positive rate. Our solution involves Level 0—Source Pre-Processing, Level 1—Object Refinement and Level 2—Situation Refinement of the Joint Directories Laboratory (JDL) Data Fusion Process Model. A description of the JDL for cyber-security is given in [15]. Information fusion is exploited in [16] to spot frauds against a mobile money transfer service by using combination rules of the Dempster-Shafer Theory.

6 Conclusions

In this paper we addressed the need for convergence of physical and logical security in order to enhance the protection of critical assets. The convergence of these two worlds brings positive effects to the general security of an organization. Integrating the efforts of physical and logical security departments allows an organization to significantly lessen security risks while also saving time and money. This paper presented a correlation system capable of collecting and processing security relevant information and events from both physical and logical domain, thus enabling the convergence of these two security areas. The proposed system has been validated in three different scenarios, where the correlation of events generated by heterogeneous probes made it possible the detection of sophisticated attacks exploiting both physical and logical security vulnerabilities.

Acknowledgments The research leading to these results has received funding from the European Commission within the context of the Seventh Framework Programme (FP7/2007–2013) under Grant Agreement No. 313034 (Situation AWare Security Operation Center, SAWSOC Project). It has been also partially supported by the TENACE PRIN Project (n. 20103P34XC) funded by the Italian Ministry of Education, University and Research, and by the Embedded Systems in critical domains POR Project (CUP B25B09000100007) funded by the Campania region in the context of the POR Campania FSE 2007–2013, Asse IV and Asse V.

References

1. Tips to reduce false security alarms with proper installation, education and training. http://www.sourcesecurity.com/news/articles/co-2173-ga.4866.html
2. Repp, N., Berbner, R., Heckmann, O., Steinmetz, R.: A cross-layer approach to performance monitoring of web services. In: Proceedings of the Workshop on Emerging Web Services Technology, CEUR-WS, December 2006
3. Yu-Sung, W., Bagchi, S., Garg, S., Singh, N.: SCIDIVE: a stateful and cross protocol intrusion detection architecture for voice-over-IP environments. In: Proceedings of Dependable Systems and Networks Conference, 28 June 2004, pp. 433–442 (2004)
4. Vigna, G., Robertson, W., Vishal, K., Kemmerer, R.A.: A stateful intrusion detection system for World-Wide Web servers. In: Proceedings of the 19th Annual Computer Security Applications Conference, 8–12 December 2003, pp. 34–43 (2003)
5. Verssimo, P., Correia, M., Neves, N., Sousa, P.: Intrusion-resilient middleware design and validation. In: Information Assurance, Security and Privacy Services (Handbooks in Information Systems, vol. 4), Emerald Group Pub. Ltd., pp. 615–678 (2009)
6. Sousa, P.: Proactive Resilience. In: Proceedings of the 6th European Dependable Computing Conference (EDCC-6) Supplemental Volume, Coimbra, Portugal, October (2006)
7. Dondossola, G., Deconinck, G., Di Giandomenico, F., Donatelli, S., Kaaniche, M., Verssimo, P.: Critical utility infrastructure resilience. In: Workshop on Security and Networking in Critical Real-Time and Embedded Systems (CRTES'06), with RTAS'06, San Jose, California, April (2006)
8. Ficco, M., Daidone, A., Coppolino, L., Bondavalli, A.: An event correlation approach for fault diagnosis in SCADA infrastructures. In: Proceedings of the 13th European Workshop on Dependable Computing (EWDC 2011), Pisa, Italy, May 2011, pp. 15–20. ACM Press (2011). doi:10.1145/1978582.1978586

9. Ficco, M., Romano, L.: A generic intrusion detection and diagnoser system based on complex event processing. In: Proceedings of the 1st International Conference on Data Compression, Communications and Processing (CCP 2011), Palinuro, Italy, June 2011, pp. 275–284. IEEE CS Press (2011). doi:10.1109/CCP.2011.43

10. Wang, W., Lu, Z.: Cyber security in the smart grid: survey and challenges. Comput. Netw. **57**(5), 1344–1371 (2013). doi:10.1016/j.comnet.2012.12.017

11. Coppolino, L., D'Antonio, S., Esposito, M., Romano, L.: Exploiting diversity and correlation to improve the performance of intrusion detection systems. In: Proceedings of the International Conference on Network and Service Security, N2S'09, Paris, June 2009, pp. 24–26 (2009)

12. Coppolino, L., D'Antonio, S., Formicola, V., Romano, L.: Enhancing SIEM Technology to Protect Critical Infrastructures. Critical Information Infrastructures Security Lecture Notes in Computer Science 7722, 10–21 (2013)

13. Rosa, L., Alves, P., Cruz, T., Simes, P., Monteiro, E.: A comparative study of correlation engines for security event management. In: Proceedings of the 10th International Conference on Cyber Warfare and Security (ICCWS-2015), Kruger National Park, South Africa (2015)

14. Myers, J., Grimaila, M.R., Mills, R.F.: Log-based distributed security event detection using simple event correlator. In: System Sciences (HICSS), 2011 44th Hawaii International Conference, Kauai, 4–7 January 2011. doi:10.1109/HICSS.2011.288

15. Giacobe, N.A.: Application of the JDL data fusion process model for cyber security. In: Proceedings of the SPIE 7710, Multisensor, Multisource Information Fusion: Architectures, Algorithms, and Applications 2010, 77100R, 28 April 2010. doi:10.1117/12.850275

16. Coppolino, L., D'Antonio, S., Formicola, V., Massei, C., Romano, L.: Use of the Dempster Shafer theory to detect account takeovers in mobile money transfer services. J. Ambient Intell. Humaniz. Comput. (April 2015). doi:10.1007/s12652-015-0276-9

17. Multi Sensor Data Fusion: Hugh Durrant-Whyte, Australian Centre for Field Robotics, The University of Sydney NSW 2006, Australia (2006)

18. Apache Storm. https://storm.apache.org/ (2015). Accessed 15 April 2015

19. EsperTech Esper: http://www.espertech.com/esper/index_redirected.php (2015). Accessed 15 April 2015

Model-Based Vulnerability Assessment of Self-Adaptive Protection Systems

Ricardo J. Rodríguez and Stefano Marrone

Abstract Security mechanisms are at the base of modern computer systems, demanded to be more and more reactive to changing environments and malicious intentions. Security policies unable to change in time are destined to be exploited and thus, system security compromised. However, the ability to properly change security policies is only possible once the most effective mechanism to adopt under specific conditions is known. To accomplish this goal, we propose to build a vulnerability model of the system by means of a model-based, layered security approach, then used to quantitatively evaluate the best protection mechanism at a given time and hence, to adapt the system to changing environments. The evaluation relies on the use of a powerful, flexible formalism such as Dynamic Bayesian Networks.

1 Introduction

Many modern computer systems are needed and used for our day-to-day living. For instance, e-mail, digital media news, or search engine websites are continuously visited by Internet users—as pointed out by Alexa's traffic top ten ranking websites. These systems are exposed to Internet users with malicious intents (i.e., attackers) who may disrupt its normal operation, thus leading to service unavailability and even financial losses.

This work was partially supported by Spanish National Cybersecurity Institute (INCIBE) according to rule 19 of the Digital Confidence Plan (Digital Agency of Spain) and the University of León under contract X43.

R.J. Rodríguez (✉)
University of León, Leon, Spain
e-mail: rj.rodriguez@unileon.es

S. Marrone
DiMat, Seconda Università di Napoli, Napoli, Italy
e-mail: stefano.marrone@unina2.it

© Springer International Publishing Switzerland 2016
P. Novais et al. (eds.), *Intelligent Distributed Computing IX*,
Studies in Computational Intelligence 616,
DOI 10.1007/978-3-319-25017-5_41

Cyber-security helps protecting against these risks first by identifying assets targeted by an attacker and then applying security measures (or policies) to protect them. However, the ever-changing nature of Internet hinders the effectiveness of these security policies, normally built upon a good design and relying on manual tuning [1]. Although security of a system can be improved in several ways (e.g., secure communication protocols, additional security devices, strict security policies), it does not come at zero cost and affects other non-functional properties, such as availability, QoS, or performance [2].

In that sense, systems able to change their behaviour and/or structure in response to the environment and the system itself have become a recent, important trend in the research community [3]. *Self-adaptive systems* can become a great solution to critical systems where failures or malfunctions may result in catastrophic consequences [4]. For instance, a system can be secured following a self-adaptive approach where its resilience, dependability, and configuration may change depending on several external conditions (e.g., a system may redirect connections to a spare server with a different OS when detects several intrusion attempts to the original server). Of course, self-adaption to different situations does not come either at zero cost [5].

In this paper, we propose a model-based approach that allows to represent critical systems with self-adaptive capabilities. Our approach relies on security layers that relate to them, being also complementary and increasing system's complexity. A security layer comprises sensors that monitor external conditions and produce an assessment, reporting when an attack attempt is discovered. When this happens, our approach counteracts adding other security layer, thus enhancing the security of the system. Vulnerability assessment that is a prime step in the definition of such systems is performed by means of formal models—in particular, using Dynamic Bayesian Networks (DBN). An example inspired by a real network intrusion case illustrates our approach.

The paper is organised as follows. Section 2 relates closest works in the literature and previous concepts. Section 3 introduces our model-based, layered security approach while Sect. 4 describes the DBN modelling guidelines. Section 5 shows the application to a network intrusion case. Section 6 states conclusions and future work.

2 Background and Related Works

DBN are a way to extend Bayesian Networks (BN) to model the probability distributions of a collections of random variables taking time into account [6]. In such variables, the Conditional Probability Table (CPT), the mean the probability distribution function of a (D)BN variable is defined, also considers the dependency from previous values of other variables (as well as from the variable itself). Figure 1 shows a sample DBN model where a variable Y is subject to other variable X inside a single slice time: this dependency is represented by the continuous arrow from X to Y. Variables may also influence the value of others in a future time slice. This is the

Fig. 1 A DBN example

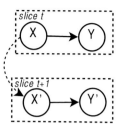

case of X that influences its own future value X', represented by the dashed line from X to X'. In this sample model, Y_t depends on X_t while X_t depends on X_{t-1}. In this sense, the term "dynamic" means we are modelling a dynamic system, not that the network changes over time.

Vulnerability evaluation represents a challenging topic especially in the field of cyber-protection: the high mutability of the knowledge on both systems and protection systems causes a never-ending process where new defence mechanisms are followed by the discovery of new exploitations in a very short time. For these reasons, security of cyber-physical infrastructures is often considered a multi-faceted, multi-disciplinary problem that requires an integrated approach [4, 7].

In security analysis, several modelling methods integrate attack and defence aspects, at different level of abstraction: Garcia addresses both of them, but attacks are described at a high level of detail [8]; Defence trees (an extension of attack trees [9]) accounts for both attacks and countermeasures [10]; Attack Response Trees incorporate both attack and response mechanism notations [11]. Another meaningful example of integration of different aspects of modelling and sensing techniques in [12] where a hierarchical approach to intrusion detection is boosted by the usage of Knowledge Engineering techniques. Other works rely on the quantitative use of other formalisms such as Generalized Stochastic Petri Nets (GPSN) [13] and Bayesian Networks (BN) [14].

Notwithstanding their flexibility, DBN have not received the right attention from the scientific community. With respect to the cited works, the proposed approach aims at exploit all the features of the DBN: the DBN formalism has been chosen to conjugate modelling expressiveness and capability to analyse models. In synthesis, DBN are more powerful with respect to BN and more simple to use and analyse than GPSN. Hence, with respect to the works in [13, 14], this paper respectively improves time to analyse the model (allowing run-time usage possible for large systems) and introduces relationships between events in time (that is not possible with plain BN).

To best of our knowledge, there are few works using DBN for vulnerability and security analysis [15, 16]. With respect to these works, this paper proposes a DBN modelling approach where, respectively, the focus is not only on attack phases but can also capture the dynamic and evolving relationships between attack actions and

Fig. 2 A *security layer*
model element

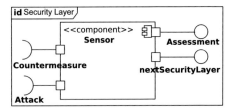

defence mechanisms; and DBN are not used as machine learning tool to infer knowledge about the attacker habits, but as a tool for applying such knowledge in protection system design. In fact, it could be possible a synergy between these approaches.

3 The Onion Security Model

Our approach relies on security layers. We define a *Security Layer* (*SL*) as a UML Component Diagram (UML-CD) which comprises a *Sensor* component as depicted in Fig. 2. A UML-CD is a UML structural diagram used to illustrate pieces of software and/or controllers that make up a system [17]. A *Sensor* defines a model of a system aimed at capturing and interacting with real-world events (e.g., a firewall, an Intrusion Detection System (IDS), or a network load-balancer). The *Sensor* has two inputs (left hand of the component) and two outputs (right hand).

The *Attack* input port describes the sequence of steps leading to a successful attack on the system. This sequence of steps can be modelled, for instance, by means of a state-chart diagram. Output ports are *Assessment*, which expresses the quantitative assessment performed by the *Sensor*, and *nextSecurityLayer*, which allows for connecting an *SL* with other *SL'* through *Countermeasure* input port. Thus, different security layers are connected through *Countermeasure/nextSecurityLayer* ports. Note that these diagrams are only used for representing the static part of the system: its dynamics is expressed by the DBN model, introduced in the next section.

Figure 3 sketches our model-based, layered approach using aforementioned security layers. Several security layers are defined in a system to be protected, where each security layer SL_{i+1} is more secure (and consequently more restrictive somehow) than SL_i. The specific moment of transition among layers SL_i, SL_{i+1} is determined by $Assessment_i$, depending on the input $Attack_i$.

Fig. 3 The *onion security*
model

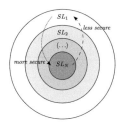

4 Modelling Guidelines for DBN

The objective of this section is to guide the reader to structure a DBN model according to the proposed approach. Figure 4 shows the proposed overall DBN model schema. The actions accomplished by the *Attacker* are modelled by an **Attack** submodel: this submodel communicates system *events* to a **Sensor** submodel whose aim is to represent real world sensing concerns (e.g., honeypots, monitoring systems, log analysers can all be modelled as sensors). Detected events (*detections*) are then communicated to an **Assessment** submodel. This last submodel is in charge to analyse detected events and to decide whether an attack is actually occurring, then assigning a proper SL as a countermeasure.

Here, we propose a modelling approach able to capture the adaptive feedback given by the **Assessment** model to the **Sensor** model: one of the simplest adaptive strategy may consist in (de)activating sensors when some operating conditions require a more accurate estimation of the suspected threat. The model also highlights proper *alarms* raised to the *Security Operators*. In the following, we focus on the single submodels showing how some recurrent situations can be modelled with "DBN patterns".

4.1 Attack Modelling

We start from the general hypothesis that *an attack is a sequence of attacking steps*: each of these is modelled by a DBN variable ranging {*Unattempted*, *Ongoing*, *Succeed*, *Failed*} values. In the following, these values are expressed by their initials.

The most simple DBN pattern is constituted by a single attack step which evolves in time and does not depend on other steps. Hence, the value of this variable is a function only of its value at previous stage. Figure 5 shows the DBN submodel representing this situation and its related CPT. The parameters used in the CPT are: the probability P_{start} of an attacker that attempts to exploit the step; the probability P_{end} of the action ending in a single time slice; and the quantification of the vulnerability (i.e., the probability P_{succ} of a successful attack to the vulnerability). Finally, $P_s = P_{end} \cdot P_{succ}$, and $P_{ns} = P_{end} \cdot (1 - P_{succ})$.

A more complex situation is represented by two attack steps that are sequentially dependent, i.e., a step A can influence a step B. Figure 6a depicts a scenario where a

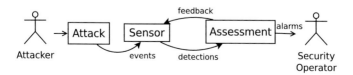

Fig. 4 DBN model structure

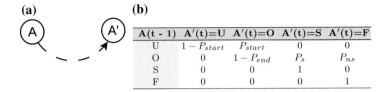

Fig. 5 DBN (**a**) model of single step and its (**b**) CPT

Fig. 6 DBN (**a**) model of a sequence of steps and its (**b**) CPT

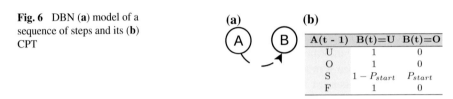

dependency from A to B exists (i.e., from A at slice t to B at slice $t+1$). In this case, the CPT in Fig. 6b shows that B is *Ongoing* when A is *Successed* with probability P_{start}; while it remains *Unattempted* with probability $(1 - P_{start})$. Otherwise, B remains *Unattempted*. Note that the CPT does not consider *Succeed* and *Failed* cases, since they can be subject of other DBN submodels.

Other scenarios may involve non-deterministic choices where an independent successful event can trigger only one of two successive events; deterministic choices, used when the value of another boolean variable is used to determine which specific event occurs; or parallelism, when two actions start in parallel after the success of a first. Furthermore, more patterns can be combined to model complex attack scenarios.

4.2 Sensor Modelling

DBN modelling of sensing activity refers to a single DBN pattern as depicted in Fig. 7a. The pattern comprises four nodes: a node S that models the sensor itself; a node D that models the detection activity; an activation node AC, which is described

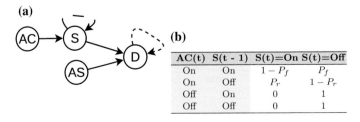

Fig. 7 DBN Sensing pattern: (**a**) DBN submodel, (**b**) CPT of the S node

later; a node *AS* that detects an attack step. The node *AC* means the activation of the sensor by means of the **Assessment** submodel, i.e., it models the effects of the "adaptive feedback" actions from the *Assessment* module as in Fig. 4.

The nodes have the following values: *S* can be {*On, Off*}, which represents the functioning of the sensor; *AS* has four values as previously described; *D* can be {*True, False*}, where *True* means the sensor detects the threat, *False* otherwise; *AS* represents an activation command to the sensor (hence, it can be represented by {*On, Off*} in its simplest form). *S* can depend on *AC* since this last node is not mandatory, representing an always-on device where it is not present. Moreover, the value of *S* also depends on its previous value since failures and repairing are possible for such devices. Figure 7b reports a typical CPT for an *S* node where P_f and P_r are respectively the probabilities of failure and repairing of the device in a time slice. Let *T* be the amount of time in a time slice, λ and μ the failure and repairing rates of the device. Hence, $P_f = \lambda \cdot T$ and $P_r = \mu \cdot T$.; *D* nodes are more complex since they involve three parent nodes. The underlying logic of a CPT for *D* nodes is described by the following statements: when the sensor is off, an attack step is never detected; when the sensor is on and it has previously raised an alarm, the device continues to produce an alarm until it is switched off; when the sensor is on and the sensor has not previously raised an alarm and the threat is unattempted or failed, the device can erroneously detect a threat with a *fpp* probability; if the sensor is on and the sensor has not previously raised an alarm and the threat is ongoing or successfully brought, the device can detect a threat with a certain probability (1 − *fnp*). The value of *fpp* (resp. *fnp*) is the false positive (resp. negative) probability, i.e., the probability that the device does (resp. does not) detect an alarm if the threat does not (resp. does) occur.

4.3 Assessment Modelling

The **Assessment** submodel represents decision mechanisms combining atomic detections of threats into complex assessment of system SLs. The combination of simple detections can be made by the algebraic operators inspired in [18]: AND, OR, NOT, and SEQ operator (i.e., an operator that is true when the inputs become true in a certain order). For the sake of space, we omit here the translation of these operators into BN models, previously reported in [19].

Figure 8a represents the most interesting case of the SEQ operator. The correlation is made between two detections *D*1 and *D*2 which are "*D*-nodes" of a DBN sensing pattern; the output of such correlation is the *D* node. *D* has the same {*True, False*} nature of *D*1 and *D*2. *F* is a node representing the feedback given to the sensor layer (i.e., the *AC* node of the Sensing pattern). Another node, *SQ*, is in charge of keeping memory of the sequence of the events; the values of this node are {*NIL, OK, KO*} that respectively stand for: none of the events has arrived, the sequence is correct, and the sequence is not correct. While the CPT of the *D* node implements an AND of the three parent nodes, the CPT of *SQ* recognises if *D*1 occurs before *D*2 (see Fig. 8b).

(a)

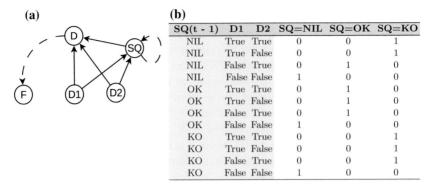

(b)

SQ(t - 1)	D1	D2	SQ=NIL	SQ=OK	SQ=KO
NIL	True	True	0	0	1
NIL	True	False	0	0	1
NIL	False	True	0	1	0
NIL	False	False	1	0	0
OK	True	True	0	1	0
OK	True	False	0	1	0
OK	False	True	0	1	0
OK	False	False	1	0	0
KO	True	True	0	0	1
KO	True	False	0	0	1
KO	False	True	0	0	1
KO	False	False	1	0	0

Fig. 8 SEQ-based DBN assessment: (**a**) submodel, (**b**) CPT of the *SQ* node

It is worth to note that more operators can be composed together in order to create complex logical expressions: the *D* nodes of a DBN submodel can be used as input node of another submodel. Moreover, the outputs of the top submodels represent the alarms raised to Security Operators.

5 Case Study

This section applies modelling approach to an example inspired by a real network intrusion case study occurred in a cancer and AIDS research organisation [20]; whose research laboratories became unavailable for several days resulting in financial losses.

As described in [20], an illegal user account was created in one of the organisation's servers by first accessing—using a stolen account—and then exploiting vulnerable services in such a server. Using this server as a base of operations, other computers were similarly compromised, while the user returned regularly to exfiltrate sensitive data from the network. Since the organisation only monitored on the Internet border, but not its internal subnets, these attacks launched from within the organisation's own network were totally in a blind spot and become unnoticed. Applying our approach in this scenario, we define four different security layers described textually as follows:

SL_1: An Internet border firewall that monitors incoming network packets, analyses them based on predefined filtering rules, and performs an assessment providing the rate of suspicious incoming packets.

SL_2: A behavioural-based network IDS focused on analysing any network packets, comparing current network flow to previous known attack-flow models. Similarly, the assessment returns a rate to express the confidence of being under attack at a given moment.

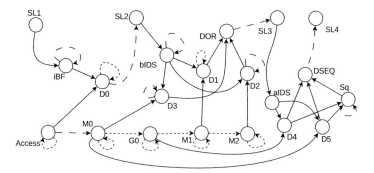

Fig. 9 DBN model of the running example.eps

SL_3: An anomaly network IDS focused on analysing packets coming from a specific
network. In this case, the assessment provides a confidence level of having indeed
an intrusion.

SL_4: A load-balance server, which redirects incoming suspicious network traffic to
an isolated machine acting as a honeypot. Thus, all suspicious traffic can be logged
for forensics analysis.

Figure 9 represents the DBN model of the case study. We suppose that the attack
starts from gaining unauthorised access (*Access*); it continues by intruding in the
machine *M0* and using this system to attack *M1* and then *M2* by means of the gateway *G0*. The first Security Level is modelled by the *SL1* node that is used as an acti-
vator of an Internet Border Firewall (modelled through the sensing pattern—*iBF*,
D0). The feedback of this pattern (considering a trivial assessment submodel) is to
trigger the *SL2*. At this level, all the networks (i.e., all the machines) are monitored
by a behavioural IDS: when any anomaly is detected (modelled by a sensing pattern, *bIDS-D1-D2-D3*, and by an OR-based assessment, *DOR*) the *SL3* is activated.
The last security level implements an anomaly network IDS focused on detecting a
sequence of non-normal packets sent from *M0* and from *G0*. In this way, a sensing
pattern and an SEQ-based operator are used: the output is the activation of the *SL4*
which cut off the threat from the network to protect and isolate it.

6 Conclusions

This paper introduces an approach for the evaluation of the vulnerability of self-
adaptive protection systems. This approach is based on the Onion Security Model, a
high-level modelling approach where higher security levels increase the security by
improving sensing and countermeasures based on lower ones. To make the approach
concrete, Dynamic Bayesian Networks are used to capture probabilistic relationships
as well as dependency in time among attacking and protecting events. First applications to a real intrusion scenario show that DBN model well these concerns. This

paper constitutes a first step in the modelling of such systems. Future works will demonstrate the effectiveness of the approach also by comparing the efficiency of DBN with related to other formalisms (e.g., BN and GSPN). Next steps will involve the completion of the available DBN patterns also by giving some examples of countermeasure modelling.

References

1. Devanbu, P.T., Stubblebine, S.: Software engineering for security: a roadmap. In: Proceedings of the Conference on The Future of Software Engineering, ICSE'00, New York, pp. 227–239. ACM (2000)
2. Rodríguez, R.J., Trubiani, C., Merseguer, J.: Fault-tolerant techniques and security mechanisms for model-based performance prediction of critical systems. In: Proceedings of the 3rd ISARCS, pp. 21–30. ACM (2012)
3. de Lemos, R., et al.: Software engineering for self-adaptive systems: a second research roadmap. Software Engineering for Self-adaptive Systems II. Volume 7475 of Lecture Notes in Computer Science, pp. 1–32. Springer, Berlin (2013)
4. NIPP 2013-Partnering for Critical Infrastructure Security and Resilience. U.S. D.H.S., Technical report (2013)
5. Perez-Palacin, D., Mirandola, R., Merseguer, J.: On the relationships between QoS and software adaptability at the architectural level. J. Syst. Softw. **87**, 1–17 (2014)
6. Dean, T., Kanazawa, K.: A model for reasoning about persistence and causation. Comput. Intell. **5**(2), 142–150 (1989)
7. Macdonald, D., Clements, S., Patrick, S., Perkins, C., Muller, G., Lancaster, M., Hutton, W.: Cyber/physical security vulnerability assessment integration. In: Innovative Smart Grid Technologies (ISGT), 2013 IEEE PES., February 2013, pp. 1–6
8. Garcia, M.L.: Vulnerability Assessment of Physical Protection Systems, 1st edn. Butterworth-Heinemann (2005)
9. Mauw, S., Oostdijk, M.: Foundations of attack trees. In: Information Security and Cryptology–ICISC 2005, 8th International Conference, Seoul, Korea, 1–2 December 2005, pp. 186–198, Revised Selected Papers (2005)
10. Bistarelli, S., Fioravanti, F., Peretti, P., Santini, F.: Evaluation of complex security scenarios using defense trees and economic indexes. J. Exp. Theor. Artif. Intell. **24**(2), 161–192 (2012)
11. Zonouz, S.A., Khurana, H., Sanders, W.H., Yardley, T.M.: RRE: a game-theoretic intrusion response and recovery engine. IEEE Trans. Parallel Distrib. Syst. **25**(2), 395–406 (2014)
12. Ficco, M.: Security event correlation approach for cloud computing. Int. J. High Perform. Comput. Netw. 7(3), 173–185 (2013). September
13. Flammini, F., Marrone, S., Mazzocca, N., Vittorini, V.: Petri net modelling of physical vulnerability. Critical Information Infrastructure Security. Volume 6983 of LNCS, pp. 128–139. Springer, Berlin (2013)
14. Xie, P., Li, J.H., Ou, X., Liu, P., Levy, R.: Using Bayesian networks for cyber security analysis. In: 2010 IEEE/IFIP International Conference on Dependable Systems and Networks (DSN), June 2010, pp. 211–220
15. Frigault, M., Wang, L., Singhal, A., Jajodia, S.: Measuring network security using dynamic Bayesian network. In: Proceedings of the 4th ACM Workshop on Quality of Protection, QoP'08, New York, pp. 23–30. ACM (2008)
16. Tang, K., Zhou, M.T., Wang, W.Y.: Insider cyber threat situational awareness framwork using dynamic Bayesian networks. In: Proceedings of the 4th International Conference on Computer Science Education (ICCSE), pp. 1146–1150, July 2009

17. OMG: Unified Modelling Language: Superstructure. Object Management Group, August 2011. Version 2.4, formal/11-08-05
18. Chakravarthy, S., Mishra, D.: Snoop: an expressive event specification language for active databases. Data Knowl. Eng. **14**(1), 1–26 (1994)
19. Flammini, F., Marrone, S., Mazzocca, N., Pappalardo, A., Pragliola, C., Vittorini, V.: Trustworthiness evaluation of multi-sensor situation recognition in transit surveillance scenarios. In: Security Engineering and Intelligence Informatics. Volume 8128 of Lecture Notes in Computer Science, pp. 442–456 (2013)
20. Casey, E.: Case study: network intrusion investigation—lessons in forensic preparation. Digit. Investig. **2**(4), 254–260 (2005)

Forensic Data Analysis Challenges in Large Scale Systems

Damien Conroy

Abstract Large-scale systems generate data, including log data, on a different scale to that typically subjected to either monitoring or forensic analysis. Where scale renders the tasks of security monitoring or investigation overwhelming this represents a vulnerability. There is significant scope to apply big-data techniques to the challenges of monitoring and forensic data analysis in large-scale systems. However, it is key that, where big-data techniques are applied, accepted standards for eDiscovery are adhered to, so that the results of any forensic analysis stand the best chance of being accepted as evidence.

1 Introduction

When applied to security, Data Analytic is also referred to as Security Analytic [10]. In [5], authors highlight the need to adopt Big Data technologies in security data processing since current security information and event management systems (SIEMs) cannot cope with the increasing variety of data sources, i.e. structured, semi-structured and unstructured data. Also, SIEMs cannot manage the increasing volumes of data and requirements in terms of low latency and timely processing. Moreover, big data technologies offer valid solutions to face low-and-slow attacks and to detect Advanced Persistent Threats.

Forensic analysis of data from any system must be carried out in the context of an acceptable framework, for example, the EDRM framework [7]. While addressing the challenges presented by the big-data nature of their logs, any approach to the forensic analysis of large-scale systems must also address the following challenges:

- the creation of evidence acceptable to a court;
- adherence to accepted processes;

D. Conroy (✉)
Espion Limited, Dublin, Ireland
e-mail: damien.conroy@espiongroup.com

© Springer International Publishing Switzerland 2016
P. Novais et al. (eds.), *Intelligent Distributed Computing IX*,
Studies in Computational Intelligence 616,
DOI 10.1007/978-3-319-25017-5_42

451

- maintaining integrity where tools are applied to deal with the big-data nature of forensic data in large-scale systems;
- building support for forensic analysis through 'forensics-by-design' in large-scale systems.

2 eDiscovery Reference Model

The EDRM is an organization comprising industry and academic partners that provides resources to support the standardization of eDiscovery processes. The eDiscovery Reference Model is a framework which practitioners may use to guide their eDiscovery activities in order to preserve the integrity and authenticity of electronically stored information (ESI).

In the context of forensic data analysis the EDRM covers the following process stages:

- *Information Governance* - The Information Governance stage of the EDRM process refers to activity carried out as a matter of course (independent of any event) to ensure that, in the event of an incident, the data pertaining to any forensic investigation is accessible and fit for purpose.
- *Identification* - "Locating potential sources of ESI and determining its scope, breadth and depth." The identification stage must be supported by suitable tools and techniques. In large scale systems this presents a particular challenge as the sources to be located may be numerous and diverse. Where the Information Governance stage has provided an up-to-date and accurate asset register this may be leveraged during the Identification stage.
- *Preservation* - "Ensuring that ESI is protected against inappropriate alteration or destruction." This stage entails storing the relevant data in such a manner as to preserve its integrity and provide assurances at a later stage in the investigation that it could not have been tampered with. Preservation is achieved by, among other measures, hashing and signing data at appropriate points in the Collection stage.
- *Collection* - "Gathering ESI for further use in the e-discovery process (processing, review, etc.)." The collection of data on large-scale distributed systems, especially SCADA systems, presents challenges that are not encountered in other information systems. Section 3.2 examines these differences in more detail and provides some suggestions as to how "big-data" technologies could contribute to collection.
- *Processing* - "Reducing the volume of ESI and converting it, if necessary, to forms more suitable for review and analysis." Processing approaches have been revolutionized by the emergence of big-data solutions. Sections 3.1 and 3.2 addresses how developments in big data processing have impacted forensic data analysis.
- *Review and Analysis* "Evaluating ESI for relevance and privilege. Evaluating ESI for content and context, including key patterns, topics, people and discussion." Overlapping with the "Processing" stage in this context, the "Review" and

"Analysis" stages may be supported by the visualization techniques that have emerged through big data work (Sect. 3.4).

- *Production and Presentation* - "Delivering ESI to others in appropriate forms and using appropriate delivery mechanisms. Displaying ESI before audiences (at depositions, hearings, trials, etc.), especially in native and near-native forms, to elicit further information, validate existing facts or positions, or persuade an audience." Data presented as evidence must be assembled correctly in a format that is acceptable as evidence. Assembling it correctly means that it must be reviewed by an expert who can testify as to its integrity and authenticity. Section 3.3 discusses veracity in big data.

3 Forensic Analysis and Big Data

Real-time forensic data analysis in large-scale systems necessitates processing data that exhibits the qualities of "big data", the four "V"s, volume, velocity, variety and (when correctly handled) veracity.

High quality trustable data, veracity, is the goal of carrying out the analysis. This is achieved by adhering to EDRM standards while dealing with volume, velocity and variety. This section examines the implications of the four 'V's for forensic data analysis. A fifth 'V', visualization, and how it can contribute to forensic data analysis is also considered.

3.1 Variety

3.1.1 Diversity in Monitoring and Control Systems

SCADA systems generate log data in a diverse range of formats and, in order to achieve efficiencies, data must be normalized based on common semantic elements before it is subjected to analysis. In order to ensure the veracity of the data it must be forensically stored as it is generated at source, before it is normalized.

Systems must be capable of quickly assimilating previously unencountered data formats. Processing tools for such data is not usually available "off-the-shelf" and techniques such as grammars may be used to deal with this and they must be capable of recovering where the format is not recognized [4].

Critical infrastructure relies on control system protocols that were designed without security in mind. As they have become more connected to growing IT infrastructure their payload is increasingly being carried over networks that expose it to vulnerabilities. The forensic analysis of data from control networks is difficult. The protocols of control systems do not lend themselves to analysis.

3.1.2 Cyber and Physical Systems

In the context of security analytic, forensic data analysis is primarily focused on information systems and has matured in that domain. While physical/control systems may have logged data the data was collected for different purposes. The question is how existing log analysis tools may be applied in emerging environments. Emerging environments: cyber/physical systems. The SAWSOC project [8] examines how the convergence of data from physical and logical/cyber systems presents challenges for forensic data analysis that are not adequately addressed by current SIEM systems. It also investigates the convergence of physical access control systems with logical access control systems. Recently some achievements have been made (e.g., SEM and SIM have merged into SIEM, and LACS and PACS have merged into IM), Security Operations Center (SOC) technology has improved significantly, but much is yet to be done. SAWSOC holistic approach and enhanced awareness technology will allow dependable (i.e. accurate, timely, and trustworthy) detection and diagnosis of attacks.

3.2 Volume and Velocity

3.2.1 Integrity in Growing Volumes

The IoT and cloud applications can scale quickly and log analysis systems must scale to handle the log (and other) data generated. Data generation increases in both volume and velocity as the number of source systems increases. The analysis infrastructure must scale to meet the demands of source systems while maintaining the integrity of the data. This has implications for the application of techniques such as hashing as signing that incur additional computational cost as volume and velocity increase.

Employing distributed systems such as Hadoop clusters may provide a means of scaling to meet some demands. However, aggregation techniques for the assembly of evidence may not be as simple as those used in indexing large bodies of text. Some forensics tasks may require orchestration between nodes of the cluster.

3.2.2 Complex Event Processing (CEP)

The "noise" and complexity of systems feeding into the SIEM could be combated with data analytic and CEP to reduce the overhead on Security Operation Centre analysts trying to make sense out of huge amounts of data and alerts. This approach has been successfully applied to intrusion detection [6]. Some common complaints about SIEM applications are that they are too complex; take a long time to deploy, are too expensive with high installation costs, create a lot of data noise and are not cloud friendly. The functions proposed above would help to alleviate these issues while also helping to pinpoint risks that current technology usually requires contextual clues to distinguish. In SAWSOC [8] the convergence of logical and physical security is dependent on the capture of physical security data by the logical security platform, where it may be correlated with logical security data. For that reason, the distinction between logical and physical security data may be unclear during some phases of the security process. The distinction will usually be apparent at the data acquisition layer where the infrastructure behind the probes used to capture data from physical systems will be quite different from that of logical security logging. The needs referred to include meeting the requirements for full convergence of physical and logical security provides easy-to-use dashboards with drill-down capabilities Supporting the user, the expert.

3.3 Veracity

3.3.1 Integrity

The veracity of data is key if that data is to be used as evidence of activity. Data acquisition is the first and most important process within digital forensics to ensure data integrity and admissibility [2]. In some cases the source system may provide log metadata such as hashes that can be used later to verify the integrity of the data. Devices in SCADA systems, however, may not have the processing power or capability to apply hashing or cryptographic techniques. Quite often when this functionality is not be available on the device the system responsible for collecting the data must also take responsibility for ensuring the veracity of the data.

The IoT must place authenticity of devices at its core [3]. Intel has done this with a secure stack on the Quark SoC (system on a chip) that ensures from boot that nothing has been tampered with. However, most devices currently deployed in control systems are far less sophisticated and lack the processing power to support such features.

3.3.2 Chain of Custody

Tracking the movement of the data maintains a chain of custody. A process used to maintain and document the chronological history of the handling of electronic evidence. A chain of custody ensures that the data presented is "as originally acquired" and has not been altered prior to admission into evidence. Some providers maintain an electronic chain-of-custody link between all electronic data and its original physical media throughout the production process.

Physically distributed environments introduce a number of challenges. Legal restrictions may prevent the movement of data outside jurisdictions. Data collection may be impacted by the communications links available to devices in the field. Critical infrastructure by its nature is often distributed geographically [1]. In a different context, cloud environments present a similar challenge in forensic data analysis.

The issues of where logs are located and how they can be accessed present the same issues in any distributed system.

3.3.3 Data Forensics in the Cloud

Cloud-based services are becoming more popular and integration with secure storage in the cloud is likely to become a more prominent consideration as this continues. Furthermore, ensuring secure transmission and storage of data to ensure confidentiality and integrity of all source data will be vital as this develops. Data analytic may be applied to almost all event logs and some other enterprise information sources. This is key in establishing baselines for network and resource utilization, event correlation and event prediction. These requirements illustrate how NoSQL databases such as Redis or Mongo DB would aid in faster searching of large and diverse data sets; coupled with Hadoop for data analytic integration to provide fast and valuable insights into the organizations infrastructure and threat environment.

3.4 Visualisation

Visualization is key in allowing security experts to analyze the data [9]. Analysts must have the facility to drill down through vast volumes of data and recognize features in the data in a timely fashion. Forensic data analytic can borrow from big-data visualization to enhance the user experience of facilities such as SOCs and SIEMs.

4 Emerging Models

System designs and productivity models are undertaking changes by taking into account increasing complexity of operative environments into business intelligence activities, as indicated in [12]. In [11], authors present an application of Big Data with hard requirements typical of mission critical infrastructures. Specifically they provide a system for automated sensor planning (or sensor management) based on shared situation awareness (SA), i.e. sensors are dynamically tasked (or re-tasked) based on the latest status of information requirements and On-Line Analytic Predictive Processing (OLAP). In the work, scalable data mining/analysis algorithms are systematically applied as well as tools and platforms for ingesting real-time sensor data for SA and predictive monitoring. This approach is applied in a multi-intelligence sensor data use case and in a man-machine crowd-sourcing use case. Authors define tactical environments as characterized by: Intelligence, Surveillance and Reconnaissance (ISR) applications, mission-driven goals and heterogeneous data sources.

5 Conclusions

Increased levels of automation in systems deployment, increasing volumes and increasing variety of data are a reality as the IoT grows and cloud computing becomes more widely adopted. It is clear that forensic data analysis must adopt the tools and techniques of 'big-data' to keep up with developments in large-scale systems. However, it is vital that, while the technical challenges of addressing these changes are tackled, the nature of evidence is recognized when gathering and processing digital forensics. Key in maintaining that recognition is ongoing integration of big-data data analysis activities with the stages of an accepted forensic analysis model such as the EDRM, ensuring that, at all stages of the EDRM model, the 'big-data' tools adopted provide analysis that can be incorporated into evidence.

References

1. Afzaal, M., Di Sarno, C., Coppolino, L., D'Antonio, S., Romano, L.: A resilient architecture for forensic storage of events in critical infrastructures. In: 2012 IEEE 14th International Symposium on High-Assurance Systems Engineering (HASE), pp. 48–55. IEEE, (2012)
2. Alqahtany, S., Clarke, N., Furnell, S., Reich, C.: A forensically-enabled iaas cloud computing architecture. In: Australian Digital Forensics Conference (2014)
3. Basnight, Z., Butts, J., Lopez, J., Dube, T.: Analysis of programmable logic controller firmware for threat assessment and forensic investigation. In: Proceedings of the 8th International Conference on Information Warfare and Security: ICIW 2013, pp. 9. Academic Conferences Limited, (2013)
4. Campanile, F., Cilardo, A., Coppolino, L., Romano, L.: Adaptable parsing of real-time data streams. In: 15th EUROMICRO International Conference on Parallel, Distributed and Network-Based Processing, PDP'07, pp. 412–418. IEEE, (2007)
5. Cárdenas, A.A., Manadhata, P.K., Rajan, S.P.: Big data analytics for security. IEEE Secur. Priv. **11**(6), 74–76 (2013)
6. Ficco, M., Romano, L.: A generic intrusion detection and diagnoser system based on complex event processing. In: 2011 First International Conference on Data Compression, Communications and Processing (CCP), pp. 275–284, (2011)
7. E.: ediscovery reference model. http://www.edrm.net (2014)
8. E.: A situation aware security operations centre. http://www.sawsoc.eu (2015)
9. Krasser, S., Conti, G., Grizzard, J., Gribschaw, J., Owen, H.: Real-time and forensic network data analysis using animated and coordinated visualization. In: Proceedings from the Sixth Annual IEEE SMC Information Assurance Workshop, 2005. IAW'05, pp. 42–49. IEEE, (2005)
10. Mahmood, T., Afzal, U.: Security analytics: big data analytics for cybersecurity: a review of trends, techniques and tools. In: 2013 2nd National Conference on Information Assurance (NCIA), pp. 129–134. IEEE, (2013)
11. Savas, O., Sagduyu, Y., Deng, J., Li, J.: Tactical big data analytics: challenges, use cases, and solutions. ACM SIGMETRICS Perform. Eval. Rev. **41**(4), 86–89 (2014)
12. Yu, E., Lapouchnian, A.: Architecting the enterprise to leverage a confluence of emerging technologies. In: Proceedings of the 2013 Conference of the Center for Advanced Studies on Collaborative Research, pp. 408–414. IBM Corp. (2013)

Part XI
International Workshop on Future Internet and Smart Networks (FI&SN'2015)

International Workshop on Future Internet and Smart Networks

Alexandre Santos, Pascal Lorenz and António Costa

Abstract The *International Workshop on Future Internet and Smart Networks*, *FI&SN'2015*, has been organized in conjunction by the Computer Communications and Networks Group, Centro ALGORITMI, University of Minho, Portugal and University of Haute Alsace, France.

Future Internet is a very meaningful name for the real evolution of Internet technology nowadays. Internet should no longer mean the concept of having a desktop or portable computer, or even any smart device, connected to plenty other of such systems, servers, or to the cloud.

Future Internet is to be the place where smart objects, smart phones, smart vehicles, computers, grids, clouds, cities, etc, are to be interconnected using dynamic and evolving intelligent network solutions. Underlying the possibility of having everything interconnected—computers, devices, things and people - and being able to ubiquitously exchange useful information, even in mobility, there should be high-speed broadband links to enable *Smart Networks*.

Smart Networks solutions here discussed range from Software Defined Networks to Named Data Networking and also from Network Resilience Optimization to Automated Network Management. As an important sum up, an insight into the challenges and key enabling technologies for a People-Centric Internet of Things, leading the way to a people-centric society in the near future, is presented and discussed.

A. Santos (✉) · A. Costa
Centro ALGORITMI, University of Minho, Braga, Portugal
e-mail: alex@di.uminho.pt

A. Costa
e-mail: costa@di.uminho.pt

P. Lorenz
University of Haute Alsace, Mulhouse Cedex, France
e-mail: pascal.lorenz@uha.fr

© Springer International Publishing Switzerland 2016
P. Novais et al. (eds.), *Intelligent Distributed Computing IX*,
Studies in Computational Intelligence 616,
DOI 10.1007/978-3-319-25017-5_43

People-Centric Internet of Things—Challenges, Approach, and Enabling Technologies

Fernando Boavida, Andreas Kliem, Thomas Renner,
Jukka Riekki, Christophe Jouvray, Michal Jacovi, Stepan Ivanov,
Fiorella Guadagni, Paulo Gil and Alicia Triviño

Abstract Technology now offers the possibility of delivering a vast range of low-cost people-centric services to citizens. Internet of Things (IoT) supporting technologies are becoming robust, viable and cheaper. Mobile phones are increasingly more powerful and disseminated. On the other hand, social networks and virtual worlds are experiencing an exploding popularity and have millions of users. These low-cost technologies can now be used to create an Internet of People (IoP), a dynamically configurable integration platform of connected smart objects that allows enhanced, people-centric applications. As opposed to things-centric ones, IoP combines the real, sensory world with the virtual world for the benefit of people while it also enables the development of sensing applications in contexts

---

This is an invited paper.

F. Boavida (✉)
CISUC, DEI, Universidade de Coimbra, Coimbra, Portugal
e-mail: boavida@uc.pt

A. Kliem · T. Renner
Technische Universitaet Berlin, Berlin, Germany
e-mail: andreas.kliem@tu-berlin.de

T. Renner
e-mail: thomas.renner@tu-berlin.de

J. Riekki
University of Oulu, Oulu, Finland
e-mail: jpr@ee.oulu.fi

C. Jouvray
Trialog, Paris, France
e-mail: christophe.jouvray@trialog.com

M. Jacovi
IBM Israel Science and Technology, Haifa, Israel
e-mail: JACOVI@il.ibm.com

S. Ivanov
Waterford Institute of Technology, TSSG, Waterford, Ireland
e-mail: sivanov@tssg.org

© Springer International Publishing Switzerland 2016
P. Novais et al. (eds.), *Intelligent Distributed Computing IX*,
Studies in Computational Intelligence 616,
DOI 10.1007/978-3-319-25017-5_44

such as e-health, sustainable mobility, social networks enhancement or fulfilling people's special needs. This paper identifies the main challenges, a possible approach, and key enabling technologies for a people-centric society based on the Internet of Things.

Keywords Internet of people · People-centric IoT · Smart systems integration

1 Introduction

Although considerable work has been done in the recent past regarding the Internet of Things (IoT) [4], most technologies and solutions for accessing real-world information are vertical, i.e., they are either closed, platform-specific, or application-specific. Recent efforts to define IoT reference architectures, such as IoT-A [7], OpenIoT [8], SENSEI [9], or FIWARE [10], are important steps in the right direction, but still they lack adaptability, intuitiveness, and integration features that are crucial for people-centric applications. So, on one hand, there is need to define an IoT architecture that goes beyond vertical solutions by integrating all required technologies and components into a common, open and multi-application platform. On the other hand, there is need to develop a set of common building blocks, middleware and services that can be used to construct people-oriented applications in an open, dynamic and more effective way into smart environments including but not restricted to smart cities, businesses, education and e-health. We call it an Internet of People (IoP) [5].

One important, overall challenge for IoP is to define a generic version of an architecture that can be used for supporting specific solutions for each particular people-centric application domain. Naturally, this will require identifying specific challenges regarding several key aspects, such as interoperability, reliable communications, self-management and adaptability, human-machine interaction, security and privacy, ontologies, and big data analytics.

Subsequently, in addition to the IoP architecture definition, it is important to develop and make generally available several easy-to-use tools, namely middleware

F. Guadagni
San Raffaele S.p.A, Milan, Italy
e-mail: guadagnifiorella@gmail.com

P. Gil
UNINOVA—Instituto de Desenvolvimento de Novas Tecnologias, Setúbal, Portugal
e-mail: psg@fct.unl.pt

A. Triviño
Universidad de Málaga, Málaga, Spain
e-mail: atc@uma.es

and services, on which people-centric applications can be built. These will build on technological solutions such as wireless sensor networks, wireless mesh networks, mobility, and ubiquity. Moreover, these tools must be context-aware, so they can be used to build applications in a variety of contexts, such as smart cities, e-learning, and e-health contexts, thus enhancing the autonomy and quality of life of citizens.

Following this general identification of motivations and overall approach, the remainder of this paper is organised as follows. Section 2 identifies the main challenges for developing people-centric Internet of Things solutions. Section 3 details a possible approach, by addressing the vision, infrastructural needs and design principles. Section 4 identifies enabling technologies, including relevant related work. Section 5 provides the conclusions and identifics guidelines for further work.

2 Challenges

Several challenges can be identified in what concerns developing people-centric Internet of Things platforms and applications. This section identifies two overall challenges and several related and/or complementary challenges. All of them are key to the success of the IoP paradigm that will be presented in Sect. 3.

Open, Smart Platform The IoP concept requires an open, smart platform that will support People2People and People2Thing interactions and can be used to develop a variety of people-centric applications. Moreover, IoP does not limit itself to a technology-oriented approach nor to an application-oriented view. IoP provides a comprehensive approach that brings together actors along the value chain, from suppliers of components and customized computing systems to system integrators and end users, going from reference architectures to applications, from application-specific approaches to an open application-development framework, from an individual devices view to resource virtualization, and from the Internet of Things to the Internet of People.

Sharing in IoT Environments Sharing physical devices leads to a paradigm shift in how IoT-related applications can be designed. Paradigms like Infrastructure as a Service (IaaS), On-Demand Resource Provisioning or Pay-As-You-Go (PAYG) pricing models became very popular along with the proliferation of cloud computing and its applications. However, looking at IoT-related applications, a completely different picture of how these applications work and are designed can be observed. IoT applications are often built upon a gateway-based approach. This can be briefly described as a single system (e.g. a router or a smart phone), that integrates available sensors, collects data from them and forwards the resulting data streams to application layer components (e.g. a computer centre hosting data analysis applications). The IoP approach aims at broadening our understanding of the term cloud. By introducing concepts for provisioning of sensors and embedded devices on a PAYG basis, the cloud turns from an endless remote resource to an overall resource surrounding us constantly.

Connectivity, Mobility and Ubiquity As we witness an unprecedented increase in the deployment and use of wireless technologies (mobility management, pervasive sensing, automated object-to-object and object-to-person communications, the Internet of Things, etc.), it is becoming important to guarantee universal connectivity, using a variety of communication technologies, including 4 G, 5 G, IEEE 802.11ad, wireless mesh networks (WMN), mobile and vehicular ad hoc networks (MANET/VANET), and devise new and more efficient ways for their operation. WMNs and MANET/VANET may play an important role in generalised use of IoT. Nevertheless, despite considerable work done in the past in the area of routing in WMNs [1], the fact is that several challenges persist and there is need to go beyond traditional proactive or reactive routing algorithms and protocols.

Adaptive, Dynamic and Mobile Configuration Capabilities There is need for tools and methods to cope with moving or disappearing nodes while keeping the transparency constraint. Device integration platforms should enable integrating sensors into any smart devices capable of doing so. This properly reflects the mobility of devices and users, because devices with limited communication range may need to be integrated at different locations (e.g. medical sensors moving with a patient in case of an emergency). In addition, service might have to be migrated or the data routing probably has to be adapted, which may have significant impact on the overall network structure. New nodes can introduce new features, which again may require adapting significant parts of the service deployment, routing and network structure.

Effective Device Integration Novel device integration and management platforms able to handle large amounts of devices (proprietary and standard based) are needed, assuring device integration and platform adaptation at runtime (online-reconfiguration), and providing device abstraction to expose uniform interfaces of heterogeneous devices to applications. For this purpose, platforms will need to understand the devices (e.g. capabilities, data structures they produce, device configurations) or at least need to be able to gather integration knowledge if required (e.g., when a new device joins the platform). This may demand for sensor markup languages (SensorML) and sensor ontologies.

Scalability and Expandability There is need for dynamic expandability of network components (things), services, applications and users. These capabilities are fundamental for an effective device integration and adaptive, dynamic and mobile configuration capabilities. Scalable and expandable systems for a large amount of heterogeneous devices and data streams, as well as ability to establish billions of different IoT connections between devices and objects, are an important challenge.

High Availability, Dependability and Fault Tolerance Adaptive and dynamic functionalities are needed for monitoring and managing the infrastructure in a self-manageable mode at runtime. This will allow platforms to be permanently available and have the ability to quickly recover from faults, as well as dynamic access and network management for a large number of robust and dynamic connections. There is also the need for integrating online adjustment technologies from other domains, like Software-Defined Networking (SDN) and Data-Centric infrastructures.

Quality of Service and Non-Functional Requirements There is the need for functionalities to manage and differentiate between critical (e.g. e-Health) and non-critical (e.g. entertainment) applications and their data streams from different domains on the same commodity transport and infrastructure. Therefore, platforms should consider and allow for quality-of-service and non-functional requirements such as reliability, determinism, or performance to transmit and deliver data in real-time.

Interoperability, Data-Models and Nomenclatures One important challenge is the ability for independent devices to cooperate and exchange information with each other. Therefore, there is need to efficiently provide knowledge repositories, which allow handling heterogeneous (probably unknown) incoming data streams in a protocol agnostic fashion. This will be a key enabler to provide technologies like context-awareness, content-based routing or quality of service, and integrate different IoT domains with each other.

User-Centred Requirements Analysis Nowadays, IoT systems are mainly focused on the technical level, like performance, interoperability, integration, etc. However, whenever use-cases are targeting human users the focus must not be solely on these aspects, as the human factor must be also considered. It is thus essential to apply a user-centred approach based on the use of the repertory grid method as well as the application of personalized and interactive e-assessment technology. This will allow identifying application-specific user features and understanding the users' needs, motivations and beliefs.

Big Data (Graph) Analysis People-centric IoT architectures must be used for modelling the Internet of People and the things they interact with, i.e., the relationships of people-to-people and people-to-things. In this respect, there is clear need for research and progress beyond state-of-the-art in at least the following three areas: efficiently and scalably streaming data into the graph; real-time discovery of effected patterns; and discovering trends based on social and temporal proximity.

Security and Privacy Secure granting and withdrawal of device access tokens is required to allow for device sharing. Issues related to trusted nodes, authentication, security and, privacy are crucial for the implementation, deployment and success of any people-centric application platform. Final users must be able to define privacy preferences in order customize policies according to their demand. Legal aspects and regulations must be completely met.

3   Approach

This section presents an overall IoP vision, a possible supporting infrastructure, and some design principles.

3.1 Vision

The emerging Internet of Things (IoT) concept and the availability of a multitude of
sensors, smart devices and applications for use by individuals and communities
point to the need for defining a people-centric IoT architecture—which we name
Internet of People, IoP (Boavida 2013)—that can go beyond devices, technologies,
services and passive entities and can be used in people-oriented IoT applications.
The IoP paradigm can be considered as a specialization of the IoT paradigm, in
which humans and their interactions can simultaneously be viewed as data sources
and sinks, in a network of connected embedded devices, bridging the gap between
IoT and the beneficiaries of technologies.

The basic IoT assumption is that people are no longer supported by a single
monolithic computing system, such as a PC, but rather use all the small embedded
systems (smart devices/objects) surrounding them to fulfill their needs. Currently,
most of these smart devices act like closed "boxes" and barely interconnect or
collaborate with each other. Moreover, usually an application domain oriented
segmentation of IoT related solutions can be observed (i.e. one box for entertainment, one box for smart home control, one box for e-health services). Hence, in
order to allow for both efficient resource utilization and smarter applications, an
application domain independent solution serving multiple applications within an
openly designed and integrated IoP infrastructure is required. The infrastructure
needs to open and connect so far isolated heterogeneous devices with each other,
provide sufficient enablers for spontaneous interoperability, and offer open APIs
that allow people and services to utilize the infrastructure in an application domain
independent and technology agnostic manner.

3.2 Infrastructure

As shown in Fig. 1, several basic components can be identified as elements of a
possible IoP infrastructure, on which services and applications can be built. The
complexity introduced by the IoP vision and its infrastructure design requires
establishing a uniform notion of abstraction throughout the whole architecture.
Neither sensors or devices nor applications or users should have to care about the
heterogeneity of the corresponding spaces (i.e. which nodes integrate or execute
them). To hide the complexity of the IoP infrastructure, it is separated into three
spaces, namely Physical Space, IoP Runtime Space and Social Space.

The Physical Space includes all physical devices, systems and networks collaborating in the IoP domain. In order to allow for seamless integration of all
available nodes in the physical space, the following segmentation is defined:
(i) Device Nodes (DN) refer to data resources (i.e. sensors); techniques such as
sensor virtualization or on-demand provisioning of the physical device itself can be
used to enable them for IoP operation; (ii) Aggregation Nodes (AN) refer to smart

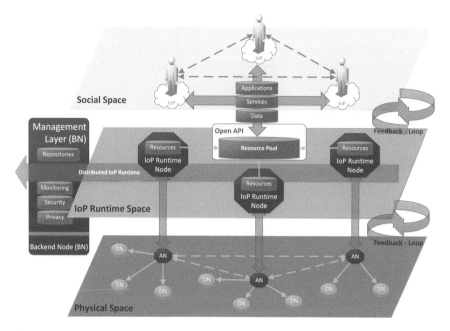

Fig. 1 IoP infrastructure components

devices that additionally provide computational or storage resources and allow hosting the IoP Runtime Nodes (i.e. a software node of the distributed IoP runtime middleware corresponding to the IoP Runtime Space); (iii) Backend Nodes (BN) refer to regular servers (e.g. IaaS Cloud) and provide management and monitoring components used for the IoP runtime.

The IoP Runtime Space will provide capabilities to uniformly access the different types of nodes. Its distributed runtime environment abstracts from the technical details of the underlying physical space and therefore hides the complexity of the physical infrastructure from the application layer. The following components are defined: (i) IoP Runtime Nodes (IRN) refer to software nodes that constitute the distributed IoP Runtime Middleware; (ii) Distributed IoP Runtime is the core middleware solution composed out of the connected IRNs; it is used by the IoP to integrate physical resources and expose them to applications and services; (iii) Management Layer, which provides repositories, like a device directory, offering knowledge about devices and data streams (e.g. nomenclatures, data-models) and therefore allowing IRNs to reconfigure themselves at runtime in order to properly handle unknown devices and incoming data streams in an application domain independent manner; (iv) the Resource Pool abstracts from all individual resources of the underlying physical space and centralizes all resources of the platform in federated and virtual resource pool.

The Social Space models the people and their things as nodes in a Big Graph along with their applications collaborating in the IoP domain. These people have access to the shared resource pool and can use the data sensed by different DNs and the computational resources of all ANs for their applications and services.

3.3   Design Principles

A possible IoP platform addressing the identified challenges and implementing the presented vision and infrastructure will benefit from several design principles identified in the following paragraphs.

Technology and Protocol Agnostic The IoP infrastructure shall be designed in a technology and protocol independent manner. Based on modularity features, knowledge required to integrate and operate newly developed devices and communication protocols shall be addable at runtime without the requirement to manually change or adapt core components of the infrastructure.

Platform Independence The IoP Runtime has to ensure platform independence. Each software module provided by the knowledge base needs to be compliant with the IRN specifications and is therefore executable on each instance, regardless of the underlying platform.

Adaptability and Openness The infrastructure, in particular the IoP Runtime, should run in a highly dynamic environment, in which changes occur very frequently, creating the need for adaptation support for changes in environment and changes that are imposed by the users themselves. The IoP Runtime shall be able to autonomously adapt itself to the requirements of the current environment, (e.g., changes in device, network, service, application, and user requirements). This means that the distributed IoP middleware should act as a general device integrator and service executor, not statically related to any pre-defined set of devices, application domains or vendors.

Peer to Peer Collaboration Because of the federated shape of the resource pool and the possibly huge amount of participating actors that can contribute and consume resources, a peer-to-peer style of interaction is required. This interaction happens both locally with nearby people and resources, and system-wide.

Abstraction and Spontaneous Interoperability The IoP infrastructure must be highly dynamic regarding the resources available in the pool and the communication links established between the participants or between the spaces (i.e. between applications and devices). A related requirement is providing appropriate measures for abstraction that allow hiding functional details like device control or protocol logic from applications.

Cloud Computing Paradigms As mentioned in Sect. 2, one of the main technical requirements for the IoP runtime is to map resource virtualization and provisioning concepts into the IoT world. This goes along with several upcoming approaches like sensor virtualization or cyber physical cloud computing. From the perspective of a user, the platform that serves his/her needs is no longer a set of statically bound physical devices and sensors.

Context Awareness The IoP architecture needs to provide dynamic and adaptive capabilities to support a great variety of smart environments, services, business and persons. Context awareness can be a key driver to enable these capabilities, because it allows applications to adapt its behaviour automatically to the current user context.

Quality of Service Given the huge amount of sensors and smart devices, the rapidly increasing amount of data, and the dynamic IoP infrastructure, efficiently applying and monitoring Quality of Service will become a major issue in IoP. Services can only be delivered efficiently if the required data are available at the required location at the required time.

Security and Privacy The IoP infrastructure shall provide necessary security and privacy features. This can include code signing mechanisms to ensure the integrity of software modules, mechanisms to ensure integrity and confidentiality for exchanged data, authentication and authorization mechanisms, or anonymization techniques. Users shall have the possibility to define different levels of confidentiality or integrity.

4 Enabling Technologies

IoP in particular and IoT in general have a strong relationship to and partially rely on other technologies and paradigms known from the distributed systems and computing domain. Some of these technologies and paradigms, which contribute foundations necessary to set up the IoP approach, will be introduced and put into a contextual relationship.

Machine-to-Machine Communication M2M describes the exchange of information between devices like machines, cars, sensors or, actuators usually performed in an automated manner and without human interaction [16]. Thus, M2M is often referred to as the building block of IoT, because the virtual representations of things made available by IoT can also be described as the service endpoints to an M2M system. M2M has a high relevance to the IoP approach, since it deals with similar challenges like heterogeneity of devices and communication networks, device manageability or scalability in general that altogether lead to the overall problem of device integration.

Mobile Grid and Mobile Cloud Computing Mobile computing evolved out of the dissemination of small, mobile and wirelessly connected devices like smart phones that offer computing capabilities. The term mobile grid covers both, the demand for users with mobile devices to access resources offered by the grid and the utilization and integration of the resources offered by the mobile devices themselves. Thus, the mobile grid can be defined as an extension to the regular grid providing capabilities to support mobile users and resources in a seamless, transparent, secure and efficient way [13].

Sensor Networks and Cloud Integration Wireless Sensor Networks (WSN) may consist of several up to thousands of resource-constrained nodes, and are often designed towards the specific requirements of the application domain. In contrast to cloud computing, data consumers usually have to be aware of the actual location, the resource constraints and the infrastructure management requirements of the sensor nodes in order to properly access and utilize them. This often limits the set of consumers being able to access the WSN. As a remedy, concepts like sensor-cloud integration [3] or sensor virtualization [2] were introduced. These approaches basically aim at overcoming the resource constraints of traditional WSNs by integrating cloud resources and providing access of multiple users to physical sensors.

Cognitive Services A number of models of selective attention have been proposed in Cognitive Science (e.g., [11]). Particularly related with these models is the issue of measuring the value of information. Most of those measures rely on assessing the utility or the informativeness of information (e.g., [14]). However, little attention has been given to the surprising and motive congruence value of information, given the beliefs and desires/goals of a user or of an agent acting on his/her behalf. Cognitive models for ordinary or creative reasoning are of high importance in the IoP architecture.

Big Graph Technologies In the era of Big Data, Big Graphs have a special place, by modelling not only the objects, but the relationships between them. The proliferation of social networks is the main driver behind the evolution of Big Graph technologies, as the interactions of people over social network map naturally into a graph. Social networks often model not only people (as nodes), but also objects that they interact with (e.g., online documents, posts, comments).

In the following paragraphs, some architectures/frameworks, resulting mostly from R&D activities under public funding (EU-FP7), are also mentioned, as they are related to and can be used in the development of the IoP vision.

Future Internet Architectures Future Internet Architectures is a generic term for several research projects and initiatives, like FI-PPP [12], FIWARE [10] or, FI-STAR [15], that investigate in the improvement or redesign of the aging IP-based infrastructure in order to cope with challenges like ubiquitous network access, mobility, or integrated security. It is assumed, that the increasing amount of users and the demand for future applications require a paradigm shift from machine-centered and packet delivery based infrastructures towards data, content and, user-centered ones.

SOCIETIES [6] Open scalable service architecture and platform for pervasive computing was developed during a European funded research project. The project expands the concept of pervasive computing from the scope of an individual user to a community. Relevance, similarity of contextual information and social networking history are used to connect users and organize them into communities. The communities are formed in an intelligent manner to ensure their ability for self-organization, self-orchestration, self-healing. The communities are further used for information exchange and resource sharing between users and their devices.

SENSEI In the SENSEI project [9] the focus has been drawn on the realization of ambient intelligence in a future network and service environment. In this environment, heterogeneous wireless sensor and actuator networks (WSAN) are integrated into a common framework of global scale and make it available to services and applications via universal service interfaces. In this pursuit, SENSEI intended to create an open business driven architecture that fundamentally could address inherent scalability problems for a large number of globally distributed wireless sensor and actuator nodes.

IoT-A IoT-A [7] technical objective was to create the architectural foundations of the Future Internet of Things, allowing seamless integration of heterogeneous IoT technologies into a coherent architecture and their federation with other systems of the Future Internet. In this context an architectural reference model for the interoperability of IoT systems was introduced. The project also focused on other technological issues, such as scalability, mobility, management, reliability, security and privacy.

5 Conclusion

Current low-cost sensing technologies and IoT-related developments make it now possible to go from simple sensing and actuating applications to people-centric applications. Nevertheless, despite considerable advancement of the state of the art, most emerging systems and applications are still platform-specific and/or application-specific. In the current paper we identified the main challenges, a possible approach and the key enabling technologies for open, platform- and application-independent, people-centric systems.

The main overall challenges are the development of an open, smart platform able to support people-to-people and people-to-thing interactions, and the virtualisation and sharing of physical and logical devices. Complementary challenges include connectivity, mobility and ubiquity, dynamic configuration and provisioning, device integration, scalability and expandability, dependability and fault tolerance, quality of service, data models and nomenclatures, user-centred analysis big data analysis and, last but not least, security and privacy.

A possible approach to the implementation of the IoP vision was briefly presented, by identifying the infrastructure components and main design principles. Lastly, several enabling and supporting technologies were identified in order to provide the reader with relevant information on related work.

Acknowledgments The authors would like to thank the participants in the IoP consortium for their insights and contributions.

References

1. Akyildiz, I.F., Wang, X., Wang, W.: Wireless mesh networks: a survey. Comput. Netw. J. **47** (4), 445–487 (2005)
2. Alam, S., Chowdhury, M.M., Noll, J.: (2010) Senaas: an event-driven sensor virtualization approach for internet of things cloud. In: Paper presented at NESEA 2010—IEEE International Conference on Networked Embedded Systems for Enterprise Applications, Xi'an Jiaotong-Liverpool University International Conference Center, Suzhou, China, 25–26 Nov 2010
3. Alamri, A., Ansari, W.S., Hassan, M.M., Hossain, M.S., Alelaiwi, A., Hossain, M.A.: (2013) A survey on sensor-cloud: architecture, applications, and approaches. Int. J. Distrib. Sensor Netw. 2013: Article ID 917923, http://dx.doi.org/10.1155/2013/917923
4. Atzori, L., Iera, A., Morabito, G.: The internet of things: a survey. Comput. Netw. J. **54**(15), 2787–2805 (2010)
5. Boavida, F., Silva, J.S.: IoP—Internet of People. Future Internet Networking Session. ICT 2013, Vilnius, Lithuania. http://ec.europa.eu/digital-agenda/events/cf/ict2013/item-display.cfm?id=10400 (2013). Accessed 1 June 2015
6. Doolin, K.: SOCIETIES—Self Orchestrating Community Ambient Intelligence Spaces. http://www.ict-societies.eu/ (2014). Accessed 1 June 2015
7. Günter, K.: IoT-A—Internet of Things Architecture. http://www.iot-a.eu/public (2013). Accessed 1 June 2015
8. Hauswirth, M.: OpenIoT—Open Source cloud solution for the Internet of Things. http://openiot.eu/ (2014). Accessed 1 June 2015
9. Hérault, L.: SENSEI—Integrating the Physical with the Digital World of the Network of the Future—http://www.sensei-project.eu/ (2010). Accessed 1 June 2015
10. Hierro, J.: FIWARE—Core platform of the Future Internet. http://www.fiware.org/ (2014). Accessed 1 June 2015
11. Horvitz, E., Jacobs, A., Hovel, D.: (1999) Attention-sensitive alerting. In: Paper Presented at the Fifteenth Conference on Uncertainty and Artificial Intelligence. Morgan Kaufmann, p. 305–313, Stockholm Sweden, 30 July—1 August (1999)
12. Lakaniemi, I.: FI-PPP—Future Internet Public-Private Partnership. Internet-Enabled Innovation in Europe. http://www.fi-ppp.eu/ (2013). Accessed 1 June 2015
13. Litke, A., Skoutas, D., Varvarigou, T.: Mobile grid computing: changes and challenges of resource management in a mobile grid environment. In: Paper presented at the 5th International Conference on Practical Aspects of Knowledge Management (PAKM 2004), Vienna, Austria, 2–3 Dec 2004
14. MacKay, D.: (1992) Information-based objective functions for active data selection. J Neural Comput. **4**(4), 590–604 (1992)
15. Usländer, T., Berre, A.J., Granell, C., Havlik, D., Lorenzo, J., Sabeur, Z., Modafferi, S.: The future internet enablement of the environment information space. Environmental Software Systems, Fostering Information Sharing, pp. 109–120. Springer, Berlin Heidelberg (2013)
16. Wu, G., Talwar, S., Johnsson, K., Himayat, N., Johnson, K.D.: M2M: From mobile to embedded internet. Commun. Mag. IEEE **49**(4), 36–43 (2011)

SDN-Based Service Delivery in Smart Environments

Lucas Mendes Ribeiro Arbiza, Liane Margarida Rockenbach Tarouco,
Leandro Márcio Bertholdo and Lisandro Zambenedetti Granville

Abstract Internet of Things scenarios demand adaptability to hold the heterogeneity of systems and devices employed; gateway-based solutions are a common answer to the issues of smart environments. The development of software for gateways, using a middleware to handle the different devices demands, is possible in our working scenario, but the maintenance cost of such a solution is high because of software development constraints imposed by hardware and system limitations of the gateway. This paper depicts an SDN approach for smart environments resulted from a refactoring of a previous middleware proposal. Through an instantiation of the refactored middleware in a home network to deliver a health monitoring service the benefits are demonstrated; the benefits regard the digital representation of the physical realm, deployment and maintenance of services, and management of devices and networks.

1 Introduction

The rapid growth of population impose to cities a major challenge to their sustainability, as well as a threat to the infrastructure of main services provided to the popu-

L.M.R. Arbiza (✉) · L.M.R. Tarouco · L.M. Bertholdo · L.Z. Granville
Informatics Institute, Federal University of Rio Grande Do Sul (UFRGS),
Porto Alegre, Brazil
e-mail: lmrarbiza@inf.ufrgs.br

L.M.R. Tarouco
e-mail: liane@penta.ufrgs.br

L.M. Bertholdo
e-mail: berthold@penta.ufrgs.br

L.Z. Granville
e-mail: granville@inf.ufrgs.br

© Springer International Publishing Switzerland 2016
P. Novais et al. (eds.), *Intelligent Distributed Computing IX*,
Studies in Computational Intelligence 616,
DOI 10.1007/978-3-319-25017-5_45

475

lation. A solution is expected to show up with the advent of smart cities. As discussed in [1], more than 150 cities can be documented around the world as smart.

Smart cities require IT services to capture, integrate, analyze, plan, inform, and act intelligently on city activities, resulting in a better place to live. This implies services to make life easier for people and businesses. A smart community is a community that carries out conscious efforts to use information technology to transform significantly and fundamentally the way of life and work within its territory, instead of following an incremental way.

From this perspective, a smart city refers to a physical environment in which communication and information technology, and sensor systems that are part of it, disappear as they become embedded into physical objects and the environments in which people live, travel, and work [2]. Smart cities should use smart computing technologies to build critical infrastructure components and services of a city more intelligent, interconnected, and efficient [3].

Homes have becoming increasingly smarter embedding many devices able to sense their surroundings; usually those devices are part of a solution for home automation, security, healthcare, among other purposes. Smart solutions employed in citizens homes may be part of broad system, such as a smart city solution for intelligent use and distribution of energy. Smartness in home environments bring some issues for network management, security and orchestration due to the demands to handle the heterogeneity of smart devices when they are employed in services delivering.

Software-Defined Networking (SDN) is a paradigm where network intelligence and control are taken from forwarding devices and are deployed in central controllers where network logic behavior is defined by software, developed or customized to fit the needs of each network environment. SDN provides to the network the flexibility of software, allowing the development of features not available in network hardware; using SDN enables the development or the use of applications for network orchestration and management, suitable to deal with heterogeneity of smart environments.

This paper presents the use of a SDN-based middleware [4] in a health monitoring environment aiming to achieve a simpler and easier to manage solution to monitor patients in their own homes, when compared to a previous middleware where SDN was not used. In the same scenario, we also exploit the use of SDN resources for network management for smart environments. The presented approach also enables the delivering of multiple services for smart environments, sharing the same network infrastructure.

The rest of this paper is organized as follow. Section 2 presents some concepts of smart environment, focusing specially in healthcare. Section 3 details what is SDN paradigm, how it works, and its resources. In Sect. 4 we discuss how SDN can be used to empower smart environments, we present the SDN-based middleware used detailing its architecture, implementation and workflow. Section 5 demonstrates the use of the SDN-based middleware for health monitoring and benefits achieved by using the SDN approach. In Sect. 6 we present our final considerations.

2 Smart Environments

Smart cities are composed of smart devices. Currently, a large number of smart objects and different types of devices are interconnected and communicate via the Internet Protocol, which creates a worldwide ubiquitous and pervasive network referred to as the Internet of Things (IoT) [5, 6]. With the inception of IoT, the Internet is further extended to connect things, such as power meters, heartbeat monitors, temperature meters, and many powerful operations, such as health care units, green energy services, and smart farming utilities, that can be made available to people for enhanced quality of life.

IoT is becoming the Internet of Everything (IoE) [7]. Most of smart devices, used in the context of Internet of Things (IoT), do not employ generic/open standards when communicating. One of the challenges in this context derives from the need to find a form of automated communication and to integrate the various devices and protocols, making it possible to obtain information about the scenario being monitored.

Data collected by devices on a given environment are combined to provide services (e.g., smart homes, and healthcare) [8]. The combination can also be made through mashups [9], where data from different sources and using various resources are handled to make possible creating a vision of a whole.

3 Software-Defined Networks

SDN is mainly known for decoupling network control plane from the forwarding devices, usually combined in the same device, such as routers. Decoupling enables the network logic to be defined at the software level and to implement features that may not be available in the network hardware in use [10]. In general, the network logic behavior is defined configuring every network device individually, using vendor specific syntax and limited to the features available according to the licenses acquired. SDN allows to centralize network intelligence in a controller that sends forwarding rules, called *flow entries*, to the switches and routers defining logic behavior of the network [11].

Figure 1 illustrates the three layers SDN architecture: (1) Infrastructure layer: comprised of forwarding devices (switches and routers) enabled with an SDN standard, such as OpenFlow; (2) Control layer: one or more devices running a controller software enabling the communication between application and infrastructure layers; and (3) Application layer: one or more applications by which network intelligence is implemented; this layer makes use of the control layer to configure forwarding devices. Control layer communicates with infrastructure layer through OpenFlow messages or other SDN standard.

When a forwarding device, also called OpenFlow switch, receives a packet the switch looks up in its *flow tables* for a *flow entry* matching the received packet. If

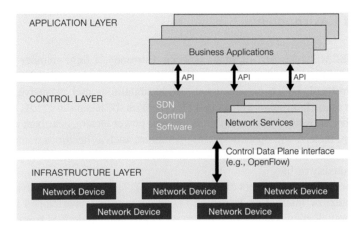

Fig. 1 Software-defined networking architecture [10]

none of the existing *flow entries* match the packet, the switch sends a *packet-in* to the controller. *packet-in* is processed at the application layer and a *flow entry* is sent to the switch through the controller containing forwarding rules for the packet.

The controller has an entire view of the network; it is aware of every switch connected to it and of every device connected to the switches. That privileged view is a valuable resource exploited in some works that propose SDN-based approaches in different network related fields, for example: network virtualization: SDN is used to create isolated slices in the network over the same physical hardware [12–14]; routing: routes established by BGP and OSPF are reflected to low cost switches not enabled with routing protocols [15]; QoS: network paths are dynamically allocated assuring the best performance and availability for differentiated traffic [16]; mobility: aiming to prevent Wi-Fi handover from occurring when users are moving the traffic is sent to more than one access point simultaneously [17]; network management: SDN is used combined with SNMP and others protocols and security and authentication mechanisms existing in a campus, events arising from different sources and network policies are translated to *flow entries* to be installed on the forwarding devices [18].

Taking network control to the application level empowers a low cost network with capabilities not available in the existing hardware, as seen in [15]. By using SDN, network administrators do not dependent on features provided by vendors anymore, since new features can be developed using high level abstraction programming languages and management solutions may be combined with control applications to improve network intelligence, management, and automation.

4 Empowering Smart Environments with SDN

Recently, SDN started to be exploited in smart environments. In research efforts such as [19, 20], the authors made use of SDN in home networks, as we do in our work as well. In [19], the authors provide a simplified interface by which non technical users are able to configure their home networks properly. Those users usually expose themselves to security and privacy risks just because they do not know technical information required to configure most of access points. OpenFlow is used to translate network settings configured by users to *flow entries* to be installed in the SDN empowered access point. In [20], SDN is used to slice the network in isolated virtual networks allowing different service providers to share the same network infrastructure to deliver their services at users' homes. Each provider have control over its slice to deliver the service as needed.

Based on the works mentioned above, in [4] we used SDN to refactor an IoT middleware designed to enable health monitoring of patients with chronic illnesses in their own homes [21]. In an environment empowered with SDN, we are able to provide a wider view of a sensed environment. In [22], IoT is defined as a digital representation of the physical realm built from data collected by devices enabled to sense the environment they are in. Although off-the-self sensing devices usually employ a vertical communication called *silos* [23–25], in this case retrieved data are sent to proprietary servers and cannot be retrieved directly from devices. OpenFlow provides resources by which one can build a representation of an environment based on the communication of all connected devices in a home network. SDN also benefits IoT in network orchestration and management. The following subsections present our approach to achieve our goals in IoT environments through SDN.

4.1 Architecture and Workflow

The architecture depicted in Fig. 2 is split into three layers: the access points, the controller called Derailleur, and the application called ThingsFlow. APs act as Open-Flow switches forwarding packets according to the *flow entries* received. Derailleur controller listens to AP connections; when an AP establishes a connection, Derailleur uses OpenFlow messages to retrieve information from the AP to build an abstraction as a switch object. Each AP may be identified and accessed individually. Derailleur owns and manages switches objects that are created or destroyed reflecting APs status. The controller triggers events to be handled by the application when APs connect, disconnect, or send OpenFlow messages. ThingsFlow is the application where network intelligence is implemented. Each AP may provide a different set of services, so ThingsFlow installs in a given AP only the *flow tables* of services that AP provides. The ThingsFlow also collects counters from APs and make it available to the services, thus to be used for different purposes, for example, management and services features.

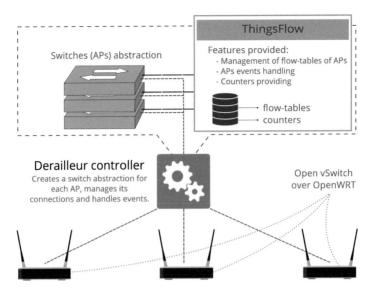

Fig. 2 Architecture overview and components roles [4]

4.2 Implementation Details

APs run OpenWRT firmware, a Linux-based operating system for embedded devices, built containing the virtual switch Open vSwitch which provides OpenFlow capabilities to APs. APs can be replaced by any other hardware and operating system able to run Open vSwitch.

The Derailleur controller was developed in C++ using libfluid [26], the winner of Open Networking Foundation OpenFlow Driver Competition. libfluid is composed of two libraries: *libfluid_base*, that provides server features such as listening loop and events handling; and *libfluid_msg*, that provides mechanisms to build and parse OpenFlow messages.

ThingsFlow was developed inheriting an abstract application class provided by Derailleur. Through the abstract application class, the controller shares and provides access to the switches objects to ThingsFlow.

5 SND-based Healthcare

To demonstrate the benefits achieved employing our SDN-based approach in a smart environment, we use the same health monitoring example that we were working when we designed the previous middleware, in the REMOA[1] project. REMOA is a

---

[1] *Rede Cidadã de Monitoramento do Ambiente Baseado no Conceito de Internet das Coisas* (Citizen Network for Environment Monitoring Based on the Concept of Internet of Things).

project that targets home solutions for care/telemonitoring of patients with chronic illnesses. The project also encompasses the design and implementation of a middleware to address issues found in monitoring devices used in the project, enabling interoperability and security needed in the context of IoT for healthcare. The issues addressed are the following:

- Interoperability: used devices do not reply requests; they just send the data they read to vendor servers employing proprietary standards to communicate, what characterizes a *silo*. Settings, such as destination server or protocols to be used when transmitting data, cannot be changed.
- Security and privacy: data transmission is neither encrypted nor authenticated.
- Management: traditional management mechanisms, such as those based on ICMP or SNMP, cannot be used because of the sleeping scheduled mechanism employed by devices to save battery.
- Data structure: proprietary standards are also used to structure the data sent. It forces every data received to be parsed according to vendor standards. Usually vendors of devices designed for end-users do not provide the documentation required for parsing. The scheme used to parse the messages results from the analysis of the communication of the devices.

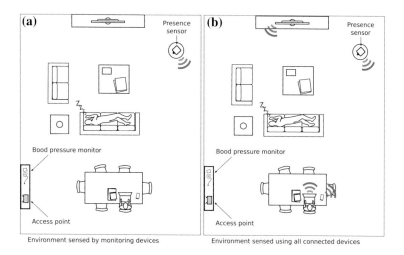

Fig. 3 Comparison between digital representations of the same environment with and without using OpenFlow counters. **a** Environment sensed by monitoring devices. **b** Environment sensed using all connected devices

Figure 3 illustrates two different views of the same environment at the same moment. The hypertensive patient has fallen asleep while was watching TV; the patient is late with blood pressure measurement. Movements of patient are not captured by presence sensor because do no exceed thresholds. The left side illustrates the view provided by the REMOA middleware, where digital representation of the environment is based only on data provided bymonitoring devices. The right side illustrates a view built combining data from monitoring devices and all other devices transmitting at that moment.

Monitoring based on health devices and presence sensor would trigger an emergence event if data from OpenFlow counters were not taken into account. The view illustrated in the right side of the figure is built using OpenFlow counters from *flow entries* installed on APs to forward traffic of devices used in the house of the monitored patient. These counters allow monitoring staff to know that the patient is not alone because a smartphone and a notebook belonging to one of the residents are in use at that moment. It is also possible to know that the TV is on. Monitoring staff can suppose that the patient is watching TV, making short movements and forgetting to do measurement. The action to be taken could be a phone call to remember the patient about the missing measurements.

In [21], the management of health monitoring devices was based in SNMP. Information about devices communication were stored in a Management Information Base (MIB) in the AP by a middleware module. In the SDN-based approach, this information is provided by OpenFlow counters, collected from APs by ThingsFlow, that makes counters available to the service provider that implements management mechanisms suitable for each monitoring device. Counters are combined with other data sources, such as the battery level transmitted by monitoring devices along collected data. Another important benefit achieved is the scalability regarding deployment of APs and monitoring devices. The configuration of the network behavior in the patient's residence is completely automated by OpenFlow; ThingsFlow provides the suitable *flow entries* for each AP. Through the orchestration of the network provided by SDN the network complexity was moved from APs to remote servers allowing developers to use of any development resources available in services features development instead of getting limited by hardware and system constraints of the APs. OpenFlow also rewrites packets headers to prevent health data of patients from being sent to proprietary serves; those packets are so forwarded to the monitoring server where patient data are processed.

6 Final Considerations

Designing a network solution for an smart environment is not an easy task because of the different demands of the smart devices. Solutions are usually based on local gateways that not only are in charge of related communication tasks but also must provide security, compatibility, and network management. To keep a gateway-based solution simple and less expensive, it is necessary to use an approach that enables the flexibility and functionality demanded, without high complexity and cost. In this sense, transferring more complex functions to a remote server and leaving the gateway to take care of only the task of switching and routing is the proposed solution presented in this paper.

Moving network control to the application level empowers a low cost network with capabilities usually not available in the gateway hardware. By using SDN, network administrators are neither dependent nor limited to the features provided by gateway vendors anymore. New features can be developed using high level abstraction programming languages and management solutions can be combined with control applications to improve network intelligence and automation.

The environment presented in this work provides health care service at home level to patients with chronic illness, but can also be used in other smart city environments as well. This work demonstrated how SDN can be used to build a wider view of an smart environment combining data from sensing devices with counters of network communication of all connected devices. SDN is also a good choice for smart environments because it enables the development of more appropriate management mechanisms and improves flexibility in network orchestration to fit the diversity of smart devices. As future work, we will deploy more complex scenarios with a wider range of services, aiming to exploit deeper SDN resources and its benefits for this kind of environment.

References

1. Anthopoulos, L., Fitsilis, P.: 16th International Conference on Advanced Communication Technology (ICACT), pp. 190–195 (2014). doi:10.1109/ICACT.2014.6778947
2. Steventon, A., Wright, S. (eds.): Intelligent Spaces: The Application of Pervasive ICT, 1st edn. Springer, London (2006). doi:10.1007/978-1-84628-429-8
3. Washburn, D., Sindhu, U., Balaouras, S., Dines, R.A., Hayes, N., Nelson, L.E.: Helping CIOs Understand "Smart City" Initiatives. Technical Report, Forrester Research Inc (2010)
4. Arbiza, L.M.R., Bertholdo, L.M., Santos, C.R.d.P., Granville, L.G., Tarouco, L.M.R.: 30th ACM/SIGAPP Symposium On Applied Computing. ACM, Salamanca, pp. 640–645 (2015). doi:10.1145/2695664.2695861
5. Gubbi, J., Buyya, R., Marusic, S., Palaniswami, M.: Future Gener. Comput. Syst. **29**(7), 1645 (2013). doi:10.1016/j.future.2013.01.010
6. Mashal, I., Alsaryrah, O., Chung, T.Y., Yang, C.Z., Kuo, W.H., Agrawal, D.P.: Ad Hoc Networks **28**, 68 (2015). doi:10.1016/j.adhoc.2014.12.006
7. Yeo, K.S., Chian, M., Ng, T., Tuan, D.A.: 14th International Symposium on Integrated Circuits (ISIC), pp. 568–571 (2014). doi:10.1109/ISICIR.2014.7029523
8. Blasco, R., Marco, A., Casas, R., Cirujano, D., Picking, R.: Sensors **14**(1), 1629 (2014). doi:10.3390/s140101629
9. Santos, C.R.P.d., Bezerra, R.S., Ceron, J.M., Granville, L.Z., Tarouco, L.M.R.: IEEE Commun. Mag. **48**(12), 112 (2010)
10. Software-Defined Networking: The New Norm for Networks (2012). https://www.opennetworking.org/images/stories/downloads/sdn-resources/white-papers/wp-sdn-newnorm.pdf
11. Wickboldt, J.A.: Jesus, W.P.d., Iolani, P.H., Both, C.B., Rochol, J., Granville, L.Z. IEEE Commun. Mag. **53**(1), 278 (2015)
12. Mckeown, N., Anderson, T., Peterson, L., Rexford, J., Shenker, S., Louis, S.: (2008)
13. Sherwood, R., Gibb, G., Yap, K.K., Appenzeller, G., Casado, M., McKeown, N., Parulkar, G.: (2009). http://archive.openflow.org/downloads/technicalreports/openflow-tr-2009-1-flowvisor.pdf

14. Rafael, R.B.: Pereira Esteves. Lisandro Zambenedetti Granville. IEEE Commun. Mag. **51**(7), 80 (2013)
15. RouteFlow Project. https://sites.google.com/site/routeflow/home
16. Egilmez, H.E., Dane, S.T., Bagci, K.T., Tekalp, A.M.: Signal and Information Processing Association Annual Summit and Conference (APSIPA ASC), Asia-Pacific, pp. 1–8. IEEE, Koc University, Istanbul (2012)
17. Yap, K.K., Kobayashi, M., Sherwood, R., Huang, T.Y., Chan, M., Handigol, N., McKeown, N.: ACM SIGCOMM Comput. Commun. Rev. **40**(1), 125 (2010)
18. Kim, H., Feamster, N.: IEEE Commun. Mag. **51**(2), 114 (2013). doi:10.1109/MCOM.2013.6461195
19. Chetty, M., Feamster, N.: SIGCOMM Comput. Commun. Rev. **42**(3), 54 (2012). doi:10.1145/2317307.2317318
20. Yiakoumis, Y., Yap, K.K., Katti, S., Parulkar, G., McKeown, N.: Proceedings of the 2ndACM SIGCOMM Workshop on Home Networks. HomeNets'11. ACM, New York, pp. 1–6 (2011). doi:10.1145/2018567.2018569
21. Tarouco, L.M.R., Bertholdo, L.M., Granville, L.Z., Arbiza, L.M.R., Carbone, F., Marotta, M., de Santanna, J.J.C.: IEEE International Conference on Communications, International Workshop on Mobile Consumer Health Care Networks, Systems and Services. Ottawa, pp. 6121–6125 (2012). doi:10.1109/ICC.2012.6364830
22. Miorandi, D., Sicari, S., Pellegrini, F.D., Chlamtac, I.: Ad Hoc Netw. **10**(7), 1497 (2012). doi:10.1016/j.adhoc.2012.02.016
23. IOT-A: Internet of Things Architecture. http://www.iot-a.eu/public
24. Wu, G., Talwar, S., Johnsson, K., Himayat, N., Johnson, K.D.: IEEE Commun. Mag. **49**(4), 36 (2011). doi:10.1109/MCOM.2011.5741144
25. Coetzee, L., Eksteen, J.: IST-Africa Conference Proceedings, pp. 1–9 (2011)
26. libfluid—The ONF OpenFlow Driver. http://opennetworkingfoundation.github.io/libfluid/index.html

Automated Network Resilience Optimization Using Computational Intelligence Methods

Vitor Pereira, Miguel Rocha and Pedro Sousa

Abstract This paper presents an automated optimization framework able to provide network administrators with resilient routing configurations for link-state protocols, such as OSPF or IS-IS. In order to deal with the formulated NP-hard optimization problems, the devised framework is underpinned by the use of computational intelligence optimization engines, such as Multi-objective Evolutionary Algorithms (MOEAs). With the objective of demonstrating the framework capabilities, two illustrative Traffic Engineering methods are described, allowing to attain routing configurations robust to changes in the traffic demands and maintaining the network stable even in the presence of link failure events. The presented illustrative results clearly corroborate the usefulness of the proposed automated framework along with the devised optimization methods.

1 Introduction

Nowadays, IP based network infrastructures have to support a myriad of applications and services generating high volumes of traffic and many of them with strict operational and availability requirements. In this perspective, actual network infrastructures should present high levels of resilience in order to correctly behave under a wide set of operational conditions [6]. As is well known, routing protocols are key elements of IP converged networks, thus having a major influence in the operational conditions of such communication infrastructures. In this specific field, link-state

V. Pereira (✉) · P. Sousa
Centro Algoritmi/Department of Informatics, University of Minho, Braga, Portugal
e-mail: Vitor.Pereira@algoritmi.uminho.pt

P. Sousa
e-mail: pns@di.uminho.pt

M. Rocha
Centre Biological Engineering/Department of Informatics, University of Minho,
Braga, Portugal
e-mail: mrocha@di.uminho.pt

© Springer International Publishing Switzerland 2016
P. Novais et al. (eds.), *Intelligent Distributed Computing IX*,
Studies in Computational Intelligence 616,
DOI 10.1007/978-3-319-25017-5_46

approaches such as Intermediate System to Intermediate System (IS-IS) or Open Shortest Path First (OSPF) [7] protocols are very popular, being often used by Internet Service Providers (ISPs) administrators to deliver connectivity between all network equipment.

The area of Traffic Engineering (TE) usually deals with performance evaluation and performance optimization of operational IP networks. In particular, relevant research in this area focused on the objective of achieving an efficient traffic distribution in the networking infrastructure, taking into account the expected traffic demands (e.g. [1, 2]). This translates to a NP-hard optimization problem that seeks to find a set of routing weights that are able to optimize the congestion levels of the network, considering the aggregated traffic demands specified for each source-destination pair (usually expressed by traffic matrices [8]). The use of computational intelligence methods to solve this TE related problem has presented encouraging results, namely with the use of Evolutionary Algorithms to solve congestion based formulations, or other variants involving multi-constrained optimization approaches (e.g. [3–5]).

Considering the above mentioned, this work aims to foster this research field by presenting a contribution specifically focused on devising mechanisms able to provide resilient aware routing configurations, using now Multi-objective Evolutionary Algorithms (MOEAs) as the optimization engines. As result, this paper presents an automated and intelligent optimization framework assisting network administrators in the configuration of resilient network infrastructures, providing them with a set of configuration alternatives expressing distinct trade-offs between the considered objectives. In order to illustrate the capabilities of the devised solution two optimization methods are presented being able to provide resilient routing configurations capable to deal with multiple demand matrices and with possible link failure events that may occur in the network infra-structure.

The paper proceeds with Sect. 2 which describes an optimization framework for routing configurations, the adopted mathematical formulation and the used multi-objective optimization engines; Sect. 3 presents an optimization method assuring the correct network behavior for multiple demand matrices assumptions; Sect. 4 focus on a resilient mechanism providing adequate configurations even in the case of network link failures; Sect. 5 presents the conclusions.

2 A Computational Intelligence Aware Optimization Framework

Figure 1 provides a high level description of the devised optimization framework and its main components. The framework has two main core components: the routing simulator and the computational intelligence optimization module. In a simplified perspective, the former computes shortest paths in the same manner as the link-state OSPF protocol (using the Dijkstra algorithm [10]) and distributes the traffic aggre-

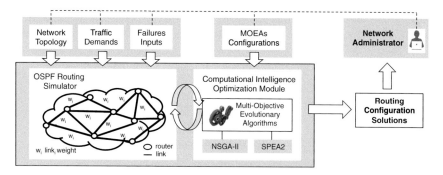

Fig. 1 High level description of the computational intelligence optimization framework

gates along the network links. The latter is responsible for performing complex routing configuration optimization tasks. This module integrates two Multi-Objective Evolutionary Algorithms (MOEAs), the NSGA-II [13] and the SPEA2 [12], provided by a Java-based library entitled JEColi [11]. Such computational intelligence engines, along with the optimization approaches discussed along this paper will be publicly available in a new release of the NetOpt tool (http://darwin.di.uminho.pt/netopt).

As also depicted in Fig. 1, the framework receives a description of the network topology to be optimized, along with the traffic matrix representing the estimated traffic traversing the network [9]. Furthermore, and in the context of one of the methods proposed in this work, network failure related inputs might also be provided to the framework. Additional input parameters tuning the MOEAs operation during the optimization processes are also allowed to be configured by the user. As output, the framework provides the network administrator with near-optimal routing configurations for the addressed optimization problem(s). More technical details about the configuration outputs provided by the framework are discussed in Sect. 2.2.

2.1 Mathematical Formulation

The resilience aware methods proposed in this work aim to attain an efficient distribution of the traffic aggregates in the links of the network domain avoiding, as much as possible, the existence of congested links. The framework represents the network topology as a direct graph $G(N, A)$, with N representing a set of nodes (network routers), and A representing a set of arcs (network links), with a capacity of c_a for each $a \in A$.

For a specific routing configuration, and considering the given traffic matrix, $f_a^{(s,t)}$ expresses the amount of traffic routed over the arc a having source s and destination t. Thus, the utilization of an arc a can be defined as in Eq. (1) with ℓ_a being the sum of all flows $f_a^{(s,t)}$ that travel over it. Considering the utilization degree of an arc we

adopt the cost function, Φ_a, proposed by Fortz and Thorup [14] and which derivate is
presented by Eq. (2), as a linear cost function which penalizes high congested links.

$$u_a = \frac{\ell_a}{c_a} \tag{1}$$

$$\Phi'_a = \begin{cases} 1 & for \ 0 \leq u_a < \frac{1}{3} \\ 3 & for \ \frac{1}{3} \leq u_a < \frac{2}{3} \\ 10 & for \ \frac{2}{3} \leq u_a < \frac{9}{10} \\ 70 & for \ \frac{9}{10} \leq u_a < 1 \\ 500 & for \ 1 \leq u_a < \frac{11}{10} \\ 5000 & for \ u_a \geq \frac{11}{10} \end{cases} \tag{2}$$

$$\Phi = \sum_{a \in A} \Phi_a \tag{3}$$

Given this formulation, a possible optimization objective consists in distributing
traffic demands in the network in order to minimize the sum of all costs, as expressed
by Eq. (3). A normalized congestion measure, Φ^*, is also defined in order to enable
results comparison between distinct topologies. It is important to note that when Φ^*
equals 1, all loads are below 1/3 of the link capacity, while when all arcs are exactly
full the value of Φ^* is 10 2/3. In the results presentation, this value is considered as
a threshold that bounds the acceptable working region of the network.

The resilient aware methods proposed in this work deal with multi-objective optimization problems, which means that they target the simultaneous minimization of
several Φ^* functions under distinct operational conditions. More details regarding
such approaches are given in Sects. 3 and 4.

2.2 Multi-objective Optimization

Since the mid-1980s, MOEAs are being used to solve several multiple-criterion
problems, being one of the most competitive approaches in this field [15]. The
MOEAs integrated in the devised optimization framework are two popular algorithms, the SPEA2 and the NSGA-II, widely accepted as two of the algorithms with
best performance.

In the used MOEAs each individual encodes a routing solution as a vector of
integer values, where each value (gene) corresponds to the weight of a link (arc) in
the network, and therefore the size of the individual equals the number of links in
the network. Although OSPF link weights are integers valued from 1 to 65535, only
values in range [1; 20] were considered, allowing to reduce the search space and,

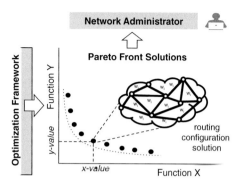

Fig. 2 Conceptual view of the solutions (pareto front) returned by the optimization framework assuming an optimization process trying to minimize two objective functions. In this case the quality of a specific solution is measured using *Function X* and *Function Y*, obtaining fitness values of *x-value* and *y-value*, respectively for each one of the objectives

simultaneously, increasing the probability of finding equal cost multipaths which benefits effective bandwidth use by load-balancing traffic over multiple paths. The individuals that populate the initial populations were randomly generated, with arc weights taken from a uniform distribution within the reduced range. The MOEAs resort to the following reproduction operators for solutions combination and genetic diversity: (i) Random mutation, replaces a given gene by a random value, within the allowed range; (ii) Incremental/decremental mutation, replaces a given gene by the next or by the previous integer value, with equal probabilities, within the allowed range and (iii) Uniform crossover, this operator works by taking two parents as input and generating two offspring. For each position in the genome, a binary variable is randomly generated: if its value is 1, the first offspring takes the gene from the first parent in that position, while the second offspring takes the gene from the second parent; if the random value is 0, the roles of the parents are reversed.

MOEAs are appropriate to deal with the multi-objective nature of the optimization problems discussed in the following sections. In fact, MOEAs return a set of solutions with distinct trade-offs between the considered objectives, allowing the network administrator decide which solution to implement. This is illustrated in Fig. 2, which provides a simplified view of the framework optimization outputs for a generic multi-objective problem involving the minimization of two objective functions. As observed in Fig. 2, the framework computes a Pareto front (i.e. a set of non-dominated[1] solutions) containing a distributed set of solutions covering the whole set of possible trade-offs between the optimization aims of the addressed problem. The administrator is then allowed to selected the most adequate solution according with the desired tradeoff. In this case, each one of the Pareto front solutions (i.e. each black filled dot represented in Fig. 2) represents a set of routing weights to be configured in the network.

---

[1]When a solution is dominated by another one, it means that it is worse than the second in at least one of the objectives and it is not better in none.

3 Optimization Methods—Demand Matrices

The functional conditions of a network environment are not static, they rather change over time. Traffic volume and behavior, for example, suffer alterations over particular periods. Although some of the fluctuations on traffic can be unpredictable, others, such as global variations over specific periods of time (e.g. night and day variability) can be foreseen and translated into distinct traffic demands estimations. Those estimations, represented in the form of matrices, frequently have uncorrelated source-destination individual entries or different overall levels of traffic. Traditional TE methods are not well suited to address those variations, as they usually assume fixed traffic volumes between each source-destination pair. A routing configuration may be appropriate to warrant a good performance of the network regarding a specific traffic estimation, but can fail in respect to another. Finding a configuration that is adequate for both can be addressed as a multi-objective problem (MOP) that summarizes as follows.

For a given network topology and two traffic demand matrices, the cost function Φ^* is used to define two functions, Φ_1^* and Φ_2^*, that evaluate the congestion of the network associated to each of the traffic demand matrices respectively. The aim consists in finding a weights configuration (w) that simultaneously minimizes both objectives, Φ_1^* and Φ_2^*.

To evaluate this mutli-objective (MO) approach, three synthetic network topologies were used, with different sizes (30 and 50 nodes) and distinct average in/out degree of each node (2 and 4). For each of these topologies, a set of traffic demands matrices D_i were randomly generated with different levels of traffic amount. The i variable reflects the expected mean of congestion on all links of the network and takes values in (0.3, 0.4, 0.5). The correlation between two matrices in the same scenario is also kept under control with an approximated value $r = 0.5$. For comparison purposes, two traditional and commonly used weights configuration schemes are included in the presented results: Unit (unitary weights are assigned to each link) and InvCap (with weight inversely proportional to the link capacity). Furthermore, to highlight the benefits of the MO approach, two single-objective optimizations, that solely minimize one of the two objectives, are also included. The single-objective optimizations were performed resorting to a Single-objective EA (SOEA), while NSGA-II was used on MO optimizations. The obtained results are presented in Table 1. Values above the threshold of acceptable congestion on the network are identified with a grey filled background, meaning that in such cases the network was unable to accommodate the considered traffic demands. The results for the case regarding the optimization for two $D_{0.4}$ traffic matrices on the 30_4 and 50_2 networks are not presented as all values were above the threshold of acceptable congestion cost.

The results show that, in most cases, by resorting to multi-objective optimization, it is possible to obtain a routing configuration that enables the network to perform well even if two disruptive traffic demands matrices need to be considered. Since the MO optimization provides a set of equally good solutions with distinct trade-offs

Table 1 Congestion optimization for two traffic demand matrices (minimum values)

Algorithm	Demands #1	Demands #2	30_2 Φ_1^*	30_2 Φ_2^*	30_4 Φ_1^*	30_4 Φ_2^*	50_2 Φ_1^*	50_2 Φ_2^*
Unit	0.3	0.3	83.80	130.68	255.27	160.54	339.96	313.34
InvCap			31.68	15.60	263.04	75.66	437.70	434.55
Single (#1)			1.48	5.34	3.63	132.24	2.36	19.88
Single (#2)			6.45	1.56	78.23	2.03	24.91	2.10
Multi-obj			1.45	1.53	3.48	2.22	1.78	1.88
Unit	0.3	0.4	83.806	227.74	255.27	426.75	339.96	430.14
InvCap			31.68	203.53	263.04	717.95	437.70	812.01
Single (#1)			1.48	92.78	4.99	278.04	2.36	84.47
Single (#2)			1.89	2.53	9.50	23.81	13.84	15.05
Multi-obj			1.40	2.51	2.79	17.59	2.05	15.80
Unit	0.4	0.4	157.00	227.74				
InvCap			221.10	203.53				
Single (#1)			2.61	47.54				
Single (#2)			28.91	2.53				
Multi-obj			2.17	2.24				

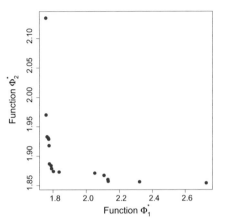

Fig. 3 Pareto front for two demands optimization. *(Scenario: Topology 50_2 and two traffic demands $D_{0.3}$)*

between the objectives, as can be seen in the non-dominated solution set representation in Fig. 3, the results presented in Table 1 are the solutions where both objectives are just as important. Nevertheless, it would be possible for a network administrator to choose from the Pareto solution set a configuration that more accurately reflects its needs. Also presented in Table 1, are the congestion values provided by the traditional configuration schemes, Unit and InvCap, where they totally fail to warrant a good performance level for heterogeneous traffic demands. When the two traffic matrices are divergent, single objective optimizations are also often unable to deliver suitable solutions. Although they provide a good level of congestion for the optimized objective, they do not grant the same level of performance for the other, being the network unable to accommodate the traffic for the unconsidered demand matrix. For most of the cases, the MO optimization algorithm was the only one capable to achieve weights configurations that enable a satisfactory network behavior for the two demands matrices by concurrently minimizing both objectives. Another important aspect of MOEA is that the optimization mechanisms included in the algorithms, such as the NSGA-II algorithm, allow to attain better congestion levels even in the scope of a single traffic demand matrix weights optimization. This can be observed in the 30_2 topology scenario, with demands level 0.3, where the MO optimization provides a better ranked solution in both objectives. This is mainly due to the diversity of solutions kept within the population during the optimization process.

4 Optimization Methods—Link Failures

Other types of events, which have severe impact in the network performance, can also successfully be tackled by MO optimization. When a link fails, the network traffic, that previously flowed through it, is shifted to other shortest paths which have meanwhile been recalculated by the routing protocol. This relocation of traffic can lead to

Fig. 4 Pareto front for link
failure optimization.
(Scenario: Topology 30_4 *and*
traffic demands $D_{0.3}$)

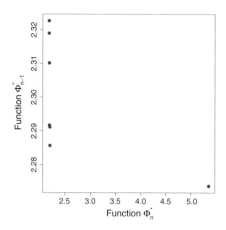

congestion in some parts of the network, and undermine its performance. Although
it is not possible to predict such events, a network administrator can identify which
link failure can cause for some reason (e.g. link capacity, network architecture, failure
probability, etc.) a significant impact on the network functional conditions. It would
therefore be adequate to protect this link against such event, whilst ensuring the con-
tinuity of an acceptable congestion level on the network. This new MO problem, that
aims to minimize the network congestion before and after the failure of a single link,
can be defined as follows. For a given network topology, a traffic demand matrix and
a previously selected topology link, the target consists in finding a set of weights (w)
that minimize simultaneously the objective functions Φ_n^* and Φ_{n-1}^*, which, respectively, assess the congestion cost of the network in a normal state (n), and when the
selected link fails ($n - 1$).

The optimization results for the link failure MO problem are shown in Table 2,
and, as above, congestion costs obtained by applying traditional weights configurations schemes (Unit and InvCap) and a single objective optimization were also
included for comparison. Again, the included MO results are those where the solutions give equal importance to both objectives. The considered single objective optimization only minimizes the congestion for the network normal state, as it would
make no sense to optimize the congestion considering only the failing state, disre-
garding the performance of the network before a link failure that may never take
place. In the experiments, the link that belongs to the largest number of shortest
paths, when a InvCap weights configuration is applied, was used as criteria to select
the failing link. The NetOpt framework, however, offers a broader set of selection
criteria, such as the link with higher load or the link whose failure has the greatest
impact on the network congestion cost.

Within this set of experiments, the MOEA was also the algorithm that showed
the best overall results. Even in cases where the threshold is surpassed, such as in
the 50_2 topology experiments with demands level 0.4, the MOEA offers a solution
set that can assure a near acceptable congestion performance, where all other meth-

Table 2 Link failure congestion optimization (minimum values)

Algorithm	Demands	30_2		30_4		50_2	
		Φ_n^*	Φ_{n-1}^*	Φ_n^*	Φ_{n-1}^*	Φ_n^*	Φ_{n-1}^*
Unit	0.3	130.68	165.30	198.96	234.23	339.96	373.56
InvCap		15.60	88.13	323.76	269.23	437.70	565.62
Single (n)		1.56	30.89	2.00	58.65	1.98	18.09
Multi-obj		1.44	1.48	2.30	2.25	1.77	1.78
Unit	0.4	160.95	165.30	426.75	499.26	339.96	478.33
InvCap		108.28	271.60	717.95	723.39	812.00	919.68
Single (n)		2.03	94.68	7.88	107.13	19.63	97.77
Multi-obj		1.75	1.80	18.66	10.13	11.52	11.49

ods clearly failed. A Pareto front of the solutions provided by the MOEA, for the 30_4 topology with $D_{0.3}$ demands specific case, is shown in Fig. 4, and illustrates the broad choice of solutions available to the administrator. It is important to remark that small penalties on the congestion cost for the network in its normal state are entirely justified by the gain on the failing state congestion, when the MO optimization algorithm is used. This can be observed in particular by comparing some single objective optimization results with those provided by the MOEA.

As an example of a possible choice given to an administrator, in the experiments with the network topology 30_4 and demands $D_{0.4}$, the congestion cost value pair (18.66, 10.13) is presented in Table 2 as a representative result. There are, nonetheless, other options such as the cost value pair (9.22, 34.29) that keeps the congestion on the normal state under the acceptable threshold, but, although better than those provided by the other algorithms, with a worst congestion in a failed state. An administrator should choose which is the most adequate. The NetOpt framework, for that matter, provides a set of tools that can help the decision making, with several informations regarding individual links usage within each weights configuration solution. Nevertheless, the results obtained in all the demand instances and topologies clearly indicate the obvious advantages for an administrators to resort to this preventive multi-objective link failure optimization method.

5 Conclusions

This paper presented an optimization framework for routing configurations based on computational intelligence methods. In particular, Multi-objective Evolutionary Algorithms are used to solve complex optimization problems pursuing near-optimal network configurations able to improve the resilience levels of network infrastructures. As a proof-of-concept two illustrative TE methods were described along with illustrative optimization results. The first method allows to achieve network routing configurations that are robust to changes in the traffic demands traversing the infrastructure, which are expressed by traffic matrices. The second proposed mechanism ensures that the network continues to operate with an appropriate level of quality even in the presence of fault situations of certain infrastructure links. In both cases, the network operator is able to select a specific solution, from a computed Pareto front, representing the most appropriate trade-off between the considered objectives.

The presented optimization results clearly corroborate the effectiveness of the used optimization engines on solving complex network optimization problems. Thus, the devised optimization framework is a valuable tool for network administrator allowing for automated optimization processes of network resilience levels.

Acknowledgments This work has been partially supported by FCT - Fundação para a Ciência e Tecnologia Portugal in the scope of the project: UID/CEC/00319/2013.

References

1. Altin, A., Fortz, B., Thorup, M., Umit, H.: Intra-domain traffic engineering with shortest path routing protocols. Ann. Oper. Res. **204**(1), 56–95 (2013)
2. Fortz, B., Thorup, M.: Optimizing OSPF/IS-IS weights in a changing world. IEEE J. Sel. Areas Commun. **20**(4), 756–767 (2002)
3. Rocha, M., Sousa, P., Cortez, P., Rio, M.: Quality of service constrained routing optimization using evolutionary computation. Appl. Soft Comput. **11**(1), 356–364 (2011)
4. Sousa, P., Rocha, M., Rio, M., Cortez, P.: Efficient OSPF weight allocation for intra-domain QoS optimization. In: Parr, G., Malone, D., Foghl, M. (eds.) IPOM 2006—6th IEEE International Workshop on IP Operations and Management, Dublin, Ireland, LNCS 4268, October 2006, pp. 37–48. Springer (2006)
5. Pereira, V., Rocha, M., Cortez, P., Rio, M., Sousa, P.: A framework for robust traffic engineering using evolutionary computation. In: 7th International Conference on Autonomous Infrastructure, Management and Security (AIMS 2013), Barcelona, Spain, LNCS 7943, pp. 2–13. Springer (2013)
6. Lee, K., Lim, F., Ong, B.: Building Resilient IP Networks. Cisco Press (2012)
7. Moy, J.: OSPF Version 2. RFC 2328 (Standard), April (1998). Updated by RFC 5709
8. Cariden Technologies, Building Traffic Matrices: Introduction to MATE Flow Collection. White Paper—Version 2. October (2012)
9. Tune, P., Roughan, M.: Network design sensitivity analysis. In: The 2014 ACM International Conference on Measurement and Modeling of Computer Systems, SIGMETRICS 14, pp. 449–461. ACM (2014)
10. Dijkstra, E.: A note on two problems in connexion with graphs. Numerische Mathematik **1**(1), 269–271 (1959)
11. Evangelista, P., Maia, P., Rocha, M.: Implementing metaheuristic optimization algorithms with jecoli. In: Proceedings of the 2009 9th International Conference on Intelligent Systems Design and Applications, ISDA'09, Washington, pp. 505–510 (2009)
12. Zitzler, E., Laumanns, M., Thiele, L.: Spea2: Improving the strength pareto evolutionary algorithm. Technical report (2001)
13. Deb, K., Agrawal, S., Pratap, A., Meyarivan, T.: A fast and elitist multiobjective genetic algorithm: NSGA-II. IEEE Trans. Evolut. Comput. **6**(2), 182–197 (2002)
14. Fortz, B.: Internet traffic engineering by optimizing OSPF Weights. In: Proceedings of IEEE INFOCOM, pp. 519–528 (2000)
15. Coello, C.: A comprehensive survey of evolutionary-based multiobjective optimization techniques. Knowl. Inf. Syst. **1**(3), 129–156 (1999)

An Automated Framework for the Management of P2P Traffic in ISP Infrastructures

Pedro Sousa

Abstract Peer-to-Peer (P2P) is nowadays a widely used paradigm underpinning the deployment of several Internet services and applications. However, the management of P2P traffic aggregates is not an easy task for Internet Service Providers (ISPs). In this perspective, and considering an expectable proliferation in the use of such applications, future networks require the development of smart mechanisms fostering an easier coexistence between P2P applications and ISP infrastructures. This paper aims to contribute for such research efforts presenting a framework incorporating useful mechanisms to be activated by network administrators, being also able to operate as an automated management tool dealing with P2P traffic aggregates.

1 Introduction

P2P overlay networks [1] are becoming omnipresent in current networking infrastructures and it is expected that many future Internet applications may increasingly rely on this network communication paradigm. However, some P2P applications, as Bit-Torrent [2], are responsible by a relevant portion of the Internet traffic [3] and their behavior is many times unpredictable, generating high volumes of traffic traversing network infrastructures and leading to coexistence problems with ISPs. As a consequence, several efforts have been made in order to attain ISP-friendly P2P solutions (e.g. [4, 5]). Aligned with such efforts there is also the need for efficient and automated management mechanisms allowing ISP administrators to better deal with P2P traffic aggregates in their infrastructures, in place of being only restricted to use traditional bandwidth throttling mechanisms [6].

In this context, this work presents the rationale of an automated framework able to contribute for a better coexistence between ISPs and P2P applications. The framework is sustained by a BitTorrent-like collaborative P2P system integrating configurable P2P trackers also with the ability to exchange valuable information with the

P. Sousa (✉)
Centro Algoritmi/Department of Informatics, University of Minho, Braga, Portugal
e-mail: pns@di.uminho.pt

© Springer International Publishing Switzerland 2016   497
P. Novais et al. (eds.), *Intelligent Distributed Computing IX*,
Studies in Computational Intelligence 616,
DOI 10.1007/978-3-319-25017-5_47

ISP level (as also proposed by other works, e.g. [7]). Based on the devised framework some illustrative capabilities are described, focusing on some methods that can be useful from the ISP point of view, namely: the capability to estimate the traffic impact that a given P2P swarm will have on the ISP infrastructure; the ISP ability to divert P2P traffic from specific network components of the network topology; the inclusion of mechanisms allowing for P2P service quality differentiation. With the proposed solution, network administrators may explicitly trigger the described mechanisms whenever required, or use the framework as an automated tool to implement specific policies controlling the P2P traffic aggregates in the ISP domain.

Section 2 presents the rationale of the proposed framework and Sect. 3 explains some of the supported methods. The simulation platform is described in Sect. 4 along with illustrative results. Finally, Sect. 5 concludes the paper.

2 Framework Architecture

Figure 1 presents the main components of the devised framework: (i) illustrative network management and optimization tasks usually required to manage and improve the ISP infrastructures; (ii) the ISP infrastructure integrating several links and routers, some of which providing access to ISP end-users/customers; (iii) the P2P tracker internal components. The framework assumes the scenario where P2P applications and the ISP assume collaborative behaviors. Furthermore, the framework assumes

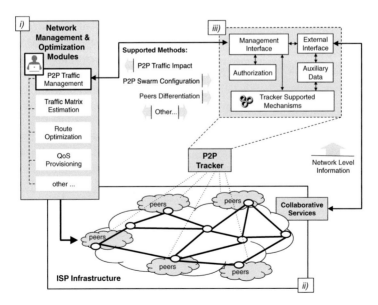

Fig. 1 High level description of the P2P management framework and associated components

the specific case of BitTorrent [2, 8] like applications, here with the tracker being the unique entity able to provide peering information, returning for this purpose a random sample of peers participating in the swarm to contacting peers. As depicted in Fig. 1 the ability to manage P2P traffic aggregates in a given infrastructure has also some relevance for other management/optimization tasks (e.g. traffic matrices estimation [9], routing optimization [10], QoS provisioning, etc.). The proposed framework assumes the existence of a P2P traffic management module (which may assume an automated behavior or be directly controlled by an administrator) able to interact with a configurable P2P tracker(s) (e.g. [11]) controlling the P2P swarm(s) behavior. The internal modules of the configurable P2P tracker are also depicted in Fig. 1, where several mechanisms are available to be activated/programmed by the P2P traffic management module (using the tracker management interface).

The devised framework assumes a collaborative perspective between the ISP and P2P levels. This is materialized by the existence of network level ISP collaborative services able to interact with the P2P tracker (using the tracker external interface), as depicted in Fig. 1. Using this interface the P2P tracker is able to access several network level information useful in the context of some specific tracker configurations (e.g. network topology, routing paths, network location of specific peers, etc.). As a reward for assuming a collaborative perspective the traffic generated by P2P applications using the proposed framework are positively discriminated by the ISP.

3 Examples of Methods Supported by the P2P Tracker

3.1 P2P Impact Estimation

This mechanism allows the P2P tracker to inform the P2P Traffic Management module (or the administrator) about the traffic impact that a given pre-scheduled P2P swarm, involving a considerable number of peers, will have in the network links of the ISP. Based on this feedback, and depending on the particular objectives in place, the ISP is able to influence the P2P swarm composition in order to protect specific elements from the underlying infrastructure (method described in Sect. 3.2).

The behavior of a P2P system as the assumed here is influenced by a large number of factors, as network level factors (e.g. network topology, peers locations, network paths, etc.) and data transfer protocol level factors (e.g. rules used by peers to exchange data pieces, etc.). Such large number of factors affecting the P2P overlay, along with the fact that some of those are extremely hard to foresee, make very difficult to define a highly accurate model to estimate the P2P traffic impact. The presented method centers the estimation efforts on the particular case of large BitTorrent P2P swarms and focus on specific network level factors that have major influence on the P2P traffic distribution. To evaluate the P2P traffic impact on the network links the P2P tracker models the network ISP infrastructure as a graph $G = (N, L)$. Furthermore, the tracker will receive from ISP level collaborating services other

Table 1 Syntax of the symbols used to compute the P2P link impact values

Symbols	Description
$G = (N, L)$	Graph expressing a network infrastructure (e.g. an ISP)
L	Set of network links of the ISP
N	Set of network nodes/routers of the ISP
A	Set of end-user areas where peers are located (each area is denoted by the corresponding network router, a, with $a \in A$ and $A \subseteq N$)
$paths_{i,j}$	Number of shortest paths between end-user areas i and j
$paths_{i,j}(l)$	Number of shortest paths between end-user areas i and j that include link l
$lif_{i,j}(l)$	Link inclusion factor for link l considering areas i,j, with $lif_{i,j}(l) = \frac{paths_{i,j}(l)}{paths_{i,j}}$
$w_{i,j}$	Ratio between the number of peers involved in possible peering adjacencies involving areas i,j and the number of peers involved in possible adjacencies involving all areas
$p_{i \leftarrow j}$	Factor denoting how close are areas j and i, with $p_{i \leftarrow j} \in [0, 1]$ and $\sum_{j \in A, j \neq i} p_{i \leftarrow j} = 1$

associated information, such as: network peers location (peers are located on end-users areas), network topology, routing information, etc. In a simplified perspective, the model extends and adapts to this P2P approach the concept of betweenness centrality that is one of several graph measures [12, 13]. The model integrates several factors used to estimate the P2P traffic impact in the network links (see Table 1 for a detailed description of the used mathematical symbols): (i) a link inclusion factor, $lif_{i,j}(l) \in [0, 1]$, is evaluated for each link $l \in L$ considering all the available end-user areas pairs. If all the available shortest paths between areas i,j include link l then $lif_{i,j}(l) = 1$; (ii) a weighting factor, $w_{i,j}$, dealing with unbalanced distribution of peers in the network, increasing the importance of shortest paths connecting areas involving higher number of peers; (iii) a preference value, $p_{i \leftarrow j}$, favoring near end-user areas pairs, as BitTorrent peers often have a higher probability to establish peering adjacencies with nearest peers in the network favoring TCP connections with lower RTTs. Equation 1 presents the devised normalized P2P impact metric (I_{P2P}) for each link l (which assigns impact values in the interval [0, 1]). This method is used by the tracker to inform the P2P Traffic Management module (of Fig. 1) about the estimated impact, where links that are assigned with higher $I_{P2P}(l)$ values are expected to be traversed by higher volumes of P2P traffic.

$$I_{P2P}(l) = \sum_{i,j \in A, \ i \neq j} [(|A| - 1) \cdot p_{i \leftarrow j}] \cdot lif_{i,j}(l) \cdot w_{i,j} \quad l \in L \quad (1)$$

3.2 ISP-controlled P2P Swarms

The framework also allows the ISP to influence the P2P swarms operation. The methods might be triggered by the administrator or integrate an automated approach programmed in the P2P Traffic Management module, e.g. allowing to react to the traffic impact values provided by the tracker or other events. Figure 2 depicts some supported methods: *link/router protection,* the tracker is informed that a given network link/router equipment should be protected from P2P traffic; *overlay minimization,* the tracker should minimize the number of routers/links traversed by P2P traffic. As depicted in Fig. 2, the P2P tracker computes the best peer sample to be returned to a given peer based on: the activated method imposing a given selection criteria, the contacting peer id, the available peers of a swarm and the collaborative information provided by the network level. Algorithm 1 presents a pseudo-code of the *router protection* method that can be activated by the P2P Traffic Management module or the administrator. The algorithm assumes a P_s set with all end-user areas pairs having swarm s peers (line 2). Each pair (a_i, a_j) indicates that peers from area i may receive peer samples with peers from area j. Next, the set J is defined to contain all links that are connected the router n that the ISP wants to protect (line 3). For each link in J an auxiliary set Z is defined containing area pairs connected by network paths traversing such link (i.e. $lif_{i,j}(l) > 0$, line 5). Next, it is verified if each area pair of Z can be removed from P_s in order to avoid that such P2P traffic aggregates between the areas traverses router n. The pair (a_i, a_j) is only removed (line 8) if the swarm does not get partitioned, i.e. possible connections established between peers of areas i and j are not necessary to guarantee that all peers of the swarm have access to all the pieces upload by the seeds of the swarm. After all the iterations, Algorithm 1 computes the allowed peering adjacencies that can be formed between swarm s peers, expressed by the P_s set. If none of Fig. 2 methods is triggered the P_s set will contain all the available area pairs. Thus, when contacted by a given peer the tracker returns a random sample (*random_peer_sample()* in Fig. 2) selected from all the available peers not violating the restrictions expressed by P_s set.

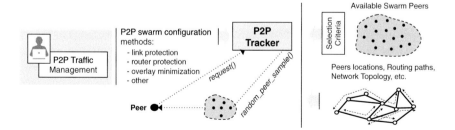

Fig. 2   High level description of methods of the framework allowing ISP-controlled P2P swarms

---

Algorithm 1 *router_protection (swarm s, router n, data info)*

1: {*s*: swarm identification; *n*: protected router; *info*: auxiliary data provided by the network}
2: $P_s \leftarrow$ Set with all (a_i, a_j) area pairs having peers from swarm s, $a_i, a_j \in A$
3: $J \leftarrow$ Set with all links $l \in L$ that are connect to router $n \in N$
4: **for all** $l \in J$ **do**
5:    $Z \leftarrow$ decreasingly ordered subset of P_s with all (a_i, a_j) area pairs having $lif_{i,j}(l) > 0$
      {Z is a $w_{i,j} * p_{i \leftarrow j}$ ordered set}
6:    **for all** $(a_i, a_j) \in Z$ **do**
7:       **if** *swarm_partitioned*$(s, P_s \setminus \{(a_i, a_j)\}) = FALSE$ **then**
8:          $P_s \leftarrow P_s \setminus \{(a_i, a_j)\}$
9:       **end if**
10:   **end for**
11: **end for**
12: *update_allowed_pairs*(s, P_s)

3.3 Peers Differentiation Strategies

This section addresses the framework capabilities in order to attain the differentiation of the P2P service offered to the peers. The objective is to enforce the ISP ability to benefit or penalize a given set of peers participating in a specific P2P swarm.

In this context, two method are defined in the framework allowing that the P2P tracker benefits or penalizes a given set of peers of a particular swarm (*penalize_peers()* and *benefit_peers()*, respectively). These methods are able to be used in a wide set of scenarios. As merely illustrative examples, the ISP may explicitly request the P2P tracker to activate such penalizing methods to punish peers which P2P behavior is contributing to the degradation of the network service quality or, alternatively, benefit specific peers of the P2P swarm as a reward mechanism for their past behavior. Independently of their particular use, the methods might be activated on-the-fly by the network administrator or integrate an automated approach where the P2P Traffic Management module of Fig. 1 is programmed to automatically activate such differentiation strategies in the tracker when a given event occur (a specific network condition event, a specific time period during the day, etc.).

Algorithms 2 and 3 present the pseudo-code of the *penalize_peers()* and *benefit_peers()* methods implemented at the tracker. As illustrated in Algorithm 2, the *penalize_peers()* method will firstly verify if the contacting peer belongs to the set of peers that should be penalized. In this case, the adopted strategy is to return to such peers a peer sample with a reduced number of peers (defined by *peer_limit*) and that can only be renewed after a given time (defined by *time_limit*) as observed in lines 2,3 of the algorithm. As consequence, such peers will be limited in the aim of discovering other peers in the swarm, thus experiencing lower service quality levels comparatively to non penalized peers receiving normal samples (line 9).

Algorithm 3 presents the pseudo-code of a tracker strategy benefiting some peers of the swarm. Here, benefited peers will form a privileged sub-swarm that will receive a given incentive which is controlled by the parameter *decision_rule* (lines 2, 3). The other peers will form a normal swarm with no access to such privileges

---

Algorithm 2 *penalize_peers(peer p, swarm s)*

1: **if** *action*(*p*, *s*) == *PENALIZE* **then**
2:     **if** first_request(*p*, *s*) **or** (current_timer() - last_request_timer(*p*, *s*)) ≥ *time_limit*) **then**
3:         *peer_sample* ← reduced_peer_sample(*s*, *peer_limit*)
4:     **else**
5:         *peer_sample* ← null
6:     **end if**
7:     last_request_timer(*p*, *s*) ← current_timer()
8: **else**
9:     *peer_sample* ← random_peer_sample(*s*)
10: **end if**
11: update_swarm_info(*p*, *s*)
12: return(*peer_sample*)

Algorithm 3 *benefit_peers(peer p, swarm s)*

1: **if** *action*(*p*, *s*) == *BENEFIT* **then**
2:     *peer_sample* ← privileged_peer_sample(*s*)
3:     *peer_sample* ← add_additional_incentives(*peer_sample*, *decision_rule*)
4: **else**
5:     *peer_sample* ← random_peer_sample(*s*)
6:     *peer_sample* ← exclude_privileged_peers(*peer_sample*)
7: **end if**
8: update_swarm_info(*p*, *s*)
9: return(*peer_sample*)

---

neither to the peers included in the privileged sub-swarm (lines 5, 6). In Sect. 4 experiments the decision rule for the privileged sub-swarm is to include in the peer sample two seeds with high upload capacity that are hidden from unprivileged peers.

4 Simulation Testbed and Illustrative Results

The main components of the framework were implemented at the ns-2 simulator [14] (Fig. 3). In order to present some illustrative results the network topology mentioned in Fig. 3 was used integrating 300 peers distributed along six end-user areas that participate in a P2P swarm exchanging a 50 MB file. In the presented experiments, one seed is assumed to exist in end-user area 1. The network uses the minimum number of hops as the criteria to compute the network routes.

4.1 P2P Impact Estimation

This example assumes the tracker programmed to inform the P2P Management module about the impact estimation of a given pre-scheduled P2P swarm. Several

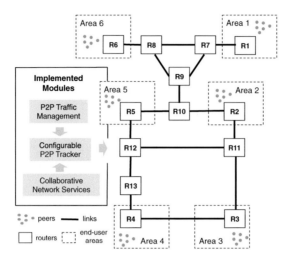

Fig. 3 Modules implemented in ns-2 and a topology with six end-users areas integrating 300 peers. P2P swarm exchanges a 50 MB file with chunks of 256 KB. Peers have upload/download capacities of 1 and 8 Mbps and propagation delays of access links vary within [1, 50] ms. The collaborative scenario assumes 50 Mbps of ISP links reserved for P2P traffic, with propagation delays two times higher than end users access links. By default, the peer sample returned by the tracker has 25 peer contacts

scenarios involving distinct peers and seeds distributions along the six end-user areas were tested. Due to space constraints only a small set of results are presented, but representative of the mechanism overall performance. Figure 4 presents the comparison between the estimated $I_{P2P}(l)$ metrics[1] and the cumulative traffic values that traversed the ISP links at the end of the simulation time, considering three distinct peers distributions (P_D) along the six end-user areas. As observed, the P2P impact metrics follow a similar trend to the traffic aggregates effectively traversing the links, thus providing a valuable information for network administrators.

4.2 ISP-controlled P2P Swarms

This section presents illustrative results obtained when the P2P Management module of the ISP (or the administrator) instructs the P2P tracker to protect some elements of the topology from P2P traffic (mechanism detailed in Algorithm 1).

Figure 5a compares the P2P traffic aggregates that traverses the routers of the ISP when the P2P tracker behaves in the normal configuration mode (white filled bars) and when the tracker is configured by the ISP in order to protect the router *R11* from the topology of Fig. 3 (black filled bars). As observed in Fig. 5a the P2P tracker forced that none of the traffic generated by the P2P swarm traversed the router *R11*

---

[1]$p_{i \leftarrow j}$ was set to 0.4 for nearest areas, the remaining areas were assigned with values of 0.15.

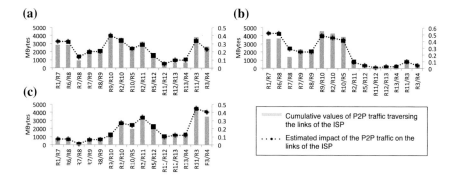

Fig. 4 P2P traffic versus Estimated $I_{P2P}(l)$ values for each ISP link in scenarios with peers distributions of **a** $P_D = (50, 50, 50, 50, 50, 50)$; **b** $P_D = (70, 70, 10, 10, 70, 70)$; **c** $P_D = (10, 70, 70, 70, 70, 10)$

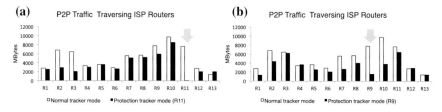

Fig. 5 P2P traffic traversing each ISP router with the tracker in the normal configuration mode and **a** programmed to protect Router 11; **b** programmed to protect Router 9

of the ISP. A slightly distinct scenario is presented Fig. 5b. Here, the ISP informs the tracker to try to protected router *R9*. As in this specific scenario only one seed is assume to exist in end-user area 1, it is not possible to completely avoid P2P traffic from traversing all *R9* links (otherwise the P2P swarm will become partitioned). Nevertheless, using the logic of Algorithm 1 the P2P tracker achieves a configuration that allows to substantially reduce the P2P traffic crossing such network element (cumulative amount of P2P traffic traversing *R9* is reduced from 7817 MB to 1575 MB, a decrease of nearly 80 % of P2P traffic traversing the equipment).

4.3 Peers Differentiation Strategies

Figure 6 results were obtained during a time period where the P2P Traffic Management module is programmed by the administrator to inform the P2P tracker that when managing new P2P swarms it should penalize/benefit specific network peers. In the first scenario, the tracker penalizes three groups of peers in end-user areas 2, 4 and 6 using the mechanism explained in Algorithm 2, returning a reduced peer sample to those peers (Fig. 6a). In the second scenario the tracker benefit two specific peer

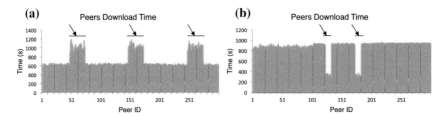

Fig. 6 Peers download times with tracker programmed to **a** penalize $peer_{ids}$ in $(50, 75)$, $(150, 175)$ and $(250, 275)$; **b** benefit $peer_{ids}$ within the intervals $(125, 135)$ and $(175, 185)$

groups from end-user areas 3 and 4 (Fig. 6b) which form a privileged sub-swarm having access to high upload capacity seeds that are hidden from the other peers of the swarm (using the configuration of Algorithm 3). In both cases there is a clear differentiation in the file download times obtained by distinct peers using the ISP network infrastructure. This confirms that the ISP was able to induce an effective P2P service quality differentiation among the selected peers.

5 Conclusions

This paper proposes a P2P management framework based on a BitTorrent-like P2P collaborative system. The solution integrates useful management methods allowing ISPs to better manage P2P traffic aggregates in their network infrastructures. Several illustrative methods were described allowing to automate some important ISP tasks in the context of P2P traffic aggregates management: (i) the possibility to estimate the traffic impact that a given pre-scheduled P2P swarm will have on the ISP topology; (ii) the protection of specific network elements from P2P traffic aggregates and (iii) the capability of the ISP to influence the P2P service quality obtained by the peers.

The devised framework was implemented and tested resorting to simulation. Several examples of the supported methods were presented and corresponding results discussed, clearly corroborating the feasibility of the proposed mechanisms.

Acknowledgments This work has been partially supported by FCT - Fundação para a Ciência e Tecnologia Portugal in the scope of the project: UID/CEC/00319/2013.

References

1. Lua, K., et al.: A survey and comparison of Peer-to-Peer overlay network schemes. IEEE Commun. Surv. Tutor. **7**(2), 72–93 (2005)
2. Choen, B.: Incentives build robustness in BitTorrent. In: Proceedings 1st Workshop on Economics of Peer-to-Peer Systems, Berkeley June 2003

3. Schulze, H., Mochalski, K.: Internet Study 2007: The Impact of P2P File Sharing, Voice over IP, Skype, Joost, Instant Messaging. One-Click Hosting and Media Streaming such as YouTube on the Internet, Technical Report (2007)
4. Liu, M. et al.: An ISP-friendly hierarchical overlay for P2P live streaming. In: Proceedings of 14th IEEE International Conference on Peer-to-Peer Computing (2014)
5. Yang, P., Xu, L.: An ISP-friendly inter-overlay coordination framework for multiple coexisting P2P systems. Peer-to-Peer Netw. Appl. **7**, 396409 (2014)
6. Wang, W., Wang, N., Howarth, M., Pavlou, G.: A dynamic Peer-to-Peer traffic limiting policy for ISP networks, pp. 317–324. In Proceedings of NOMS, IEEE/IFIP (2010)
7. Xie, H. et al.: P4P: Provider portal for applications. In: Proceedings of ACM SIGCOMM. Seattle 17–22 August 2008
8. Legout, A. et al.: Clustering and sharing incentives in bittorrent systems. In: Proceedings of ACM SIGMETRICS. San Diego, 12–16 June 2007
9. Tune, P., Roughan, M.: Network-design sensitivity analysis. Proc. ACM SIGMETRICS **14**, 449–461 (2014)
10. Pereira, V., Rocha, M., Cortez, P., Rio, M., Sousa, P.: A Framework for Robust Traffic Engineering Using Evolutionary Computation. In G.D. et al. (Ed.), AIMS Conference. Springer, LNCS, vol. 7943, 1–12 (2013)
11. Sousa, P.: Flexible peer selection mechanisms for future internet applications. In: Proceedings of BROADNETS 2009 Conference, Madrid (2009)
12. Opsahl, T., Agneessens, F., Skvoretz, J.: Node centrality in weighted networks: generalizing degree and shortest paths. Soc. Netw. **32**(3), 245–251 (2010)
13. Narayanan, S.: The betweenness centrality of biological networks, MSc thesis. Faculty of the Virginia Polytechnic Institute and State University (2005)
14. ns2: The Network Simulator NS-2 (2011). http://www.isi.edu/nsnam/ns/

A Geographic Opportunistic Forwarding Strategy for Vehicular Named Data Networking

Xuejie Liu, M. João Nicolau, António Costa, Joaquim Macedo
and Alexandre Santos

Abstract Recent advanced intelligent devices enable vehicles to retrieve information while they are traveling along a road. The store-carry-and-forward paradigm has a better performance than traditional communication due to the tolerance to intermittent connectivity in vehicular networks. Named Data Networking is an alternative to IP-based networks for data retrieval. On account of most vehicular applications taking interest in geographic location related information, this paper propose a Geographical Opportunistic Forwarding Protocol (GOFP) to support geo-tagged name based information retrieval in Vehicle Named Data Networking (V-NDN). The proposed protocol adopts the opportunistic forwarding strategy, and the position of interest and trajectories of vehicles are used in forwarding decision. Then the ONE simulator is extended to support GOFP and simulation results show that GOFP has a better performance when compared to other similar protocols in V-NDN.

Keywords Vehicular networks · Opportunistic forwarding · Named data networking · Geographic routing

X. Liu
Jilin University, Changchun, China
e-mail: xuejie@jlu.edu.cn

X. Liu · M. João Nicolau (✉) · A. Costa · J. Macedo · A. Santos
Centro ALGORITMI, Universidade Do Minho, Braga, Portugal
e-mail: joao@dsi.uminho.pt

A. Costa
e-mail: costa@di.uminho.pt

J. Macedo
e-mail: macedo@di.uminho.pt

A. Santos
e-mail: alex@di.uminho.pt

© Springer International Publishing Switzerland 2016
P. Novais et al. (eds.), *Intelligent Distributed Computing IX*,
Studies in Computational Intelligence 616,
DOI 10.1007/978-3-319-25017-5_48

509

1 Introduction

Recent advanced intelligent devices enable vehicles to retrieve information while travelling. Message transmitting normally follows multi-hop routing that use moving vehicles as intermediate nodes. However, efficient routing is challenged by high mobility of vehicles. To overcome intermittent connectivity, the store-carry-and-forward paradigm of Delay-Tolerant Networks (DTNs) is proposed, in which appropriate relay selection is the key problem. On the other hand, Named Data Networking (NDN) [1] is developed as an effective content-centric model for information retrieval, in which each node maintains three structures: Forwarding Information Base (FIB), Pending Interest Table (PIT) and Content Store (CS), to process Interest Packet (IntPkt) and Data Packet (DatPkt). A node (called consumer) sends out an IntPkt with a name to retrieve desired data. After receiving an IntPkt, nodes check local CS. If desired data is found, this IntPkt is satisfied and a DatPkt containing name and data is generated and is transmitted back along IntPkt's reverse path to consumer. Otherwise IntPkt is added in PIT and is forwarded based on FIB until desired data is found or IntPkt TTL expires. Though DTN and NDN are developed for different purposes, they have some similarities: flexible routing and network packet storage. Thus, they can be combined to improve data delivery.

Most vehicular applications are interested in location related information. Though some studies proposed to encode Position of Interest (POI) into data names to identify data, packet forwarding still adopts broadcasting or location-independent routing in traditional NDN. Since a node would not store data unless it has corresponding pending interest in its PIT [2], IntPkt is hardly to satisfy at any place far from POI and broadcasting IntPkt results in poor performance. Besides, forwarding DatPkt to consumer along IntPkt's reverse paths is impractical in V-NDN because of dynamic topology. As a possible solution to these problems, this paper presents a Geographical Opportunistic Forwarding Protocol (GOFP) for V-NDN. To the best of our knowledge, this work is the first that applies geographic information to routing named data. The store-carry-and-forward paradigm is supported in GOFP, and geographic location of POI and the trajectories of vehicles are used to select better next relay nodes. To evaluate GOFP, the ONE simulator is extended and simulation results show that GOFP has better performance, when compared to similar forwarding strategies in V-NDN.

The rest of the paper is organized as follows. Section 2 briefly summarizes related work in DTN and NDN. The proposed application scenario is discussed in Sect. 3 and the design of GOFP is described in Sect. 4. Section 5 shows simulation results and related analysis and Sect. 6 concludes presenting suggestions for future work.

2   Related Work

As most vehicular applications need to disseminate information to specific geographic areas, many geographic routing protocols are available. A position-based greedy forwarding approach and a repair strategy to choose the next hop are proposed in [3]. Distance based routing protocol [4] selects next hop base on in-vehicular distance and connectivity duration. Geographical opportunistic routing [5] and GeoSpray [6] follow the store-carry-and-forward paradigm and minimum estimated time of delivery is used as a utility function to make routing decisions. These protocols are designed for the scenario where the destination is stationary and forward process is one-way, so they do not apply in situation that destination (consumer) is mobile and communication process is a query-reply mode.

In terms of research about NDN, Grassi et al. [7] applied named data to networking running vehicles and described a prototype implementation of V-NDN. A named data based traffic information dissemination application was developed in [8], which shows that data names can greatly facilitate the dissemination process. But using broadcast to propagate packets potentially leads to poor performance. Furthermore, Kuai et al. evaluated IntPkt broadcast in [9] and indicated that it incurs to increased loss ratio at high density scenarios in V-NDN. Pesavento et al. in [10] pointed that most vehicular applications focus on getting POI related information and proposed an approach to map bi-dimensional geographic areas into a uni-dimensional naming scheme to identify geographic areas related data. Yu et al. in [11] proposed a Neighborhood-Aware Interest Forwarding (NAIF) routing protocol to improve the NDN Forwarding protocol [2]. Instead of indiscriminate flooding, NAIF selects cooperative nodes to forward IntPkt fractions. Lu et al. [12] presented a social-tie based content retrieval algorithm, where K-mean clustering algorithm is used to structure an hierarchical architecture among nodes, but it needs a process to build the social-ties.

3   Context and Application Scenario

We present the relevant application scenario to help explaining this research work motivation. Some vehicular applications require information about specific geographic areas. For instance, the parking application may need to know any available parking spaces around the vehicle's destination in order to direct the driver to the most convenient one. The service platform of parking lots broadcasts information of parking fees, current capacity and estimated available spots in several hours later. The vehicles moving near the parking lots can receive the information. We cannot assume constant connectivity between the vehicles and the service platform because of sparse vehicle density or high vehicles mobility. There are two potential processes in this scenario: (1) Consumer vehicle sends IntPkts with geo-tagged name to request the information about POI; (2) Once IntPkt reaches a vehicle carrying desired data, corresponding DatPkt is generated and is forwarded back to the moving consumer. GOFP is proposed for both these two processes.

4 GOFP Protocol Design

4.1 Data Naming Scheme and Packets Structure

The application should have a good data naming scheme that lets data providers to describe what they have and consumers to express what they want. Our data naming scheme is proposed as: */application/geo-reference/temporal-field/nonce/*. The field *application* indicates different application-dependent data. The *geo-reference* presents ID of POI, which can be converted to geographic coordinate (x, y) by GPS devices. The *temporal-field* is represented as *start-time/end-time* in IntPkt, and is set as data published time in DatPkt. The start and end time designate the time interval of desired data, e.g. the user may indicate he want the parking information from 10 AM to 11 AM. If all other fields are matched and published time of data is within the time interval of IntPkt, IntPkt is satisfied. The *nonce* is a random number used to distinguish different data providers.

The proposed structures of IntPkt and DatPkt are shown in Fig. 1. To differentiate IntPkt and DatPkt, the "type" field is reserved. The TTL (Time-To-Live) defines live time of packets in seconds and packet is discarded once it expires. The trajectory info indicates the consumer's trajectory till TTL expires. Outdated items will be deleted from the trajectory every time the packet is forwarded. DatPkt has the field content to store data.

4.2 Forwarding Strategies for Interest and Data Packets

Under assumption of equipping vehicles with GPS devices, vehicles can obtain their current position and future trajectory. GOFP is based on the opportunistic forwarding, so the most important issue involves the selection of relay nodes. Since

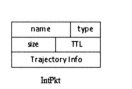

Fig. 1 Structure of interest packet (IntPkt) and data packet (DatPkt)

it is possible to obtain desired data from carried nodes, IntPkt is just forwarded to the vicinity of POI to increase satisfied probability, rather than to data provider. Conversely, DatPkts must be delivered to certain moving consumer. Thus, GOFP adopts different forwarding strategies for the IntPkt and DatPkt respectively.

4.2.1 Interest Packets Forwarding

To propagate IntPkt close to POI quickly, wireless channel should be used as much as possible because its transfer speed is faster than moving speed of vehicles. The GPS equipped vehicle knows its own trajectory and can convert place ID in IntPkt name to specific geographic position (x_a, y_a). Thus, when carrier vehicle meets a vehicle which is nearer or moves faster toward POI in a future period, the IntPkt will be forwarded to it. The position of vehicle i at time t is denoted by $P_i(t)$ and the time is slotted with customizable interval. The trajectory of the vehicle is defined as Definition 1.

Definition 1 (Manifestation of Vehicle's Trajectory) The trajectory of vehicle i is a sequence of positions in a given time span $[t_a, t_b]$, denoted by

$$T_i = \langle (t_a, P_i(t_a)), (t_{a+1}, P_i(t_{a+1})), \ldots (t_{a+k}, P_i(t_{a+k})), \ldots (t_b, P_i(t_b)) \rangle$$

Each vehicle can calculate the nearest distance to (x_a, y_a) on its trajectory in time period ($Current\ Time \leq t \leq Current\ Time + \delta$), where δ is the defined time interval.

The Euclidean distance of two points (x_1, y_1) and (x_2, y_2) is calculated by formula (1).

$$d_{i,j} = d((x_1, y_1), (x_2, y_2)) = \sqrt{(x_1 - x_2)^2 + (y_1 - y_2)^2} \qquad (1)$$

Definition 2 (Near Degree to POI): The near degree of vehicle i to POI is defined as the shortest distance of its trajectory to POI in a given interval δ. It is indicated as formula (2):

$$d_{\min}(i, poi) = mind_{pi,POI}(t) \quad t \in [Current\ Time \leq t \leq Current\ Time + \delta] \qquad (2)$$

The Near Degree of vehicle i to POI is calculated as shown in Fig. 2. If the $d_{\min}(i, poi)$ is smaller than communication range, vehicle i can get data from POI server directly.

The better relay vehicle is the one can potentially satisfy IntPkt earlier, which means a vehicle is chosen as next relay node if either it holds desired data in their CS or it will move closer to POI. In GOFP, vehicles announce the cached data name digest to their neighbors. If there is a neighbor that holds the desired data, the carrier will forward IntPkt to it. Otherwise, $d_{\min}(i, poi)$ is the metric to select the vehicle that meets the second condition. When δ is set a proper value, e.g. it is enough to let

```
Input : Positon of POI; Trajectory of vehicle i; interval δ
Output: The near degreedmin (i, poi)
1. dmin(i, poi) = dpi,POI(CurrentTime);
2. if (dmin(i, poi) <Communicaiton Range) return;
3. for(each CurrentTime< t<CurrentTime+ δ ) {
4.   if(dpi,POI(t)<dmin(i, poi))  dmin(i, poi) = dpi,POI(t);
5.   if (dmin(i, poi) <Communicaiton Range) return;
6. }
```

Fig. 2 Calculating the near degree to POI (pseudo-code)

```
Input :IntPkt bundles, neighbor vehicles set
Output: forward list
1. check and remove satisfied IntPkt();
2. for (each IntPkt pi in bundles buffer){
3.   get poi of IntPkt; read dmin(i, poi);
4.   for( each neighbor vehicle j){
5.     if( CS of j contains pi's desired named data) insert (pi,j) to forwarding list;
6.     else if (dmin(j, poi) is minimal dmin to POI) insert or update (pi,j) in forward list;
7.   }
8. }
9 . transfer IntPkts according by forwarding list
```

Fig. 3 IntPkt forwarding process (pseudo-code)

two contacted vehicle passing each other's current position using current speed, $d_{min}(i, poi)$ is able to reflect if vehicle i is moving closer to POI in future period. The smaller d_{min} the vehicle has, the closer it is to POI. If no vehicle meets these two conditions, current vehicle continues to carry IntPkt. The pseudo-code in Fig. 3 presents that vehicle i forwards IntPkt process. The forwarding list indicates subsequent relay node for each IntPkt. If there is desired data in neighbor j's CS or vehicle j has smallest near degree $d_{min}(i, poi)$ for IntPkt pi, an item (pi, j) will be added into forwarding list. Finally, all IntPkts are transferred based on this forwarding list.

4.2.2 Data Packets Forwarding

The better relay nodes would route DatPkt from the vehicle where IntPkt is satisfied, nearer or quicker to the moving consumer, which means that any vehicle that either goes closer to, or travels faster near, consumer should be the next carrier. Two conditions are considered: (1) The proximity of the trajectories of candidate node and target consumer before DatPkt expires; (2) The instant time when these two vehicles would travel closest. Unlike IntPkt forwarding, in DatPkt forwarding we consider the nearest distance between the trajectories of candidate vehicle and consumer vehicle before DatPkt expires. The consumer's trajectory in IntPkt appends in "trajectory info" field of DatPkt when DatPkt is generated.

Definition 3 (Trajectory Nearest Distance $d_{\min}(i, c)$ to moving consumer): The vehicle i's trajectory nearest distance to the consumer c before TTL of DatPkt TTL_{DatPkt} is defined as:

$$d_{\min}(i, c) = mind_{pi, pc}(t) = min\sqrt{\left(x_{pi}(t) - x_{pc}(t)\right)^2 + \left(y_{pi}(t) - y_{pc}(t)\right)^2} \qquad (3)$$
$$t \in [Current\ Time, TTL_{DatPkt}]$$

Definition 4 (Nearest Time $t_{near}(i, c)$): The nearest time $t_{near}(i, c)$ is defined as the earliest time when vehicle i and consumer c reach nearest distance $d_{\min}(i, c)$ apart.

$d_{\min}(i, c)$ and $t_{near}(i, c)$ are calculated in the same way as $d_{\min}(i, POI)$, the only difference is that the consumer's trajectory is obtained from DatPkt fields, rather than data name. They serve as metrics to determine next relay node of DatPkt and a comprehensive metric is defined as Definition 5.

Definition 5 (Comprehensive Nearest Metric $dt_{\min}(i, c)$): The comprehensive nearest metric $dt_{\min}(i, c)$ combines the nearest distance and nearest time. It is represented as formula (4).

$$dt_{\min}(i, c) = k\frac{d_{\min}(i, c)}{D(n_{sat}, c)} + (1 - k)\frac{t_{near}(i, c)}{TTL_{DatPkt}} \qquad (4)$$

where $D(n_{sat}, c)$ is the constant distance between consumer and the DatPkt provider vehicle when DatPkt is generated. k is the impact ratio of $d_{\min}(i, c)$ and $t_{near}(i, c)$. Similar to IntPkt forwarding process, if contacting a vehicle with smaller $dt_{\min}(k, c)$ for a DatPkt, vehicle i forward this DatPkt to it, else vehicle i continues to be the carrier.

4.3 Receiving Process for Interest and Data Packets

The pseudo-code of processing an incoming IntPkt is shown in Fig. 4. Whenever an IntPkt is received, the node checks local CS. On name matching, the IntPkt is satisfied and corresponding DatPkt is generated. In case of local CS not containing desired data, unsatisfied interest is stored in PIT. Only one entry is created in PIT for the same data name. If there is no entry for this name, a new entry is created and consumer information from IntPkt is stored in request hosts list that stores all hosts requesting the same data. Otherwise, consumer information is appended or updated in the existing PIT entry of this name.

```
Input :incoming IntPkt received
Output:
1. get request name and consumer information from IntPkt and check local CS;
2. if(Local CS contains desired data){
3.   generate DatPkt containing the rest of TTL and Trajectory info from IntPkt;
4.   remove the IntPkt and add DatPkt into bundles buffer; }
5. else if (local PIT contains the name){
6.     if(consumer is not in request host list) insert consumer into Request Host List;
7.     else update consumer in Request Host List with new TTL and trajectory info; }
8. else { create new interest entry for name in PIT;
9.       add consumer into Request Host List of new entry; }
10. }
```

Fig. 4 IntPkt receiving process (pseudo-code)

After receiving a DatPkt, the node check local PIT for corresponding interest. If current node is subscriber, the interest is satisfied and the data from DatPkt is stored in repository. If there is a PIT entry for this data, this PIT entry is deleted and the data is stored in local CS, then DatPkt is sent to each requested hosts respectively. Otherwise, DatPkt is discarded.

4.4 Validity of Messages

The validity of message is also one of the important issues for GOFP. Firstly, a retention period is set for the data stored in CS. Its value is application-depended and indicates the freshness of data. After retention period has elapsed, the data is useless and will be deleted from CS. Secondly, IntPkt and DatPkt all contain TTL field: TTL of IntPkt is assigned by consumer to represent the expected latest time to obtain desired data and TTL of DatPkt is set as the rest of TTL of corresponding IntPkt. The node monitors the validity of each carried packets and discards the packet once its TTL expires. Consumer will resend the interest if it never receives desired data until TTL expires. Finally, each node maintains delivered packets information it knew. The node will delete any packet that is stored in its bundles buffer and is announced as delivered by neighbor nodes.

5 Performance Evaluation

To evaluate the performance of GOFP, we extended the Opportunistic Networking Environment (ONE) [13] simulator to support the proposed forwarding strategy and compared GOFP with two algorithms: (1) FirstContact [14], an opportunistic routing algorithm in which carrier vehicle forwards packets to the first contact vehicle. Like in GOFP, current carrier removes these packets after forwarding; (2) P-Random [14], another opportunistic routing protocol, which randomly decides whether or not to forward packets to other node by a certain probability and only one copy of every packet is retained in the network. In simulations, the probability is set to 0.2. The metrics used to evaluate GOFP for different vehicle densities include the hops of Interest Packet, the hops of Data Packet, the satisfied delay of Interest Packet and the delivery delay of Data Packet.

5.1 Simulation Scenario

The deployment scenario of simulation was the map of Helsinki city, Finland. There are four groups of vehicles: a stationary data provider with zero velocity at POI, two groups of cars with different velocity whose numbers are varied to construct different traffic density and a tram group with one tram node. The main simulation parameters are listed in Table 1.

For different traffic densities, the simulation was run 25 times with different movement random seeds. A vehicle was selected randomly as the consumer in each simulation and only a single IntPkt was generated at a preset time. Results are always presented with 90 % confidence interval. Our goal is to just evaluate forwarding performance, thus the packet size is small and message drops are not considered.

5.2 Experimental Results

Figure 5a presents the average number of hops of IntPkt resulting in different algorithms. The value of k has no effect on IntPkt forwarding in GOFP. GOFP

Table 1 The main simulation parameters

Parameter	Value
Simulation time:	21600 s
Deployment field:	4500 m × 3400 m
Transmission Range:	Cars: 50 m; Trams: 200 m
Transmission rate:	Simple Interface: 250 k; High Speed Interface: 10 M
Node Speed (m/s):	The first group of cars: 2.2–8.34; The second group of cars: 2.7–13.9; Trams: 10–30
Traffic density:	40, 60, 80, 100, 120
NDN Parameter:	Interest TTL: 180 min; Trajectory interval: 10 s; Given period δ: 360 s; Weight factor k: 0.0, 0.5, 1.0

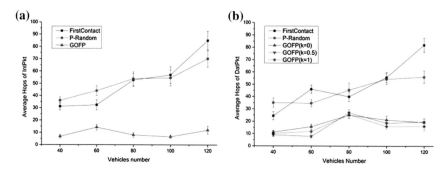

Fig. 5 Average number of Hops of IntPkt and DatPkt in different algorithms

outperforms two other algorithms to make the average hops of IntPkt holding steady under 20 nodes, which decreased respectively up to 85.8 and 69.8 % over FirstContact and P-Random algorithms whenever vehicles density is low or high. FirstContact and P-Random algorithms all show a larger number of hops which increases as vehicles number increases. This is mainly because that it has more chance to contact and transmit IntPkt to other vehicles in high density situation. GOFP choose relay vehicle with optimal forwarding metric, thus, the vehicles density does not influence the average hops of IntPkt. Figure 5b plots the average hops of DatPkt of different algorithms in varying vehicle destinies. With different values of k, GOPFs still have lower hops than FirstContact and P-Random. Though selecting next carrier vehicle at random, P-Random shows better performance than FirstContact that constantly attempts to forward the packets to first neighbors within its communication range.

Due to variation of k, GOFP presents different values for the average hops of DatPkt. When k equals 0, the nearest time metric is only considered to select next relay node; when k equals 1, only the nearest distance metric is considered; when k is 0.5, both nearest time and nearest distance are take into account. Figure 5b shows that smaller hops are required when the nearest distance is used in next carrier selection (as $k = 0.5$ or $k = 1$), while for $k = 1$, hops are higher. This is caused by the fact that the vehicle with nearest trajectory distance may carry DatPkt closer to consumer and potentially reduce the forwarding frequency.

Figure 6a presents the average satisfied delay of IntPkt, taken as the difference between the instant in time when IntPkt is generated and the instant in time when IntPkt is satisfied by a vehicle holding desired data. GOFP exhibits a 74–89.1 % delay improvement over the FirstContact and P-Random algorithms under different vehicles density. This verifies the effectiveness that the trajectory distance to POI can indicate if the vehicle is moving towards to POI in next period of time. The average success delivery delay of DatPkt is presented in Fig. 6b, which describes the difference from the time instant when the consumer sends out interest to the time instant when it receives desired data. FirstContact and P-Random present unstable performance, as their delivery delay depends much on specific situations,

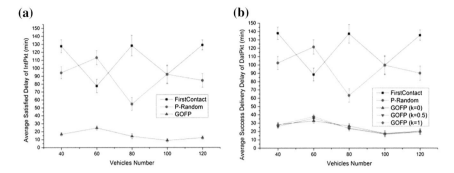

Fig. 6 Average satisfied delay of IntPkt and delivery delay of DatPkt (GOPF vs. others)

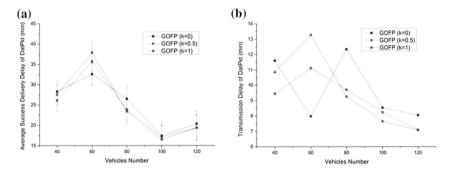

Fig. 7 Average success delivery delay and transimission delay of DatPkt in GOFP

which means that they only have better performance when the consumer is in a relatively near area. As one can see, average delivery delay of GOFP is significantly lower, no matter how many vehicles are on the road. Though only little differences between GOFPs with various value of k, it still can be seen in Fig. 7a that the delay is a little lower if the shortest time is used in next carrier selection (as $k = 0$ or $k = 0.5$). This is mainly due to the vehicle with shortest time being able to forward packets faster near to the consumer.

Figure 7b shows how GOFP performs under different values of k in the transmission delay which is the difference between the time instant when DatPkt being sent to the network and the time instant when the consumer receives it. It is obvious that the higher the vehicles density is, the less delay the GOFPs exhibit. Besides, when k equals 0 or 1, only shortest time or nearest distance is taken into account to select next carrier vehicle separately, which is incomplete and has unstable performance, while GOFP with $k = 0.5$ is better and presents satisfactory performance.

These simulation results show that GOFP has a better performance than those other two similar routing strategies, both in terms of lowering the average hops and delay.

6 Conclusions

This paper proposed GOFP, a new Geographic Opportunistic Forwarding strategy for Vehicle Named Data Networking. Through using geographic position of POI and vehicles trajectories, the different forwarding strategies are provided for IntPkt and DatPkt, respectively, in GOFP. The ONE simulator is extended to evaluate GOFP and the simulation results show that GOFP has better performance and outperforms other two similar algorithms.

As future work, the interval of vehicles' trajectory will be evaluated and the trams with fixed route, in-line with the findings in [15], will be also considered in message delivery, in order to reduce the amounts of trajectory information in packets.

Acknowledgments This work is supported in part by the Fundamental Research Funds of Jilin University, No. 450060491509 and partially supported by FCT—Fundação para a Ciência e Tecnologia Portugal in the scope of the project: UID/CEC/00319/2013.

References

1. Ahlgren, B., Dannewitz, C., Imbrenda, C., et al.: "A survey of information-centric networking"[J]. Commun. Mag. IEEE **50**(7), 26–36 (2012)
2. Jacobson, V., Smetters, D.K., Thornton, J.D., et al.: Networking named content [C]. In: Proceedings of the 5th International Conference on Emerging Networking Experiments and Technologies. ACM, pp. 1–12 (2009)
3. Karp, B., Kung, H.T.: GPSR: Greedy perimeter stateless routing for wireless networks [C]. In: Proceedings of the 6th International Conference on Mobile Computing and Networking. ACM, pp. 243–254 (2000)
4. Ramakrishna, M.: DBR: distance based routing protocol [J]. Int. J. Inf. Electron. Eng. **2**(2), 228–232 (2012)
5. Leontiadis, I., Mascolo, C.: GeOpps: geographical opportunistic routing for vehicular networks [C]. In: IEEE International Symposium on a World of Wireless, Mobile and Multimedia Networks 2007, Espoo, Finland, 18–21 June 2007, pp. 1–6 (2007)
6. Soares, V.N.G.J., Rodrigues, J.J.P.C., Farahmand, F.: GeoSpray: a geographic routing protocol for vehicular delay-tolerant networks [J]. Inf. Fusion **15**, 102–113 (2014)
7. Grassi, G., Pesavento, D., Pau, G., et al.: VANET via named data networking [C]. In: 2014 IEEE INFOCOM Workshops on Name-Oriented Mobility, pp. 410–415 (2014)
8. Wang, L., Afanasyev, A., Kuntz, R., Zhang, L.: Rapid traffic information dissemination using named data [C]. In: 1st ACM Workshop on Emerging Name-Oriented Mobile Networking Design—Architecture, Algorithms, and Applications. ACM, pp. 7–12 (2012)
9. Kuai, M., Hong, X., Flores, R.R.: Evaluating interest broadcast in vehicular named data networking [C]. In: IEEE Research and Educational Experiment Workshop, 2014 Third GENI, pp. 77–78 (2014)
10. Pesavento, D., Grassi, G., Palazzi, C.E., et al.: A naming scheme to represent geographic areas in NDN [C]. In: IEEE Wireless Days (WD), 2013 IFIP, 1–3 (2013)
11. Yu, Y.T., Dilmaghani, R.B., Calo, S., et al.: Interest propagation in named data MANETs [C]. In: Interantional Conference Computing Networking Communication IEEE, pp. 1118–1122 (2013)

12. Lu, Y., Li, X., Yu, Y.T., et al.: Information-Centric delay-tolerant mobile Ad-Hoc networks [C] In: 2014 IEEE Conference on INFOCOM 2014, WKSHPS, IEEE, pp. 428–433 (2014)
13. Keränen, A., Ott, J., Kärkkäinen, T.: The ONE simulator for DTN protocol evaluation [C]. In: Proceedings of the 2nd Int. Conf. on simulation tools and techniques. ICST (Institute Computer Sciences, Social-Informatics and Telecommunications Engineering), p. 55 (2009)
14. Jain, S., Fall, K., Patra, R.: Routing in a delay tolerant network [C]. In: Proceedings of the Conference on Applications, Technologies, Architectures, and Protocols for Computer Communications. ACM, pp. 145–157 (2004)
15. Hadiwardoyo, S., Santos, A.: Deploying Public Surface Transit to Forward Messages in DTN [C]. In: 11th International Wireless Communication Mobile Computing Conference Dubrovnick, Croatia, 24–27 Aug 2015